# Fracture Mechanics and Fatigue Damage of Materials and Structures

# Fracture Mechanics and Fatigue Damage of Materials and Structures

Editors

**Grzegorz Lesiuk**
**Dariusz Rozumek**

MDPI • Basel • Beijing • Wuhan • Barcelona • Belgrade • Manchester • Tokyo • Cluj • Tianjin

MDPI

*Editors*
Grzegorz Lesiuk
Department of Mechanics,
Materials Science and
Biomedical Engineering,
Wroclaw University of
Science and Technology,
Wroclaw, Poland

Dariusz Rozumek
Department of Mechanics
and Machine Design,
Opole University
of Technology,
Opole, Poland

*Editorial Office*
MDPI
St. Alban-Anlage 66
4052 Basel, Switzerland

This is a reprint of articles from the Special Issue published online in the open access journal *Materials* (ISSN 1996-1944) (available at: https://www.mdpi.com/journal/materials/special_issues/fracture_fatigue).

For citation purposes, cite each article independently as indicated on the article page online and as indicated below:

LastName, A.A.; LastName, B.B.; LastName, C.C. Article Title. *Journal Name* **Year**, *Volume Number*, Page Range.

**ISBN 978-3-0365-8124-8 (Hbk)**
**ISBN 978-3-0365-8125-5 (PDF)**

# Contents


**About the Editors** . . . . . . . . . . . . . . . . . . . . . . . . . . . . . . . . . . . . . . . . . . . . . . **vii**

**Preface to "Fracture Mechanics and Fatigue Damage of Materials and Structures"** . . . . . . . **ix**

**Grzegorz Lesiuk and Dariusz Rozumek**
Special Issue "Fracture Mechanics and Fatigue Damage of Materials and Structures"
Reprinted from: *Materials* **2023**, *16*, 4171, doi:10.3390/ma16114171 . . . . . . . . . . . . . . . . . . . **1**

**Volodymyr Hutsaylyuk, Yosyf Piskozub, Liubov Piskozub and Heorhiy Sulym**
Deformation and Strength Parameters of a Composite Structure with a Thin Multilayer Ribbon-like Inclusion
Reprinted from: *Materials* **2022**, *15*, 1435, doi:10.3390/ma15041435 . . . . . . . . . . . . . . . . . . . **5**

**Xinna Liu, Shuai Zhang, Yanmei Bao, Zhongran Zhang and Zhenming Yue**
Strain-Controlled Fatigue Behavior and Microevolution of 316L Stainless Steel under Cyclic Shear Path
Reprinted from: *Materials* **2022**, *15*, 5362, doi:10.3390/ma15155362 . . . . . . . . . . . . . . . . . . . **25**

**Maria Letizia Raffa, Raffaella Rizzoni and Frédéric Lebon**
Numerical Assessment of Damage Parameters for a Hard Interface Model
Reprinted from: *Materials* **2022**, *15*, 5370, doi:10.3390/ma15155370 . . . . . . . . . . . . . . . . . . . **37**

**Krzysztof Junik, Grzegorz Lesiuk, Szymon Duda, Krzysztof Jamroziak, Wojciech Błażejewski, Paweł Zielonka, et al.**
Constitutive Law Identification and Fatigue Characterization of Rigid PUR Elastomers 80 ShA and 90 ShA
Reprinted from: *Materials* **2022**, *15*, 6745, doi:10.3390/ma15196745 . . . . . . . . . . . . . . . . . . . **49**

**Antonin Bermond, Claire Roume, Jacques Stolarz, Matthieu Lenci, Jean-François Carton and Helmut Klocker**
Low Cycle Fatigue of G20Mn5 Cast Steel Relation between Microstructure and Fatigue Life
Reprinted from: *Materials* **2022**, *15*, 7072, doi:10.3390/ma15207072 . . . . . . . . . . . . . . . . . . . **65**

**Xi-Ming Yao, Yu-Chen Zhang, Qi Pei, Li-Zhu Jin, Tian-Hao Ma, Xiao-Hua He and Chang-Yu Zhou**
Empirical Solution of Stress Intensity Factors for the Inclined Inner Surface Crack of Pipe under External Pressure and Axial Compression
Reprinted from: *Materials* **2023**, *16*, 364, doi:10.3390/ma16010364 . . . . . . . . . . . . . . . . . . . . **85**

**Živilė Decker, Vitalijus Rudzinskas, Kazimierz Drozd, Jacek Caban, Jurijus Tretjakovas, Aleksander Nieoczym and Jonas Matijošius**
Analysis of the Vehicle Chassis Axle Fractures
Reprinted from: *Materials* **2023**, *16*, 806, doi:10.3390/ma16020806 . . . . . . . . . . . . . . . . . . . . **101**

**Takayuki Shiraiwa, Fabien Briffod and Manabu Enoki**
Prediction of Fatigue Crack Initiation of 7075 Aluminum Alloy by Crystal Plasticity Simulation
Reprinted from: *Materials* **2023**, *16*, 1595, doi:10.3390/ma16041595 . . . . . . . . . . . . . . . . . . . **119**

**Michał Böhm, Adam Niesłony, Szymon Derda, Robert Owsiński, Miloslav Kepka, Jr., Ivana Zetkova, et al.**
General Reference and Design S–N Curves Obtained for 1.2709 Tool Steel
Reprinted from: *Materials* **2023**, *16*, 1823, doi:10.3390/ma16051823 . . . . . . . . . . . . . . . . . . . **131**

**Faezeh Hatami and Ahmad Varvani-Farahani**
Accumulation of Plastic Strain at Notch Root of Steel Specimens Undergoing Asymmetric Fatigue Cycles: Analysis and Simulation
Reprinted from: *Materials* **2023**, *16*, 2153, doi:10.3390/ma16062153 . . . . . . . . . . . . . . . . . . . **149**

**Branko Nečemer, Franc Zupanič, Tomaž Vuherer and Srečko Glodež**
High-Cycle Fatigue Behaviour of the Aluminium Alloy 5083-H111
Reprinted from: *Materials* **2023**, *16*, 2674, doi:10.3390/ma16072674 . . . . . . . . . . . . . . . . . . . **169**

**Živilė Decker, Jurijus Tretjakovas, Kazimierz Drozd, Vitalijus Rudzinskas, Mariusz Walczak, Artūras Kilikevičius, et al.**
Material's Strength Analysis of the Coupling Node of Axle of the Truck Trailer
Reprinted from: *Materials* **2023**, *16*, 3399, doi:10.3390/ma16093399 . . . . . . . . . . . . . . . . . . . **185**

**Tong Li, Zhenting Yang, Chenghui Xu, Xinsheng Xu and Zhenhuan Zhou**
A Phase Field Approach to Two-Dimensional Quasicrystals with Mixed Mode Cracks
Reprinted from: *Materials* **2023**, *16*, 3628, doi:10.3390/ma16103628 . . . . . . . . . . . . . . . . . . . **203**

# About the Editors

**Grzegorz Lesiuk**

Grzegorz Lesiuk is employed at the Wrocław University of Technology (Faculty of Mechanical Engineering) since 2009. He received his PhD in technical sciences (mechanics) in 2013 and habilitation (in mechanical engineering) in 2020. His scientific activities are related to mechanics, fracture mechanics and the fatigue of materials and structures, as well as new materials and modeling their durability under cyclic and static loads. He has published over 109 scientific papers in national and international scientific and technical journals and over 100 conference papers and 1 patent in this field. Currently, Prof. Lesiuk is a President of the Polish Fracture Mechanics Group. He is also a member of the Polish Society of Theoretical and Applied Mechanics and the ESIS (European Structural Integrity Society). He is the co-organizer of many international and national scientific events such as XXICMFM (International Colloquium Mechanical Fatigue of Metals), XVIII KKMP (National Conference on Fracture Mechanics), and the first VCMF (Virtual Conference on Fatigue of Metals). He has been repeatedly awarded for his scientific activities: he was admitted as a member of the Academy of Young Scientists and Artists operating at the President of the City of Wroclaw (Poland); he received a nomination from the President of the City of Wroclaw (2018), Rector's Award of Wroclaw University of Technology (2018, 2020), International ESIS Award—TC12 (Merit Award) 2019 and two international TOP 1% Reviewers (Publons and Web of Science) awards (2019); and a Lierati Award (2019)—Emerald Journals Group. In 2021, Professor Grzegorz Lesiuk was listed in the TOP2% cited researchers (prepared by Stanford University in cooperation with the Elsevier publishing company).

**Dariusz Rozumek**

Since 2021, Dariusz Rozumek has been a full Professor at the Department of Mechanics and Machine Design in the Faculty of Mechanical Engineering from the Opole University of Technology, following 28 years of research and teaching activity at the Opole University of Technology, Poland. His research interests lie in the area of multiaxial fatigue and fracture, including stress and energy approaches, welding and many other topics related to the field of mechanical engineering. He has published around 282 scientific papers, 11 patents, 8 book editions and 13 book chapters. He has participated in more than 11 European and National Projects. He is also a Guest Editor of several issues in various journals and a reviewer of articles in 25 journals. He has been awarded many times with the awards of the Rector of the Opole University of Technology. He has a medal from the National Education Commission.

# Preface to "Fracture Mechanics and Fatigue Damage of Materials and Structures"

One of the most important aspects of the engineering assessment of the technical condition of structures and materials is the ability to assess the fatigue behavior of materials and structures. In addition, another important topic is the design of materials or structures that can resist fatigue and fracture. Modern science provides us with an increasing number of new materials, from superalloys of metals manufactured conventionally as well as via additive manufacturing to functionally advanced composites. Against this background, fundamental knowledge of the fatigue behavior and fracture mechanics of different material groups provides a convenient platform for communication between different interested groups and fields, from material science, numerical engineering and mathematical modeling to hybrid methods for fatigue life prediction.

This Special Issue presents the latest research findings on crack growth and fatigue damage in metallic materials. The latest achievements in experimental research and the modeling of static and cyclic loads are presented. Fundamental research insights as well as practical predictions are presented, including the influence of microstructure, defect assessment, damage analysis, mathematical modelling, numerical simulation and S-N curve parameters. For this Special Issue, 16 submissions were received and 14 papers were published, attracting submissions from 10 countries around the world.

We would also like to express our most profound appreciation to the MDPI Book staff; the editorial team of the *Materials* journal, especially Mr. Felix Guo; the Assistant Editor of this Special Issue; the talented authors; and the hardworking and professional reviewers.

**Grzegorz Lesiuk and Dariusz Rozumek**
*Editors*

*materials*

MDPI

*Editorial*

# Special Issue "Fracture Mechanics and Fatigue Damage of Materials and Structures"

**Grzegorz Lesiuk [1,*] and Dariusz Rozumek [2]**

1 Department of Mechanics, Materials Science and Biomedical Engineering, Wroclaw University of Science and Technology, Smoluchowskiego 25, PL 50-370 Wroclaw, Poland
2 Department of Mechanics and Machine Design, Opole University of Technology, Mikolajczyka 5, 45-271 Opole, Poland; d.rozumek@po.edu.pl
* Correspondence: grzegorz.lesiuk@pwr.edu.pl

**Citation:** Lesiuk, G.; Rozumek, D. Special Issue "Fracture Mechanics and Fatigue Damage of Materials and Structures". *Materials* **2023**, *16*, 4171. https://doi.org/10.3390/ma16114171

Received: 22 May 2023
Accepted: 26 May 2023
Published: 3 June 2023



## 1. Introduction and Scope

One of the most important aspects of engineering assessment of the technical condition of structures and materials is the ability to assess the fatigue behavior of materials and structures. On the other hand, an important topic is the design of materials or structures to resist fatigue and fracture. Modern science provides us with an increasing number of new materials, from superalloys of metals manufactured conventionally and by additive manufacturing to functionally advanced composites. Against this background, the fundamental knowledge of the fatigue behavior and fracture mechanics of different material groups provides a convenient platform for communication between different interested groups and fields, from material scientists, numerical engineers and mathematical modeling to hybrid methods for fatigue life prediction. This Special Issue provides such an exchange of ideas on recent developments in the field of fatigue and fracture and is especially focused on fatigue crack growth analysis, the description of fatigue damage in metals and composites, probabilistic approaches, fracture mechanics analysis, fatigue failure analysis and lifetime prediction. A summary of the articles is given in this editorial.

This Special Issue contains thirteen original research papers on fatigue and cracking in materials. The presented articles mainly concern experimental research and numerical calculations.

## 2. Contributions

In the paper by Volodymyr Hutsaylyuk et al. [1], an analytical–numerical method for determining the mechanical fields in composite structures with interphase ribbon-like deformable multilayered inhomogeneities under combined force and dislocation loading has been proposed. The values of generalized stress intensity factors for the asymptotic stress–strain fields in the vicinity of the ends of thin inhomogeneities are calculated, from which the stress concentration and local strength of the structure can be calculated. The proposed method has shown its effectiveness for solving an entire class of problems of deformation and fracture of bodies with thin deformable inclusions of finite length and can be used for mathematical modeling of the mechanical effects of thin FGM heterogeneities in composites.

Based on the twin bridge shear specimen, Xinna Liu et al. [2] performed cyclic shear experiments on 1.2 mm thin plates of 316L metastable austenitic stainless steel with different strain amplitudes from 1 to 5% at ambient temperature. The fatigue behavior of 316L stainless steel under the cyclic shear path was studied, and the microscopic evolution of the material was analyzed. The results show that the cyclic stress response of 316L stainless steel exhibits cyclic hardening, saturation and cyclic softening, and the fatigue life is negatively correlated with the strain amplitude.

The study by Maria Letizia Raffa et al. [3] applies numerical calculations. This paper proposes a numerical assessment of two model parameters, damage energy threshold and damage viscosity, of a hard interface model previously formulated by the authors. The proposed assessment protocol uses macroscopic experimental data, available in the literature, on structural adhesives under standard characterization tests. The numerical results obtained give insights into the physical interpretation of these parameters.

Furthermore, Krzysztof Junik et al. [4] present the results of a study of polyurethane rigid (PUR) elastomers in terms of constitutive law identification and analyze the effect of polyurethane elastomers' hardness on fatigue properties. The research objects were PUR materials based on 4,4′-diphenylmethane diisocyanate (MDI) with hardnesses of 80 ShA and 90 ShA, typically used in various industrial applications. Based on the experimental campaign performed under static and cyclic loading, the constitutive model proposed by Ogden is most appropriate. In addition, a hybrid numerical–experimental analysis (using FEM-DIC) of diabolo specimens' behavior is carried out in fatigue tests. Based on the fatigue test performed, it is worth noting that the energy approach describes the fatigue process synonymously compared to the displacement or strain approach.

In the article by Antonin Bermond et al. [5], the influence of micro-shrinkage porosity on a G20Mn5 cast steel was presented. G20Mn5 (normalized) ingots were cast under industrial conditions, ensuring the absence of macroporosities. Solidification leads to two very different microstructures prior to the normalization treatment: columnar dendrites beneath the surface (Skin) and equiaxed microstructures close to the center (Core). First, metallographic observations of the whole ingot revealed the same grain size in both areas. Fatigue samples were extracted by differentiating two sampling volumes corresponding to columnar (S) and equiaxed solidification (C), respectively. The distribution of microporosities was determined for all samples by micro-CT scans. Core samples exhibit microporosities with volumes 1.7 times larger than Skin samples. Low-cycle fatigue tests (three levels of fixed plastic strain) were run on both sample series (C, S). The results follow the Manson–Coffin law. Core specimens exhibit a lower fatigue life than Skin specimens. The differences in fatigue life have been successfully related to the differences in microporosity sizes.

Xi-Ming Yao et al. [6], based on the theory of fracture mechanics, used a finite element method to determine the stress intensity factors of an inclined crack on the inner surface of a pipe under axial compression load and external pressure. The effects of different influencing factors on the stress intensity factor along the crack front were systematically explored considering crack closure and were different from those under internal pressure. The effects of a high aspect ratio on $K_{II}$, the crack inclination asymmetry caused by curvature, and the effects of the friction coefficient on the stress intensity factors of the pipe with an inclined inner surface crack under axial compression load and external pressure were explored. To be suitable for defect assessment, the solutions for stress intensity factors $K_{II}$ and $K_{III}$ were derived, and new correction factors, $f_\theta$ and $f_\mu$, were proposed in the empirical solutions to accommodate the crack inclination asymmetry and the friction coefficient, respectively.

The study by Živilė Decker et al. [7] concerns the durability of suspension components in relation to the duration of the onset of fatigue. This article presents an analysis of damage to the rear axle of the semi-trailer using macroscopic observations of the damage site and dynamic FEA of stress distribution in the axle material. To identify the probable cause of the damage, eight cases of loading the semi-trailer axle were considered. Analytical solutions have shown that in various cases, the yield point is exceeded and the strength limit of the modeled semi-trailer axle is reached. The risk of damage to the vehicle's suspension system components increases on poor roads (bumps and winding road sections).

Takayuki Shiraiwar et al. [8] proposed a method for predicting fatigue crack initiation of the 7075 aluminum alloy by crystal plasticity finite element analysis considering microstructures. In order to accurately predict the total fatigue life, it is necessary to calculate the number of cycles for fatigue crack initiation, small crack growth, and long crack growth. The long crack growth life can be estimated by the Paris law, whereas fatigue crack initiation and small crack growth are sensitive to microstructures and difficult to predict.

In this work, the microstructure of 7075 aluminum alloy was reconstructed based on experimental observations in the literature, and crystal plasticity simulations were performed to calculate the elasto-plastic deformation behavior in the reconstructed polycrystalline model under cyclic deformation. The calculated local plastic strain was introduced into the crack initiation criterion (Tanaka and Mura, 1981) to predict fatigue crack initiation life. The predicted crack initiation life and crack morphology were in good agreement with the experimental results, indicating that the proposed method is effective in predicting fatigue crack initiation in aluminum alloys. Based on the obtained results, future issues regarding the prediction of fatigue crack initiation were discussed.

The research by Michał Böhm et al. [9] presented the commonly used EN 1.2709 tool steel (printing steel), which has good strength properties and high abrasion resistance and can be hardened. The research shows, however, that its fatigue strength may differ depending on the printing method and may be characterized by a wide range of fatigue life. Selected S–N curves for EN 1.2709 steel are presented after printing with the selective laser melting method. The characteristics are compared, and conclusions are presented regarding the resistance of this material to fatigue loading, especially in the tension–compression state. A combined general mean reference and design fatigue curve is presented, which incorporates our own experimental results as well as those from the literature for the tension–compression loading state. The design curve may be implemented using the finite element method by engineers and scientists to calculate the fatigue life.

Faezeh Hatami and Ahmad Varvani-Farahani [10] evaluated the ratcheting response at notch roots of 1045 steel specimens experiencing uniaxial asymmetric fatigue cycles. Local stress and strain components at the notch root were analytically evaluated using Neuber, Glinka, and Hoffman–Seeger (H-S) rules coupled with the Ahmadzadeh–Varvani (A–V) kinematic hardening model. Backstress promotion through a coupled kinematic hardening model with the Hoffman–Seeger, Neuber, and Glinka rules was studied. Relaxation in local stresses on the notched samples as hysteresis loops moved forward with plastic strain accumulation during asymmetric loading cycles was observed. Local ratcheting results were simulated through FE analysis, where the Chaboche model was employed as the material hardening rule. A consistent response of the ratcheting values was evidenced as predicted, and simulated results were compared with the measured ratcheting data.

In the paper by Branko Nečemer et al. [11], a comprehensive experimental investigation of the high-cycle fatigue (HCF) behavior of the ductile aluminium alloy AA 5083-H111 is presented. The analyzed specimens were fabricated in the rolling direction (RD) and the transverse direction (TD). The HCF tests were performed under load control (load ratio $R = 0.1$) at different loading levels under a loading frequency of 66 Hz up to the final failure of the specimen. The experimental results have shown that the S–N curves of the analyzed Al alloy consist of two linear curves with different slopes. Furthermore, RD-specimens demonstrated longer fatigue lives if compared to TD specimens. This difference was about 25% at an amplitude stress of 65 MPa, where the average fatigue life was 276,551 cycles for RD specimens and 206,727 cycles for TD specimens. Similar behavior was also found for the lower amplitude stresses and fatigue lives between 106 and 108 cycles. The difference can be caused by large Al6(Mn,Fe) particles that are elongated in the rolling direction and cause higher stress concentrations in the case of TD specimens. The micrography of the fractured surfaces has shown that the fracture characteristics were typical for the ductile materials and were similar for both specimen orientations.

Similar to [7], Živilė Decker et al. [12] focused on issues related to road transport, which is important for the national economy. Damage usually excludes the means of transport from operation, which causes disruption of supply chains. One such damage is the failure of the suspension system of the vehicle or trailer, which usually occurs when the vehicle is heavily loaded. Such a defective system has been analyzed in this publication. Mathematical apparatus and finite element method (FEM) numerical simulations were used. A dangerous axle cross-section in terms of load was indicated, and the maximum stresses in this area were calculated for two types of roads. On highways, the stress at the critical point was 199 MPa, and on uneven roads, it increased to 304 MPa, which is comparable to the yield point. It was found that the second form of vibration may cause stress in the damage area, but the excitation frequency would have to be quite high. The probability of such a load and failure event occurring under operating conditions is low.

Tong Li et al. [13] studied quasicrystals (QCs), which are representatives of a novel type of material exhibiting many remarkable specific properties. However, QCs are usually brittle, and crack propagation inevitably occurs in such materials. Therefore, it is of great significance to study the crack growth behaviors in QCs. In this work, the crack propagation of two-dimensional (2D) decagonal QCs is investigated by a fracture phase field method. In this method, a phase field variable is introduced to evaluate the damage of QCs near the crack. In the numerical examples, the crack propagation paths of 2D QCs are simulated by the proposed method, and the effects of the phason field on the crack growth behaviors of QCs are studied in detail.

## 3. Conclusions

We would like to thank all of the reviewers for their contributions and efforts in producing this Special Issue, as well as the authors for preparing the papers. We would also like to thank all the staff at the *Materials* Editorial Office, especially Felix Guo, Assistant Editor, who managed and facilitated the publication process.

**Conflicts of Interest:** The authors declare no conflict of interest.

## References

1. Hutsaylyuk, V.; Piskozub, Y.; Piskozub, L.; Sulym, H. Deformation and Strength Parameters of a Composite Structure with a Thin Multilayer Ribbon-like Inclusion. *Materials* **2022**, *15*, 1435. [CrossRef] [PubMed]
2. Liu, X.; Zhang, S.; Bao, Y.; Zhang, Z.; Yue, Z. Strain-Controlled Fatigue Behavior and Microevolution of 316L Stainless Steel under Cyclic Shear Path. *Materials* **2022**, *15*, 5362. [CrossRef] [PubMed]
3. Raffa, M.L.; Rizzoni, L.; Lebon, F. Numerical Assessment of Damage Parameters for a Hard Interface Model. *Materials* **2022**, *15*, 5370. [CrossRef]
4. Junik, K.; Lesiuk, G.; Duda, S.; Jamroziak, K.; Błażejewski, W.; Zielonka, P.; Socha, T.; Denisiewicz, A.; Kula, K.; Szczurek, A. Constitutive Law Identification and Fatigue Characterization of Rigid PUR Elastomers 80 ShA and 90 ShA. *Materials* **2022**, *15*, 6745. [CrossRef] [PubMed]
5. Bermond, A.; Roume, C.; Stolarz, J.; Lenci, M.; Carton, J.-F.; Klocker, H. Low Cycle Fatigue of G20Mn5 Cast Steel Relation between Microstructure and Fatigue Life. *Materials* **2022**, *15*, 7072. [CrossRef] [PubMed]
6. Yao, X.-M.; Zhang, Y.-C.; Pei, Q.; Jin, L.-Z.; Ma, T.-H.; He, X.-H.; Zhou, C.-Y. Empirical Solution of Stress Intensity Factors for the Inclined Inner Surface Crack of Pipe under External Pressure and Axial Compression. *Materials* **2023**, *16*, 364. [CrossRef] [PubMed]
7. Decker, Ž.; Rudzinskas, V.; Drozd, K.; Caban, J.; Tretjakovas, J.; Nieoczym, A.; Matijošius, J. Analysis of the Vehicle Chassis Axle Fractures. *Materials* **2023**, *16*, 806. [CrossRef] [PubMed]
8. Shiraiwa, T.; Briffod, F.; Enoki, M. Prediction of Fatigue Crack Initiation of 7075 Aluminum Alloy by Crystal Plasticity Simulation. *Materials* **2023**, *16*, 1595. [CrossRef] [PubMed]
9. Böhm, M.; Niesłony, A.; Derda, S.; Owsiński, R.; Kepka, M., Jr.; Zetkova, I.; Zetek, M.; Houdková, Š.; Prażmowski, M. General Reference and Design S–N Curves Obtained for 1.2709 Tool Steel. *Materials* **2023**, *16*, 1823. [CrossRef] [PubMed]
10. Hatami, F.; Varvani-Farahani, A. Accumulation of Plastic Strain at Notch Root of Steel Specimens Undergoing Asymmetric Fatigue Cycles: Analysis and Simulation. *Materials* **2023**, *16*, 2153. [CrossRef] [PubMed]
11. Nečemer, B.; Zupanič, F.; Vuherer, T.; Glodež, S. High-Cycle Fatigue Behaviour of the Aluminium Alloy 5083-H111. *Materials* **2023**, *16*, 2674. [CrossRef] [PubMed]
12. Decker, Ž.; Tretjakovas, J.; Drozd, K.; Rudzinskas, V.; Walczak, M.; Kilikevičius, A.; Matijosius, J.; Boretska, I. Material's Strength Analysis of the Coupling Node of Axle of the Truck Trailer. *Materials* **2023**, *16*, 3399. [CrossRef] [PubMed]
13. Li, T.; Yang, Z.; Xu, C.; Xu, X.; Zhou, Z. A Phase Field Approach to Two-Dimensional Quasicrystals with Mixed Mode Cracks. *Materials* **2023**, *16*, 3628. [CrossRef] [PubMed]

*materials*

MDPI

*Article*

# Deformation and Strength Parameters of a Composite Structure with a Thin Multilayer Ribbon-like Inclusion

**Volodymyr Hutsaylyuk [1,*], Yosyf Piskozub [2], Liubov Piskozub [2] and Heorhiy Sulym [3]**

[1] Institute of Robots and Machine Design, Military University of Technology, Gen. S. Kaliskiego str. 2, 00-908 Warsaw, Poland

[2] Department of Applied Mathematics and Physics, Ukrainian Academy of Printing, Pidgolosko 19, 79020 L'viv, Ukraine; piskozub@pancha.lviv.ua (Y.P.); piskozub@uad.lviv.ua (L.P.)

[3] Department of Mechanics and Applied Computer Science, Faculty of Mechanical Engineering, Bialystok University of Technology, Wiejska 45c, 15-351 Bialystok, Poland; h.sulym@pb.edu.pl

* Correspondence: volodymyr.hutsaylyuk@wat.edu.pl; Tel.: +48-22-261-839-094

**Abstract:** Within the framework of the concept of deformable solid mechanics, an analytical-numerical method to the problem of determining the mechanical fields in the composite structures with interphase ribbon-like deformable multilayered inhomogeneities under combined force and dislocation loading has been proposed. Based on the general relations of linear elasticity theory, a mathematical model of thin multilayered inclusion of finite width is constructed. The possibility of nonperfect contact along a part of the interface between the inclusion and the matrix, and between the layers of inclusion where surface energy or sliding with dry friction occurs, is envisaged. Based on the application of the theory of functions of a complex variable and the jump function method, the stress-strain field in the vicinity of the inclusion during its interaction with the concentrated forces and screw dislocations was calculated. The values of generalized stress intensity factors for the asymptotics of stress-strain fields in the vicinity of the ends of thin inhomogeneities are calculated, using which the stress concentration and local strength of the structure can be calculated. Several effects have been identified which can be used in designing the structure of layers and operation modes of such composites. The proposed method has shown its effectiveness for solving a whole class of problems of deformation and fracture of bodies with thin deformable inclusions of finite length and can be used for mathematical modeling of the mechanical effects of thin FGM heterogeneities in composites.

**Keywords:** functionally gradient material; composite; thin inhomogeneity; fracture mechanics; nonperfect contact; stress intensity factor

**Citation:** Hutsaylyuk, V.; Piskozub, Y.; Piskozub, L.; Sulym, H. Deformation and Strength Parameters of a Composite Structure with a Thin Multilayer Ribbon-like Inclusion. *Materials* **2022**, *15*, 1435. https://doi.org/10.3390/ma15041435

Academic Editor: Andrea Spagnoli

Received: 20 January 2022
Accepted: 12 February 2022
Published: 15 February 2022



## 1. Introduction

Microscopic, layered structures in fields such as microelectronics, biotechnology, energy, weaponry, etc. are gaining special attention in modern engineering and technology. Among the most important scientific projects, experts identify a significant increase in computer performance, restoration of human organs using reproduced tissues (obtained from 3D printers) and obtaining new structured materials created directly from given molecules and atoms. Quite often these inclusions are used as elements to reinforce structural parts of machines and structures or as fillers of composite materials. Thin lamellar inhomogeneities are also a characteristic phenomenon at the interphase boundaries of crystalline grains arising during crystallization [1–7]. In this regard, there is a need to provide mathematical modeling of nanostructure mechanics, which is still a pressing problem of materials science theory. At this stage of the development of mechanics, it is already possible to concentrate on the construction of the complex universal equations suitable for investigations of multiscale, including layered, structures and the development of methods for their solution.

In such structures, each layer or their combination has its functional purpose, in particular, anti-corrosion, anti-abrasion, heat protection, strengthening to inhibit and block crack growth, reduce porosity, provide a high degree of adhesion of the components [8–14]. Thanks to multilayers, it is also possible to increase the service life of structures, and their use can significantly reduce material intensity and cost and increase the endurance of products. At the same time in a structure with thin layers, there is a concentration of stresses near places of change of physical and mechanical characteristics of materials. And it is the higher, the greater the difference in their properties.

Inhomogeneous structures with optimally varying physical and mechanical properties along with the thickness, known as functionally graded materials (FGMs) [15–20], allow one to reduce such stress concentrations in the vicinity of the contact between the matrix and the interlayer by avoiding abrupt transitions in the properties of the components. A detailed review of the manufacturing techniques can be found in [21–26]. FGMs are often used in the coatings of structural elements to protect them from the harmful effects of temperature [8–11,27–35], etc. One of the frequently used variants of FGM arrangement is the combination of ceramics with metal [36,37], but this often leads to the violation of the contact between them. Due to the brittle nature of ceramics, there is a need for additional research into the applicability limits of such FGM structures [38–40]. The complexity of the geometry of structural elements and consideration of imperfections in the contact of their components stimulate the process of improving mathematical models of FGMs to ensure their qualitative design both in terms of mechanical strength [12–14,41–54] and in terms of consideration of thermal, magnetic, piezoelectric loading factors [47,48]. The use of the FGMs seems to be one of the most effective materials in the realization of sustainable development in industries.

An important aspect of strength research, including tensile strength, for such structures, is to improve their strength criteria, to determine such key parameters as stress intensity factors (SIF) in the points of singularity. Moreover, since we consider thin inhomogeneities not only in the form of classical cracks but also thin cavities filled with an arbitrary elastic or nonlinearly elastic material, it makes sense to claim that the theory of thin inclusions is an essential generalization of the crack theory and the so-called generalized SIFs, which characterize the distribution of stress and displacement fields, are analogous parameters of fracture mechanics for the theory of thin inhomogeneities [55–57].

This work aims to develop an analytical and numerical method for studying the stress-strain state and strength of composites with thin deformable multilayer ribbon-like elements that are also suitable for mathematical modeling of thin inclusions with an almost arbitrary continuous thickness variation of mechanical characteristics.

## 2. Formulation of the Problem

We consider a structure which, following the concept of deformable solid mechanics, we will further consider as a combination of two half-spaces with elastic constants $E_k, \nu_k, G_k$ $(k = 1, 2)$, at the interface of which (plane $xOz$) there is a tunnel section $L' = [-a;\ a]$ in the direction of the shear axis $Oz$ (Figure 1), in which a certain object of general thickness $2h (h \ll a)$ is inserted–a package of $M$ different thin plane-parallel layers $\{x \in L'; y \in [y_K - h_K; y_K + h_K], K = \overline{1, M}\}$ of thickness $2h_K \left(2h = 2\sum_{K=1}^{M} h_K\right), y_1 - h_1 = -h$, $y_M + h_M = h$ with orthotropic mechanical properties $G_y^{inK}, G_x^{inK}$ in the direction of two axes (Figure 2).

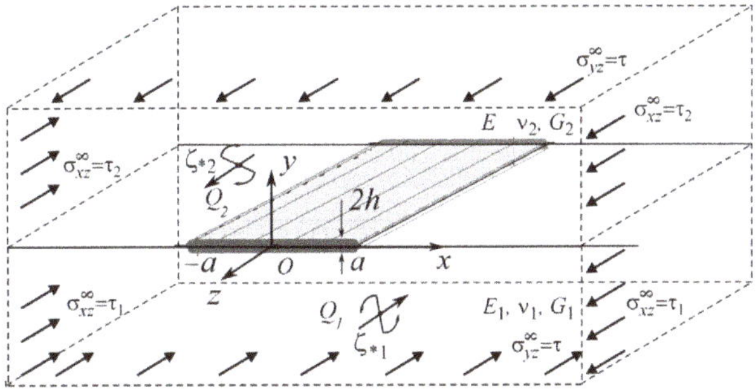


**Figure 1.** Geometry and load pattern of the problem.

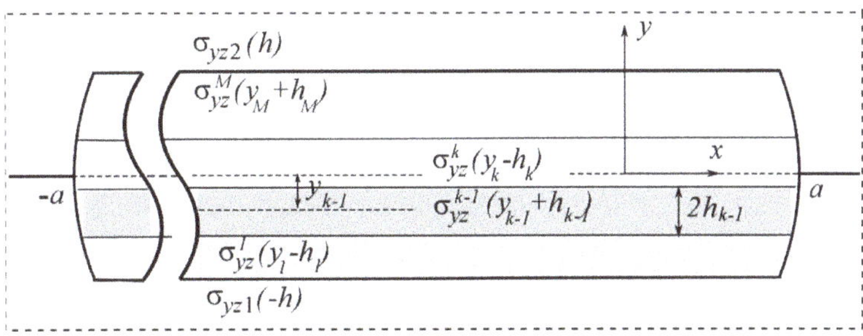


**Figure 2.** Multilayered inclusion.

The structure is loaded quasi-statically by shear factors (uniform shear at infinity $\tau, \tau_k$, concentrated forces $Q_k$, and screw dislocations $b_k$ at points $\varsigma_{k*}$), which cause longitudinal shear in the body. To ensure the straightness of the material interface at infinity, the stresses must satisfy the conditions $\sigma_{xz2}^{\infty} G_1 = \sigma_{xz1}^{\infty} G_2$, $\frac{v_2\sigma_{yy}^{\infty} - (1-v_2)\sigma_{xx2}^{\infty}}{G_2} = \frac{v_1\sigma_{yy}^{\infty} - (1-v_1)\sigma_{xx1}^{\infty}}{G_1}$.

Let us restrict ourselves to the problem of longitudinal displacement in the direction of the $z$-axis (antiplane deformation). Then, considering that the stress-strain state (SSS) of the structure in each section perpendicular to the $z$-axis is identical, we will further consider only the plane $xOy$, which consists of two planar sections of half-spaces $S_k(k = 1, 2)$ with a separation boundary between them in the form of abscissa axes $Ox$.

## 3. Materials and Methods

The construction of a mathematical model of such a layered thin inclusion-layer (internal problem) should eventually reveal the relation between the stress-strain parameters inside the inclusion and on its external surface as the influence functions $\sigma_{yz}^{in}(x, \pm h), w^{in}(x, \pm h)$, which will be used in the further solution of the problem [51–53].

Let us introduce into consideration the jumps of the stress tensor components and the displacement vector for the matrix components and individual layers on $L'$:

$$\begin{aligned} &\left[\sigma_{yz}\right]_{0,h} \cong \sigma_{yz1}(x, -h) - \sigma_{yz}(x, h) = f_3(x), \\ &\left[\frac{\partial w}{\partial x}\right]_{0,h} \cong \frac{\partial w}{\partial x}(x, -h) - \frac{\partial w}{\partial x}(x, h) = \left[\frac{\sigma_{xz}}{G}\right]_{0,h} = f_6(x),\ x \in L' \end{aligned} \tag{1}$$

$$\begin{aligned} &\left[\sigma_{yz}^{inK}\right]_{y_K,h_K} \cong \sigma_{yz}^{inK}(x, y_K - h_K) - \sigma_{yz}^{inK}(x, y_K + h_K) = f_{3K}(x),\ (K = \overline{1, M}) \\ &\left[\frac{\partial w^{inK}}{\partial x}\right]_{y_K,h_K} \cong \frac{\partial w^{inK}}{\partial x}(x, y_K - h_K) - \frac{\partial w^{inK}}{\partial x}(x, y_K + h_K) = f_{6K}(x),\ x \in L'; \end{aligned} \tag{2}$$

$f_3(x) = f_6(x) = 0$, $f_{3K}(x) = f_{6K}(x) = 0$, if $x \notin L'$.

Hereinafter marked:

$$[\bullet]_{y,h} = \bullet(x, y-h) - \bullet(x, y+h),$$

$$\langle\bullet\rangle_{y,h} = \bullet(x, y-h) + \bullet(x, y+h).$$

Let us analyze the methodology for constructing a mathematical model for the case of a multilayer package of thin inclusion layers. The basic relation for an arbitrary orthotropic elastic material with shear moduli $G_x^{inK}, G_y^{inK}$ of each of the layers given by the parameters $y_K, h_K (K = \overline{1, M})$ are the equilibrium conditions:

$$\frac{\partial \sigma_{xz}^{inK}}{\partial x} + \frac{\partial \sigma_{yz}^{inK}}{\partial y} + \rho^K F^{inK} = 0, \tag{3}$$

where $\rho^K$ denotes a density of the material, and $F^{inK}$—distribution of the mass forces, and constitutive strain-stress dependence (orthotropic linear elasticity):

$$\sigma_{xz}^{inK} = G_x^{inK} \frac{\partial w^{inK}}{\partial x}, \sigma_{yz}^{inK} = G_y^{inK} \frac{\partial w^{inK}}{\partial y}. \tag{4}$$

By integrating Equation (3) over the $x$ limits $[-a, x]$ and averaging, respectively, over the thicknesses of each of the heterogeneity layers $y \in [y_K - h_K, y_K + h_K]$, we obtain:

$$\frac{1}{2h_K} \int_{y_K - h_K}^{y_K + h_K} \sigma_{xz}^{inK}(\xi, y) dy \simeq \frac{1}{2} \left\langle \sigma_{xz}^{inK} \right\rangle_{y_K, h_K} = \frac{G_x^{inK}}{2} \left\langle \frac{\partial w^{inK}}{\partial x} \right\rangle_{y_K, h_K}, \tag{5}$$

and, accordingly, the first group of $M$ equations of mathematical models of layers:

$$\begin{gathered} \frac{G_x^{inK}}{2} \left\langle \frac{\partial w^{inK}}{\partial x} \right\rangle_{y_K, h_K} - \sigma_{xz}^{inK}(-a) - \frac{1}{2h_K} \int_{-a}^{x} \left[ \sigma_{yz}^{inK} \right]_{y_K, h_K} (\xi) d\xi + \\ + F_{aver}^{inK}(x, -h_K, h_K) = 0, \end{gathered} \tag{6}$$

where $F_{aver}^{inK}(x, -h_K, h_K) = \frac{\rho^K}{2h_K} \int_{y_K - h_K}^{y_K + h_K} \int_{-a}^{x} F^{inK}(\xi, y) d\xi dy, \ (K = \overline{1, M})$.

Considering the thin-wall ratio of the inclusion layers:

$$\begin{gathered} \frac{\partial w^{inK}}{\partial y}(x, y_K + h_K) + \frac{\partial w^{inK}}{\partial y}(x, y_K - h_K) \simeq \\ \simeq \frac{w^{inK}(x, y_K + h_K) - w^{inK}(x, y_K - h_K)}{h_K} = -\frac{\left[ w^{inK} \right]_{y_K, h_K}}{h_K}, \end{gathered} \tag{7}$$

$$w^{inK}(x, y_K \mp h_K) = w^{inK}(x, y_K \pm h_K) \mp 2h_K \frac{\partial w^{inK}}{\partial y}(x, y_K \pm h_K) (K = \overline{1, M}),$$

and constitutive relations (4), we obtain the following form of the second group of $M$ equations of the inclusion model:

$$-\frac{\left[ w^{inK} \right]_{y_K, h_K}}{h_K} = \frac{\left\langle \sigma_{yz}^{inK} \right\rangle_{y_K, h_K}}{G_y^{inK}} (K = \overline{1, M}), \tag{8}$$

which together with relations (6) fully describe the thin $M$ -layered inclusion model written in the values of the stress-strain behavior of the inclusion package materials.

Instead of the displacement jump, in many cases, it is convenient to use the formula for the jump of the strain components:

$$\left[ w^{inK} \right]_{y_K, h_K}(x) = \left[ w^{inK} \right]_{y_K, h_K}(-a) + \int_{-a}^{x} \left[ \frac{\partial w^{inK}}{\partial x} \right]_{y_K, h_K} (\xi) d\xi.$$

At each inclusion layer, the balance conditions must be satisfied:

$$\int_{-a}^{a} f_{3K}(\xi)d\xi = -N_{xzK}(-a) + N_{xzK}(a) + 2hF_{averK}^{in}(a,h), \tag{9}$$

$$\int_{-a}^{a} f_{6K}(\xi)d\xi = \left[w^{inK}\right]_{y_K,h_K}(a) - \left[w^{inK}\right]_{y_K,h_K}(-a), \tag{10}$$

where $N_{xzK}(\pm a) = 2h_K\sigma_{xzK}^{in}(\pm a)$.

The partial cases of the model (6), (8) of the form $\mu_y^{inK} = \mu_y^{in}, \mu_x^{inK} = \mu_x^{in}\,(K = \overline{1,M})$ (all layers are the same) or $h_K \to 0\ \ \mu_y^{inK} = \mu_y^{in} \to 0, \mu_x^{inK} = \mu_x^{in} \to 0(K = \overline{1,M})$ (no inclusion or crack) or $\mu_y^{inK} = \mu_y^{in} \to 0\left(G_y^{inK} = G_y^{in} \to \infty\right),\ \mu_x^{inK} = \mu_x^{in} \to \infty(G_x^{inK} = G_x^{in} \to \infty)$ (perfectly rigid homogeneous inclusion) are satisfied and coincide with those known in the literature.

The solution for the matrix as an isotropic bimaterial (external problem) is obtained by the method of the problem of conjugation of analytic functions [51–53]:

$$\begin{array}{l} \sigma_{sz}(x,y) = \sigma_{sz}^0(x,y) + \hat{\sigma}_{sz}(x,y), s = \{x,y\}, \\ w(x,y) = w^0(x,y) + \hat{w}(x,y), \end{array} \tag{11}$$

$$\begin{array}{l} \sigma_{yzk}(z) + i\sigma_{xzk}(z) = \sigma_{yzk}^0(z) + i\sigma_{xzk}^0(z) + ip_k g_3(z) - Cg_6(z) \\ (z \in S_k; r = 3,6; k = 1,2), \\ \sigma_{yzk}^{\pm}(x) = \mp p_k f_3(x) - Cg_6(x) + \sigma_{yz}^{0\pm}(x), \\ \sigma_{xzk}^{\pm}(x) = \mp Cf_6(x) + p_k g_3(x) + \sigma_{xz}^{0\pm}(x), \\ \frac{\partial w}{\partial y}^{\pm}(x) = \mp pf_3(x) - p_{3-k}g_6(x) + \frac{\sigma_{yz}^{0\pm}(x)}{G_k}, \\ \frac{\partial w}{\partial x}^{\pm}(x) = \mp p_{3-k}f_3(x) + pg_6(x) + \frac{\sigma_{xz}^{0\pm}(x)}{G_k}, \end{array} \tag{12}$$

where:

$$g_r(z) \equiv \frac{1}{\pi}\int_{L'} \frac{f_r(x)dx}{x - z},\ s_r(x) \equiv \int_{-a}^{x} f_r(x)dx, C = G_{3-k}p_k,\ p_k = pG_k, p = \frac{1}{G_1 + G_2}.$$

Here the upper indexes "+" and "−" correspond to the limit values of the functions at the upper and lower margins of the line $L$; the values marked with the index "0" on the top characterize the corresponding values in a solid body without modeled heterogeneities under the corresponding external load, and the values marked with the symbol "^" on the top, are the perturbations of the basic stress-strain field by the presence of an inclusion [53].

The following entries [47,49] shall continue to apply:

$$\begin{array}{l} \sigma_{yz}^0(z) + i\sigma_{xz}^0(z) = \tau + i\{\tau_k + D_k(z) + (p_k - p_j)\overline{D}_k(z) + 2p_k D_j(z)\}, \\ D_k(z) = -\frac{Q_k + iG_k b_k}{2\pi(z - \varsigma_{k*})}\ (z \in S_k,\ k = 1,2;\ j = 3-k). \end{array}$$

To connect the external and internal problems, one needs to use contact conditions between the components of the package. There are several variants of contact conditions between the layers and between the package and the matrix:

(1) Ideal (perfect) contact between all constituents of the package:

$$\left\{\begin{array}{l} w^{in(K-1)}(x, y_{K-1} + h_{K-1}) = w^{inK}(x, y_K - h_K)(K = \overline{2,M}), \\ \sigma_{yz}^{in(K-1)}(x, y_{K-1} + h_{K-1}) = \sigma_{yz}^{inK}(x, y_K - h_K)(x \in L'). \end{array}\right. \tag{13}$$

(2) Nonperfect contact with additional tension between layers:

$$\left\{\begin{array}{l} w^{in(K-1)}(x, y_{K-1} + h_{K-1}) = w^{inK}(x, y_K - h_K)(K = \overline{2,M}), \\ \sigma_{yz}^{inK}(x, y_K - h_K) = \sigma_{yz}^{in(K-1)}(x, y_{K-1} + h_{K-1}) - T_K. \end{array}\right. \tag{14}$$

$T_K$ are the surface stresses. When $T_K = 0$ we have the same ideal contact (13).

(3) Contact with friction between the $(K)$-th and $(K-1)$-th layers at the boundary $\{x, y_K \pm h_K\}$ in some area $x \in L_f \subset L'$

$$\sigma_{yz}^{in(K-1)}(x, y_{K-1} + h_{K-1}) = \sigma_{yz}^{inK}(x, y_K - h_K) = -sgn\left[w^{in}\right]_{y_K, h_K} \tau_{yzK}^{max}. \quad (15)$$

$\tau_{yzK}^{max}$ are the limit value of tangential stresses, at which slippage begins. When $\tau_{yzK}^{max}$ there is a smooth contact between these layers.

(4) Ideal contact between the boundary components of the package and the matrix:

$$\begin{cases} w^{in1}(x, y_1 - h_1) = w(x, y_1 - h_1) \\ \sigma_{yz}^{in1}(x, y_1 - h_1) = \sigma_{yz1}(x, y_1 - h_1), \\ w^{inM}(x, y_M + h_M) = w(x, y_M + h_M), \\ \sigma_{yz}^{inK}(x, y_M + h_M) = \sigma_{yz2}(x, y_M + h_M)(x \in L'). \end{cases} \quad (16)$$

(5) Nonperfect contact between the edge components of the package and the matrix:

$$\begin{cases} w^{in1}(x, y_1 - h_1) = w(x, y_1 - h_1) \\ \sigma_{yz}^{in1}(x, y_1 - h_1) = \sigma_{yz1}(x, y_1 - h_1) - T_1, \\ w^{inM}(x, y_M + h_M) = w(x, y_M + h_M), \\ \sigma_{yz}^{inM}(x, y_M + h_M) = \sigma_{yz2}(x, y_M + h_M) + T_{M+1}(x \in L'). \end{cases} \quad (17)$$

In the case $T_1 = T_{M+1} = 0$ we have an ideal contact (16).

(6) Contact with friction between the inclusion and the matrix within $\{x, y_1 - h_1\}$, $\{x, y_M + h_M\}$ in some area $x \in L_f \subset L'$

$$\begin{cases} \sigma_{yz1}(x, y_1 - h_1) = \sigma_{yz}^{in1}(x, y_1 - h_1) = -sgn\left[w^{in}\right]_{y_1, h_1} \tau_{yz1}^{max} \\ \sigma_{yz2}(x, y_M + h_M) = \sigma_{yz}^{inM}(x, y_M + h_M) = -sgn[w]_{y_M, h_M} \tau_{yzM}^{max}\left\{\left(x \in L_f\right)\right. \end{cases} \quad (18)$$

In the case $\tau_{yzK}^{max}$ there is a smooth contact.

One of the conditions (13)–(15) and one of the conditions (16)–(18) must be fulfilled simultaneously.

Using (2) and, for example, (14), we can obtain the expressions for the stresses and strains in the layers through the stress and strain limits for the inclusion package for the presence of interlayer tension:

$$\begin{aligned} \sigma_{yz}^{inK}(x, y_K + h_K) &= \sigma_{yz}^{in1}(x, -h) - \sum_{j=1}^{K} f_{3,j} - \sum_{j=2}^{K} T_j = \sigma_{yz}^{inM}(x, h) + \sum_{j=K+1}^{M} f_{3,j} + \sum_{j=K+1}^{M} T_j, \\ \frac{\partial w^{inK}}{\partial x}(x, y_K + h_K) &= \frac{\partial w^{in1}}{\partial x}(x, -h) - \sum_{j=1}^{K} f_{6,j} = \frac{\partial w^{inM}}{\partial x}(x, h) + \sum_{j=K+1}^{M} f_{6,j}(x \in L'). \end{aligned} \quad (19)$$

Here, the total jumps of the boundary stresses and strains for the inclusion package have the value:

$$\begin{aligned} \sigma_{yz}^{in1}(x, -h) - \sigma_{yz}^{inM}(x, h) &= \sum_{j=1}^{M} f_{3,j} + \sum_{j=2}^{M} T_j, \\ \frac{\partial w^{in1}}{\partial x}(x, -h) - \frac{\partial w^{inM}}{\partial x}(x, h) &= \sum_{j=1}^{M} f_{6,j}(x \in L'). \end{aligned} \quad (20)$$

The resulting limit stresses and strains for the inclusion package (19), the boundary values of the stresses and strains of the matrix (12), and the boundary conditions (13)–(18) form a complete system of singular integral equations (SSIE) for the solution of the problem. Note that the dissimodularity of the inclusion layers in no way affects the peculiarity of the solution of the SSIE of the problem, which allows us to obtain a large variety of effects from manipulating the properties of the layers.

To illustrate the method, let us investigate the longitudinal shear of a structure in the form of a body with a thin two-layer inclusion with layers of thickness $2h_K(K = 1, 2), 2h = 2h_1 + 2h_2$ and orthotropic mechanical properties $G_y^{inK}, G_x^{inK}$, respectively, under the condition of nonperfect mechanical contact with surface tension on the contact surfaces of the structural components under different kinds of loading (Figure 3) when $\varsigma_{k*} = x_{k*} + iy_{k*}(k = 1, 2)$.

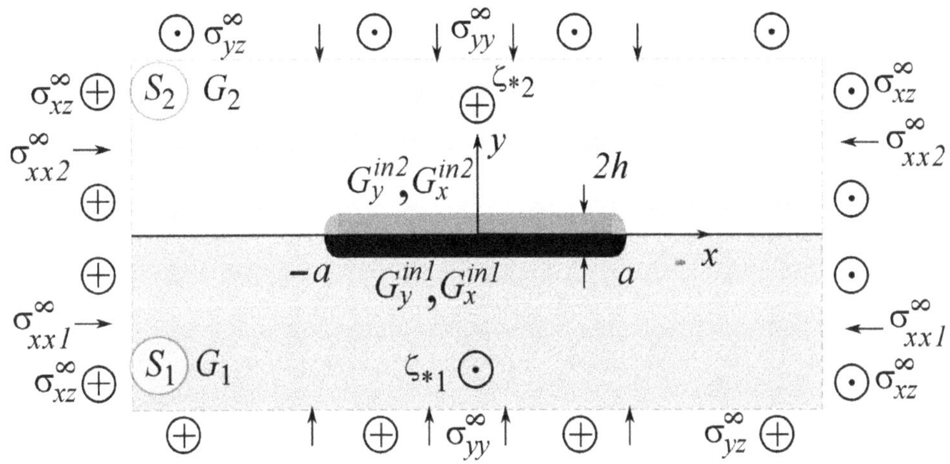


**Figure 3.** Geometry and load pattern of the problem for two-layer different-modularity thin inclusion.

According to (6), (8), the mathematical model for the two-layer inclusion is as follows:

$$\begin{cases} \mu_x^{in1}\left\langle \frac{\partial w^{in1}}{\partial x} \right\rangle_{y_1,h_1}(x) - \int_{-a}^{x} \left[\sigma_{xz}^{in1}\right]_{y_1,h_1}(\xi)d\xi = 2h_1\left(\sigma_{xz}^{in1}(-a) - F_{aver}^{in1}(x)\right), \\ \mu_x^{in2}\left\langle \frac{\partial w^{in2}}{\partial x} \right\rangle_{y_2,h_2}(x) - \int_{-a}^{x} \left[\sigma_{xz}^{in2}\right]_{y_2,h_2}(\xi)d\xi = 2h_2\left(\sigma_{xz}^{in2}(-a) - F_{aver}^{in2}(x)\right), \\ \mu_y^{in1}\left\langle \sigma_{yz}^{in1} \right\rangle_{y_1,h_1} + \int_{-a}^{x} \left[\frac{\partial w^{in1}}{\partial x}\right]_{y_1,h_1} d\xi + \left[w^{in1}\right]_{y_1,h_1}(-a) = 0, \\ \mu_y^{in2}\left\langle \sigma_{yz}^{in2} \right\rangle_{y_2,h_2} + \int_{-a}^{x} \left[\frac{\partial w^{in2}}{\partial x}\right]_{y_2,h_2} d\xi + \left[w^{in2}\right]_{y_2,h_2}(-a) = 0. \end{cases} \quad (21)$$

The boundary conditions between the surfaces of the layers and the tunnel section on $L' = [-a; a]$ take nonperfect with surface tension (14)–(17):

$$\begin{array}{l} \sigma_{yz}^{in1}(x, y_1 - h_1) = \sigma_{yz1}(x, -h) - T_1, \\ \sigma_{yz}^{in2}(x, y_2 - h_2) = \sigma_{yz1}^{in1}(x, y_1 + h_1) - T_2, \\ \sigma_{yz2}(x, h) = \sigma_{yz}^{in2}(x, y_2 + h_2) - T_3, \\ w(x, -h) = w^{in1}(x, y_1 - h_1), \\ w^{in2}(x, y_2 - h_2) = w^{in1}(x, y_1 + h_1), \\ w^{in2}(x, y_2 + h_2) = w(x, h). \end{array} \quad (22)$$

Given the boundary conditions (22), the relations between jumps (1) and (2) take the form:

$$\begin{array}{l} f_3(x) = f_{3,1}(x) + f_{3,2}(x) + T_1 + T_2 + T_3, \\ f_6(x) = f_{6,1}(x) + f_{6,2}(x). \end{array} \quad (23)$$

In addition, the integral representations of the external problem (11), (12), if the tension $T_K$ is not a function of the $Ox$ coordinate, can be written as:

$$\begin{array}{l}\sigma_{yz2}(x,h)=-p_2f_{3,1}(x)-p_2f_{3,2}(x)-Cg_{6,1}(x)-Cg_{6,2}(x)-\\-p_2(T_1+T_2+T_3)+\sigma^{0+}_{yz}(x),\\\sigma_{yz1}(x,-h)=p_1f_{3,1}(x)+p_1f_{3,2}(x)-Cg_{6,1}(x)-Cg_{6,2}(x)+\\+p_1(T_1+T_2+T_3)+\sigma^{0-}_{yz}(x),\\\frac{\partial w}{\partial x}(x,h)=-p_1f_{6,1}(x)-p_1f_{6,2}(x)+pg_{6,1}(x)+pg_{6,2}(x)+\\+\frac{p}{\pi}(T_1+T_2+T_3)ln\left|\frac{1-x}{1+x}\right|+\frac{\sigma^{0+}_{xz}(x)}{G_2},\\\frac{\partial w}{\partial x}(x,-h)=p_2f_{6,1}(x)+p_2f_{6,2}(x)+pg_{6,1}(x)+pg_{6,2}(x)+\\+\frac{p}{\pi}(T_1+T_2+T_3)ln\left|\frac{1-x}{1+x}\right|+\frac{\sigma^{0-}_{xz}(x)}{G_1},\end{array} \quad (24)$$

which agrees well with (23).

Substituting (22)–(24) into (21) considering the expressions:

$$\begin{array}{l}\sigma^{in1}_{yz1}(x,y_1-h_1)=\sigma_{yz1}(x,-h)-T_1,\\\sigma^{in2}_{yz2}(x,y_2+h_2)=\sigma_{yz2}(x,h)+T_3,\\\sigma^{in1}_{yz1}(x,y_1+h_1)=\sigma^{in2}_{yz}(x,y_2-h_2)+T_2=\\=\sigma^{in2}_{yz2}(x,y_2+h_2)+f_{3,2}(x)+T_2=\\=\sigma_{yz2}(x,h)+f_{3,2}(x)+T_2+T_3,\\\sigma^{in2}_{yz2}(x,y_2-h_2)=\sigma^{in1}_{yz1}(x,y_1+h_1)-T_2=\\=\sigma^{in1}_{yz1}(x,y_1-h_1)-f_{3,1}(x)-T_2=\\=\sigma_{yz1}(x,-h)-f_{3,1}(x)-T_2-T_1,\end{array} \quad (25)$$

generates the following kind of two-layer multi-module thin inclusion model in terms of jumps:

$$\left\{\begin{array}{l}(p_2-p_1)f_{6,1}(x)+2p_2f_{6,1}(x)+2pg_{3,1}(x)+2pg_{3,2}(x)-\frac{1}{\mu^{in1}_x}\int_{-a}^{x}f_{3,1}(\xi)d\xi=\\\quad=\frac{2h_1}{\mu^{in1}_x}(\sigma^{in1}_{xz}(-a)-F^{in1}_{aver}(x))-\left\langle\frac{\partial w^0}{\partial x}\right\rangle_h(x)-\{T_1+T_2+T_3\}\frac{2p}{\pi}ln\left|\frac{1-x}{1+x}\right|,\\-2p_1f_{6,1}(x)+(p_2-p_1)f_{6,2}(x)+2pg_{3,1}(x)+2pg_{3,2}(x)-\frac{1}{\mu^{in2}_x}\int_{-a}^{x}f_{3,2}(\xi)d\xi=\\\quad=\frac{2h_2}{\mu^{in2}_x}(\sigma^{in2}_{xz}(-a)-F^{in2}_{aver}(x))-\left\langle\frac{\partial w^0}{\partial x}\right\rangle_h(x)-\{T_1+T_2+T_3\}\frac{2p}{\pi}ln\left|\frac{1-x}{1+x}\right|,\\-(p_2-p_1)f_{3,1}(x)+2p_1f_{3,2}(x)-2Cg_{6,1}(x)-2Cg_{6,2}(x)+\frac{1}{\mu^{in1}_y}\int_{-a}^{x}f_{6,1}(\xi)d\xi=\\\quad=-\frac{1}{\mu^{in1}_y}\left[w^{in1}\right]_{y_1,h_1}(-a)+T_1-T_2-T_3-\left\langle\sigma^0_{yzk}\right\rangle_h(x),\\-2p_2f_{3,1}(x)-(p_2-p_1)f_{3,2}(x)-2Cg_{6,1}(x)-2Cg_{6,2}(x)+\frac{1}{\mu^{in2}_y}\int_{-a}^{x}f_{6,2}(\xi)d\xi=\\\quad=-\frac{1}{\mu^{in2}_y}\left[w^{in2}\right]_{y_2,h_2}(-a)+T_1+T_2-T_3-\left\langle\sigma^0_{yzk}\right\rangle_h(x).\end{array}\right. \quad (26)$$

The resulting SSIE is supplemented by additional balance conditions:

$$\begin{array}{l}\int_{-a}^{a}f_{3,1}(\xi)d\xi=N^{in1}_{xz}(a)-N^{in1}_{xz}(-a)+a(T_1+T_2),\\\int_{-a}^{a}f_{3,2}(\xi)d\xi=N^{in2}_{xz}(a)-N^{in2}_{xz}(-a)+a(T_2+T_3),\end{array} \quad (27)$$

$$\begin{array}{l}\int_{-a}^{a}f_{6,1}(\xi)d\xi=\left[w^{in1}\right]_{y_1,h_1}(a)-\left[w^{in1}\right]_{y_1,h_1}(-a),\\\int_{-a}^{a}f_{6,2}(\xi)d\xi=\left[w^{in2}\right]_{y_2,h_2}(a)-\left[w^{in2}\right]_{y_2,h_2}(-a),\end{array} \quad (28)$$

where $N^{inK}_{xz}(x)=2h_K(\sigma^{inK}_{xz}(-a)-F^{inK}_{aver}(x))(K=1,2)$.

or:

$$\begin{array}{l}\int_{-a}^{a}f_3(\xi)d\xi=N^{in}_{xz}(a)-N^{in}_{xz}(-a)+a(T_1+T_3),\\\int_{-a}^{a}f_{3,1}(\xi)d\xi+\int_{-a}^{a}f_{3,2}(\xi)d\xi+2a(T_1+T_2+T_3)=\\=N^{in1}_{xz}(a)-N^{in1}_{xz}(-a)+a(T_1+T_2)+N^{in2}_{xz}(a)-N^{in2}_{xz}(-a)+a(T_2+T_3),\end{array} \quad (29)$$

$$\int_{-a}^{a} f_{3,1}(\xi)d\xi + \int_{-a}^{a} f_{3,2}(\xi)d\xi = N_{xz}^{in1}(a) - N_{xz}^{in1}(-a) + N_{xz}^{in2}(a) - N_{xz}^{in2}(-a) - a(T_1 + T_3).$$

To preserve the quasi-static equilibrium of the considered microstructure, one should also require the fulfillment of the condition of the balance of surface forces $T_1 + T_3 = 2T_2$.

The resulting system of Equations (26)–(29) is reduced to a system of linear algebraic equations concerning the unknown coefficients of the decomposition of the desired influence functions [51–53] into a series by Jacobi-Chebyshev polynomials, described in [55].

## 4. Numerical Results and Discussion

In fracture mechanics, it is common to use the stress intensity factor (SIF) $K_3$ to describe the asymptotics of the SSS in the vicinity of the crack tip [8,38]. For the case of a thin elastic inclusion, this is not sufficient [56]. The introduction of a system of polar coordinates $(r, \theta)$ with the origin near the right or the left tip of the inclusion $z_1 = \pm rexp(i\theta) \pm a$ (Figure 3), makes it possible to obtain two-term asymptotic expressions for the distribution of the stresses and displacements in the vicinity of the tips $(|z_1| \ll 2a)$ [55] using the generalized stress intensity factors (GSIF) introduced by the expression:

$$K_{31} + iK_{32} = \lim_{r \to \infty(\theta = 0, \pi)} \sqrt{2\pi r}(\sigma_{yz} + i\sigma_{xz}).$$

Consider also the following dimensionless values, marked with a "~" at the top, which significantly reduce the number of calculations without loss in generality:

$$\begin{array}{l} \widetilde{x} = \frac{x}{a,} \widetilde{h}_K = \frac{h_K}{a,} \widetilde{y} = \frac{y}{a}, \widetilde{\tau}_k = \frac{\tau_k}{G_{gav}}, \widetilde{\tau} = \frac{\tau}{G_{gav}}, \\ \widetilde{G}_x^{inK}(\widetilde{x}) = \frac{G_x^{inK}(x)}{G_{gav},} \widetilde{G}_y^{inK}(\widetilde{x}) = \frac{G_y^{inK}(x)}{G_{gav}}, \\ \widetilde{G}_k = \frac{G_k}{G_{gav}(k=1,2),} \widetilde{p}_k = p_k, \widetilde{C} = \frac{C}{G_{gav},} \\ \widetilde{\sigma}_{xz}(\widetilde{x}) = \frac{\widetilde{\sigma}_{xz}(x)}{G_{gav},} \widetilde{\sigma}_{yz}(\widetilde{x}) = \frac{\widetilde{\sigma}_{yz}(x)}{G_{gav}}, G_{gav} = \sqrt{G_1 G_2}. \end{array} \quad (30)$$

$$\widetilde{K}_{31} = \frac{K_{31}^{+}}{2\widetilde{C}G_{gav}\sqrt{\pi a}}, \widetilde{K}_{32} = \frac{K_{32}^{+}}{2\widetilde{p}_2 G_{gav}\sqrt{\pi a}}, \quad (31)$$

where $K_{31}^{+}$, $K_{32}^{+}$—GSIF's near the tip of inclusion $x = +a$.

The use of dimensionless values will make it possible to interpret the obtained quantitative results and qualitative conclusions on any variant of specific materials of inclusion layers or matrix by simple recalculation due to the universality of the mathematical model of a thin deformable inclusion and the method of problem-solving. The investigation of the influence of the inclusion layers different modularity, external force factors at non-ideal contact with the surface tension of the structural components on the unmeasured stress-strain field parameters on the inclusion surfaces, and the dimensionless stress intensity factor $\widetilde{K}_{31}$ are illustrated in Figures 4–16. Figures 4–7 shows the results of a study of the stress distributions on the contact surfaces and displacement jumps on the inclusion as a function of the degree of dissimilarity of the inclusion layers under different external loads (Figures 4 and 5 illustrate the effect of a far-field uniform shear loading, and Figures 6 and 7 illustrate the effect of a concentrated force on similar structures) and in the absence of surface forces.

When one of the layers is significantly softer than the matrix, the effect of "unloading" (stress level reduction) of the surfaces is observed irrespective of the stiffness of the second layer. And this effect is more local than the loading by the concentrated forces located in the points $\varsigma_{k*} = x_{k*} + iy_{k*}; x_{k*} = 0, y_{2*} = -y_{1*} = d$ of order $d/a \approx O(1)$. Figures 4 and 5 reflect the known fact that the stress variation on most of the inclusion surfaces is small and changes abruptly as they approach the tips. In contrast, the applied near the inclusion concentrated force (Figures 6 and 7) essentially perturbs the character of stress distribution along the inclusion axis, its maximum value for such a loading is reached on the geometric symmetry axis of the problem. With the removal of the point of force application (increase in *d*) the character of the stresses changes approaches the characteristic of a far-field uniform shear loading (Figures 4 and 5). Figure 7d illustrates the proportionality of the displacement jumps of each layer to their stiffness.

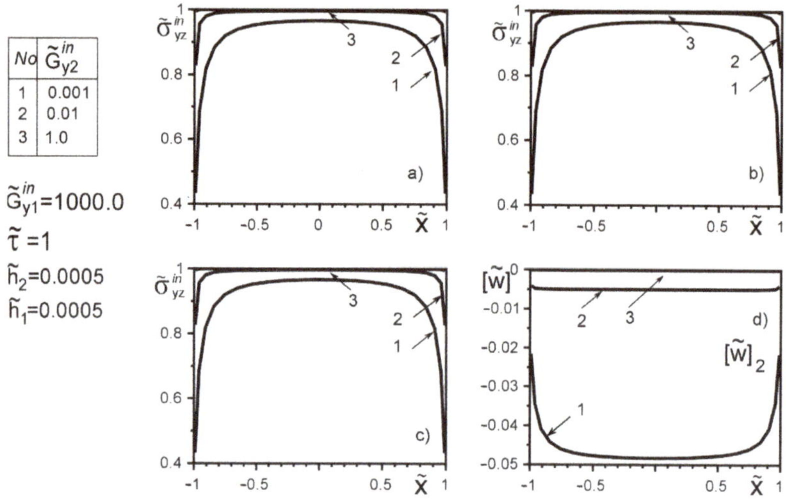


**Figure 4.** Stress distribution along with the upper interface (layer 2 of the inclusion—matrix half-space $S_2$) (**a**); the boundary between layers (layer 1–layer 2) (**b**); lower interface (layer 1—matrix half-space $S_1$ ) (**c**), and the displacement jump on the inclusion (**d**) for a layer 1 stiffer than the matrix as a function of the change in stiffness of layer 2 under the load uniformly distributed at infinity.

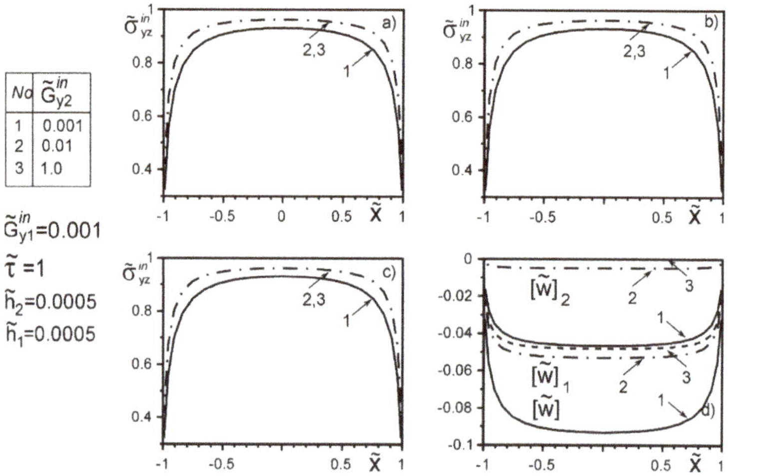


**Figure 5.** Stress distribution along with the upper interface (layer 2 of the inclusion—matrix half-space $S_2$) (**a**); the boundary between layers (layer 1–layer 2) (**b**); lower interface (layer 1—matrix half-space $S_1$ ) (**c**), and the displacement jump on the inclusion (**d**) for a layer 1 softer than the matrix as a function of the change in stiffness of layer 2 under the load uniformly distributed at infinity (1—result for the case of the same layer materials, verified by comparison with [50,53]).

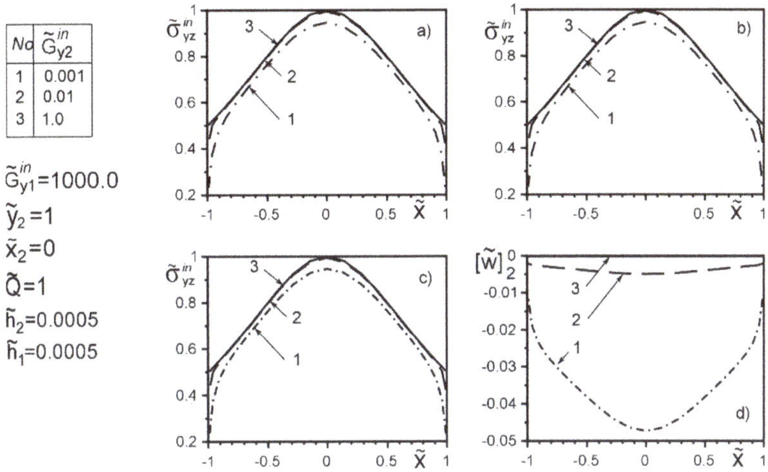


**Figure 6.** Stress distribution along with the upper interface (layer 2 of the inclusion–matrix half-space $S_2$) (**a**); the boundary between layers (layer 1–layer 2) (**b**); lower interface (layer 1—matrix half-space $S_1$ ) (**c**), and the displacement jump on the inclusion (**d**) for a layer 1 stiffer than the matrix as a function of the change in stiffness of layer 2 under the load by concentrated forces at points $\varsigma_{k*} = x_{k*} + iy_{k*}; x_{k*} = 0, y_{2*} = -y_{1*} = d$.

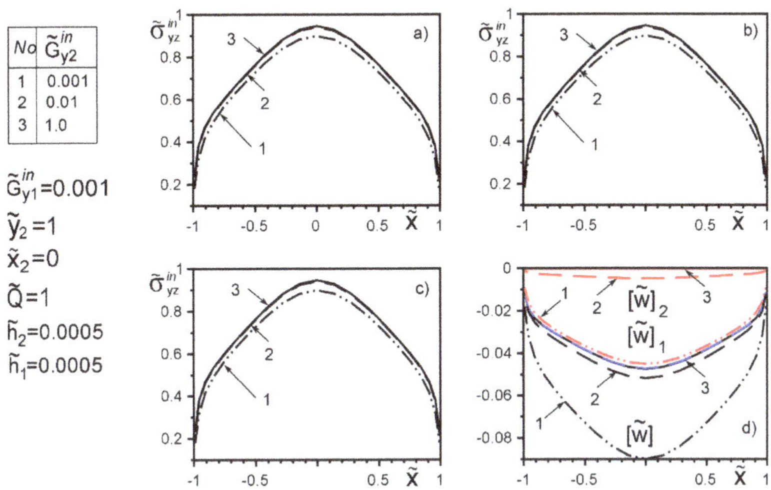


**Figure 7.** Stress distribution along with the upper interface (layer 2 of the inclusion–matrix half-space $S_2$) (**a**); the boundary between layers (layer 1–layer 2) (**b**); lower interface (layer 1—matrix half-space $S_1$ ) (**c**), and the displacement jump on the inclusion (**d**) for a layer 1 softer than the matrix as a function of the change in stiffness of layer 2 under the load by concentrated forces at points $\varsigma_{k*} = x_{k*} + iy_{k*}; x_{k*} = 0, y_{2*} = -y_{1*} = d$ (1—result for the case of the same layer materials, verified by comparison with [50,53]).

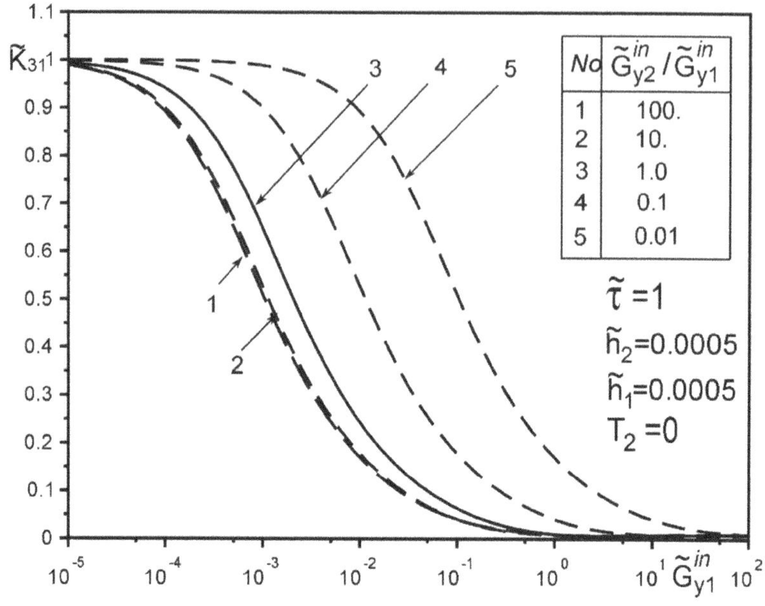


**Figure 8.** Influence of the level of dissimodularity on the GSIF $\tilde{K}_{31}$ under the load by uniformly distributed on infinity stress and absence of surface tension.

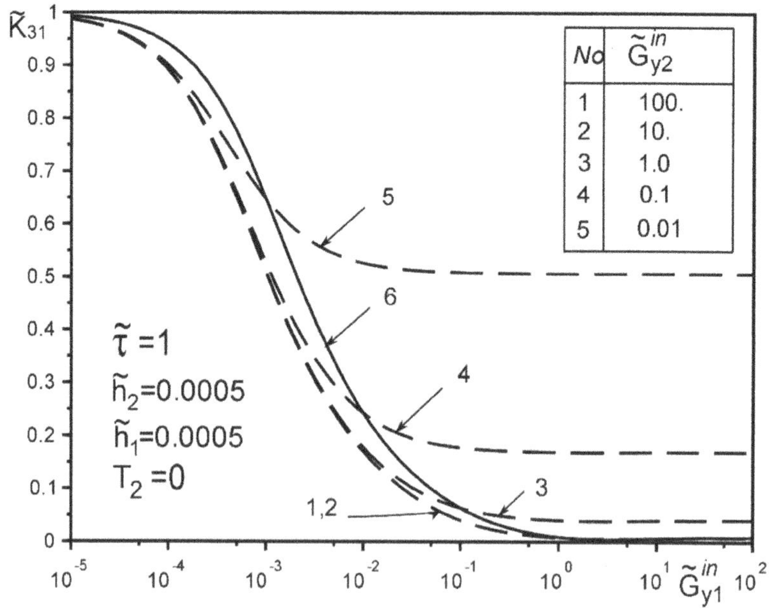


**Figure 9.** Influence of the level of dissimodularity on the GSIF $\tilde{K}_{31}$ under the load by uniformly distributed on infinity stress and absence of surface tension.

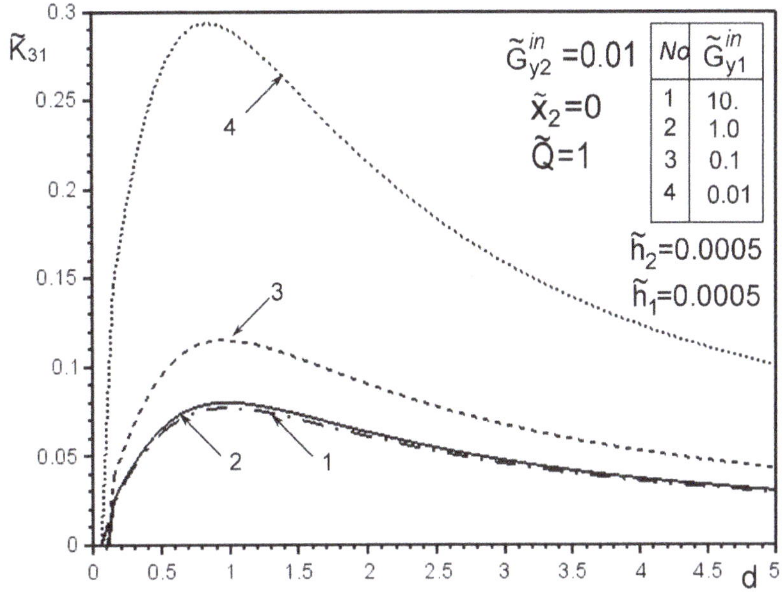


**Figure 10.** The effect of changing the distance $d$ of the point of application of concentrated forces from the inclusion and the level of dissimodularity to the GSIF $\tilde{K}_{31}$ in the absence of surface tension and layer 1 softer from the matrix (4—result for the case of the same layer materials, verified by comparison with [50,53]).

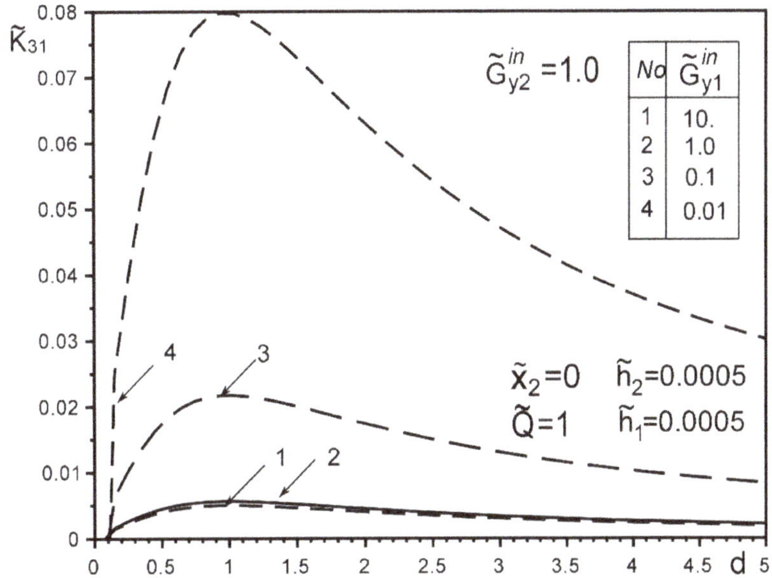


**Figure 11.** The effect of changing the distance $d$ of the point of application of concentrated force from the inclusion and the level of dissimodularity to the GSIF $\tilde{K}_{31}$ in the absence of surface tension and equivalent to the matrix material layer 1 (2—result for the case of the same layer materials, verified by comparison with [50,53]).

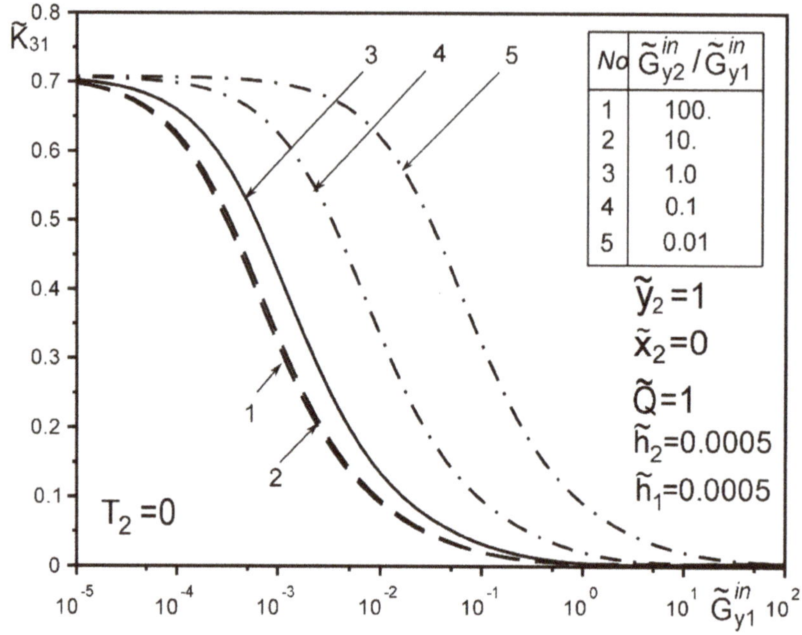


**Figure 12.** Influence of the level of dissimodularity of the inclusion layer materials on the GSIF $\tilde{K}_{31}$ in the absence of surface tension (3—result for the case of the same layer materials, verified by comparison with [50,53]).

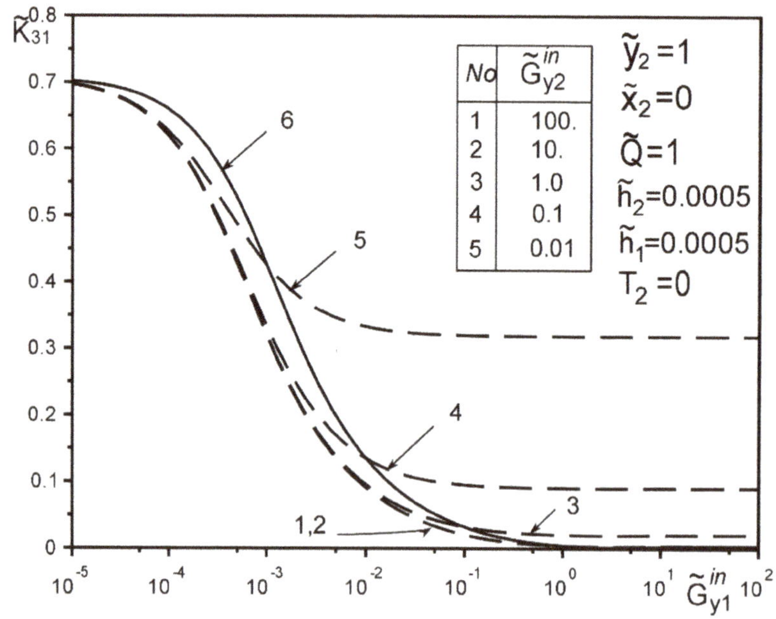


**Figure 13.** The effect of changing the level of dissimodularity of the inclusion layer materials on the GSIF $\tilde{K}_{31}$ in the absence of surface tension under a concentrated force loading.

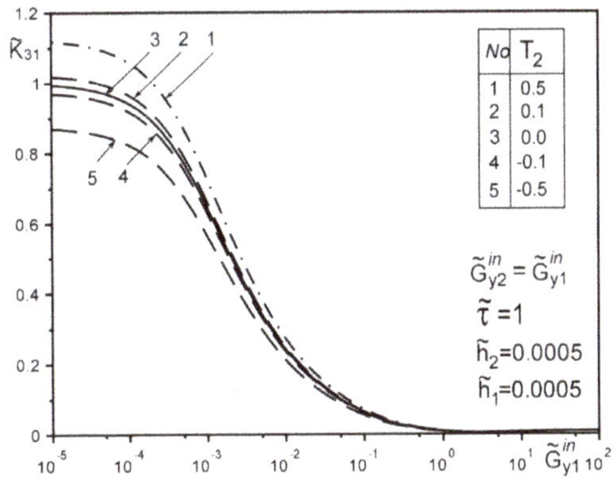


**Figure 14.** Influence of interlayer surface tension on the GSIF $\tilde{K}_{31}$ for the same inclusion layer materials when loaded by a uniformly distributed stress at infinity (3—result for the case of the same layer materials, verified by comparison with [50,53]).

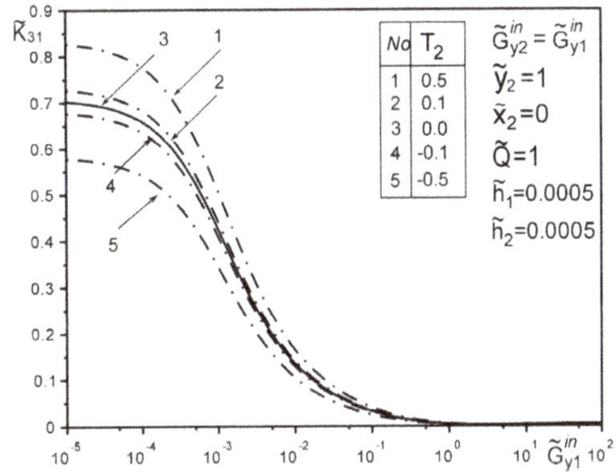


**Figure 15.** Influence of the interlayer surface tension for the same materials of the inclusion layers when loading with a concentrated force at the point $\varsigma_{2*}$.

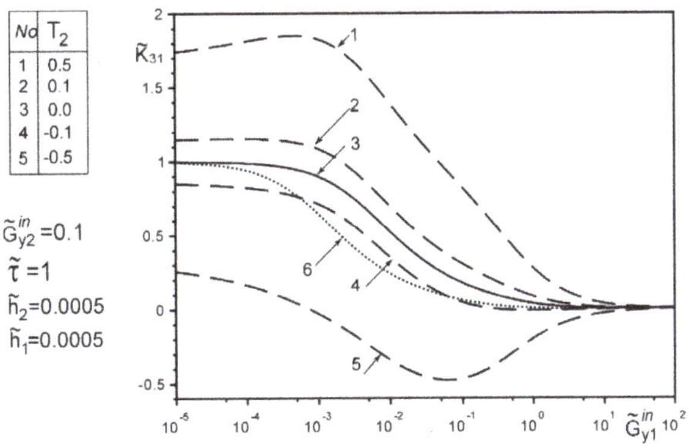


**Figure 16.** The effect of the interlayer surface tension on the GSIF $\tilde{K}_{31}$ when layer 2 is softer than the matrix, and the layer 1 is of the arbitrary material under the loading by stress uniformly distributed at infinity; (6—result for the case of the same layer materials verified by comparison with [50,53])).

Figures 8–13 show the results of the study of the effect of the level of dissimodularity on the GSIF $\widetilde{K}_{31}$ under different external loading and in the absence of surface forces. It is noteworthy that the increase in the level of dissimodularity of the materials of the inclusion layers significantly affects the GSIF when the stiffness of one of the layers is greater than that of the matrix (Figures 8, 9, 12 and 13) regardless of the type of loading.

Figures 10 and 11 confirms the known effect of the GSIF $\widetilde{K}_{31}$ maximum under the loading by a concentrated forces placed at a distance approximately $d \approx a$ from the inclusion axis, irrespective of the stiffness of the materials of the layers. However, it is more pronounced in the stiffness range of the materials of the layers softer than the matrix material. Moreover, if the material of one of the layers is equivalent to that of the matrix (Figure 11), we obtain the known results for a homogeneous elastic inclusion at the interface of the matrix materials [50–53].

Figures 14–16 show the results of the study of the influence of the level of dissimodularity of the inclusion layers and the presence of surface forces under different external loads on the GSIF $\widetilde{K}_{31}$. It was found that the presence of surface forces leads to an increase in the SIF if they are directed toward the external load, and a decrease if they are directed in the opposite direction from the external load (Figures 14 and 15). The dissimodularity of the materials of the layers significantly distorts this effect, which is especially noticeable when one of the layers is significantly softer than the matrix material. It is revealed that there are certain combinations of external load parameters, surface forces, and material properties of the layers, at which there are local SIF extremes. This effect can be useful in designing the modes of operation of structures with such a structure.

The solution method and results obtained for the two-layer inclusion have been verified by the coincidence of the numerical results with those known in literature [47,50–53] for a homogeneous thin elastic inclusion—the curves 1 in Figures 5 and 7; 3 in Figure 8; 4 in Figure 10; 2 in Figure 11; 3 in Figure 12; 3 in Figures 14 and 15; 6 in Figure 16.

## 5. Conclusions

A mathematical model of a thin multilayer inclusion of finite length with orthotropic properties of the layers is constructed, taking into account the effect of surface energy on their interfaces. On its basis, we derive a system of equations for solving the problems of antiplane deformation of a bimaterial with thin multilayer interfacial linearly elastic inclusions under arbitrary force and dislocation loading in the case where the inclusion-matrix contact may be ideal or with surface tension or sliding (smooth or frictional).

It is found that for the corresponding problems with bilayer inclusions:

(1) The growth of the level of dissimodularity of the materials of the inclusion layers significantly affects the SIF $K_{31}$ when the stiffness of one of the layers is greater than the stiffness of the matrix, regardless of the type of loading. The effect of localization of the maximum SIF $K_{31}$, when loaded by a concentrated force located at a distance approximately $d \approx a$ from the inclusion axis, is confirmed irrespective of the stiffness of the materials of the layers. However, it is more pronounced in the stiffness range of the materials of the layers softer than the matrix material. Moreover, if the material of one of the layers is equivalent to that of the matrix, then the known results for a homogeneous elastic inclusion (the second layer) at the material interface are obtained.

(2) The presence of surface forces leads to an increasing SIF if they are co-directed with the external load, and a decrease otherwise. The different modularity of the materials of the layers qualitatively changes this phenomenon, which is especially noticeable when one of the layers is significantly softer than the matrix material.

(3) There are certain combinations of external load parameters, surface forces, and material properties of the layers at which there are local SIF extremes.

All these conclusions may be useful for the design of a layered inclusion and the modes of operation of such structures. The proposed method has proved to be effective for solving a whole class of strain problems for bodies with thin deformed inclusions of finite length and may be used for the calculation of FGM inclusions. The addition of the proposed method by the homogenization method [58] will give an obvious opportunity to solve the problem of thin deformable heterogeneous inclusions in periodically layered composites.

**Author Contributions:** Conceptualization, Y.P., H.S., V.H.; methodology, Y.P., H.S.; software, Y.P., L.P.; validation, Y.P., H.S.; formal analysis, Y.P.; investigation, Y.P., H.S; resources, V.H.; data curation, Y.P.; writing—original draft preparation, Y.P.; writing—review and editing, Y.P.; visualization, Y.P, L.P.; supervision, H.S.; project administration, V.H.; funding acquisition, V.H. All authors have read and agreed to the published version of the manuscript.

**Funding:** This research received no external funding.

**Institutional Review Board Statement:** Not applicable.

**Informed Consent Statement:** Not applicable.

**Data Availability Statement:** The data presented in this study are openly available at https://doi.org/10.3390/ma14174928, reference number [51], https://doi.10.2478/ama-2018-0029, reference number [52].

**Conflicts of Interest:** The authors declare no conflict of interest.

## Nomenclature

| | |
|---|---|
| FGM | functionally graded material; |
| SIF | stress intensity factor; |
| SSIE | system of singular integral equations; |
| SSS | stress-strain state |
| $x, y, z$ | Cartesian coordinates; |
| $f_r$ | jump functions; |
| $E_k, \nu_k,\ G_k, G_x^{inK}, G_y^{inK}$ | elastic properties of the material; |
| $S_k$ | half-planes (sections of the body); |
| $a, h, h_K, y_K$ | dimensions of the inclusion layers; |
| $w, \sigma_{xz}, \sigma_{yz}, \sigma_{xx}, \sigma_{yy}$ | displacement, stresses (components of SSS); |
| $L' = [-a;\ a]$ | line, modeling the presence of thin inclusion; |
| $Q_k, b_k$ | magnitudes of concentrated forces and screw dislocations; |
| $\sigma_{yz}^{\infty}, \sigma_{xzk}^{\infty}, \sigma_{yy}^{\infty}, \sigma_{xxk}^{\infty}$ | uniformly distributed in infinity shear stresses; |

**Special denotations**

$[\bullet]_{y,h} = \bullet(x, y-h) - \bullet(x, y+h),\ \langle\bullet\rangle_{y,h} = \bullet(x, y-h) + \bullet(x, y+h);$

| | |
|---|---|
| superscripts "+" and "−" | denotes boundary values of functions on the upper and the lower to width inclusion borders accordingly; |
| superscript "in" | marks the values corresponding to inclusion; |
| superscript "○" | marks the values in the corresponding problem without any inclusion; |
| superscript "~" | marks the terms that become dimensionless; |
| subscript "k" | denotes the terms corresponding to half-plains. |

## References

1. Mura, T. *Micromechanics of Defects in Solids*, 2nd ed.; Martinus Nijhoff: Dordrecht, The Netherlands, 1987; 601p.
2. Paulo Davim, J.; Constantinos, A. (Eds.) *Nanocomposites: Materials, Manufacturing and Engineering*; Charitidis, Walter de Gruyter GmbH: Berlin, Germany; Boston, MA, USA, 2013; 211p.
3. Wang, Y.; Huang, Z.M. Analytical micromechanics models for elastoplastic behavior of long fibrous composites: A critical review and comparative study. *Materials* **2018**, *11*, 1919. [CrossRef] [PubMed]
4. Zhou, K.; Hoh, H.J.; Wang, X.; Keer, L.M.; Pang, J.H.; Song, B.; Wang, Q.J. A review of recent works on inclusions. *Mech. Mater.* **2013**, *60*, 144–158. [CrossRef]
5. Mencik, J. *Mechanics of Components with Treated or Coated Solids*; Kluwer Academic Publishing: Dordrecht, The Netherlands, 1996; 360p.
6. Nemat-Nasser, S.; Hori, M. *Micromechanics: Overall Properties of Heterogeneous Materials*; North-Holland Series in Applied Mathematics and Mechanics; Elsevier: Amsterdam, The Netherlands, 1993; Volume 37, 687p.
7. Williams, J.C. *Doctor-Blade Process, in Treatise on Materials Science and Technology*; Academic Press: New York, NY, USA, 1976; pp. 173–198.

8. Chen, J. Determination of thermal stress intensity factors for an interface crack in a graded orthotropic coating–substrate structure. *Int. J. Fract.* **2005**, *133*, 303–328. [CrossRef]
9. Chen, X.; Liu, Q. Thermal stress analysis of multi-layer thin films and coatings by an advanced boundary element method. *Comput. Model. Eng. Sci.* **2001**, *2*, 337–349.
10. Elperin, T.; Rudin, G. Thermal stresses in a coating–substrate assembly caused by internal heat source. *J. Therm. Stresses* **2016**, *39*, 90–102. [CrossRef]
11. Ding, S.H.; Li, X. Thermal stress intensity factors for an interface crack in a functionally graded layered structures. *Arch. Appl. Mech.* **2011**, *81*, 943–955. [CrossRef]
12. Naik, R.A. Simplified micromechanical equations for thermal residual stress analysis of coated fiber composites. *J. Compos. Technol. Res.* **1992**, *14*, 182–186.
13. Perkowski, D.M.; Kulchytsky-Zhyhailo, R.; Kołodziejczyk, W. On axisymmetric heat conduction problem for multilayer graded coated half-space. *J. Theor. Appl. Mech.* **2018**, *56*, 147–156. [CrossRef]
14. Shevchuk, V.A. Modeling and computation of heat transfer in a system "body–multilayer coating". *Heat Transf. Res.* **2006**, *37*, 412–423. [CrossRef]
15. Ichikawa, K. (Ed.) *Functionally Graded Materials in the 21ST Century, A Workshop on Trends and Forecasts*; Kluver Academic Publishers: Boston, MA, USA; Dordrecht, The Netherlands; London, UK, 2000; 264p.
16. Kashtalyan, M.; Menshykova, M. Three-dimensional analysis of a functionally graded coating/substrate system of finite thickness. *Phil. Trans. R. Soc. A Math. Phys. Eng. Sci.* **2008**, *366*, 1821–1826. [CrossRef]
17. Koizumi, M. The concept of FGM. *Ceram. Trans. Funct. Graded Mater.* **1993**, *34*, 3–10, NII Article ID (NAID):10014680673.
18. Miyamoto, Y.; Kaysser, W.A.; Rabin, B.H.; Kawasaki, A.; Ford, R.G. Functionally graded materials: Design, processing and applications. *Mater. Technol. Ser.* **1999**, *5*, 247–313. [CrossRef]
19. Zhang, N.; Khan, T.; Guo, H.; Shi, S.; Zhong, W.; Zhang, W. Functionally graded materials/an overview of stability, buckling, and free vibration analysis. *Adv. Mater. Sci. Eng.* **2019**, *2019*, 1354150. [CrossRef]
20. Li, Y.; Feng, Z.; Hao, L.; Huang, L.; Xin, C.; Wang, Y.; Bilotti, E.; Essa, K.; Zhang, H.; Li, Z.; et al. A Review on Functionally Graded Materials and Structures via Additive Manufacturing: From Multi-Scale Design to Versatile Functional Properties. *Adv. Mater. Technol.* **2020**, *5*, 1900981. [CrossRef]
21. Bishop, A.; Lin, C.Y.; Navaratnam, M.; Rawlings, R.D.; McShane, H.B. A functionally gradient material produced by a powder metallurgical process. *J. Mater. Sci. Lett.* **1993**, *12*, 1516–1518. [CrossRef]
22. Boch, P.; Chartier, T.; Huttepain, M. Tape casting of $Al_2O_3/ZrO_2$ laminated Composites. *J. Am. Ceram. Soc.* **1987**, *69*, 191–192. [CrossRef]
23. Kieback, B.; Neubrand, A.; Riedel, H. Processing techniques for functionally graded materials. *Mater. Sci. Eng. A.* **2003**, *362*, 81–105. [CrossRef]
24. Mistler, R.E. High strength alumina substrates produced by a multiple-layer casting technique. *Am. Ceram. Soc. Bull.* **1973**, *52*, 850–854.
25. Takahashi, M.; Itoh, Y.; Kashiwaya, H. Fabrication and evaluation of W/Cu gradient material by sintering and infiltration technique. In *Proceeding of the First International Symposium on Functionally Gradient Materials FGM*; Yamanouchi, M., Ed.; Functionally Gradient Materials Forum, The Society of Non-Traditional Technology: Tokyo, Japan, 1990; pp. 129–134.
26. Uchida, Y. Properties of functionally graded materials, Manufactured by progressive lamination method for applications. *Aichi Inst. Technol. Res. Rep.* **2004**, *39–B*, 39–51.
27. Jin, Z.H. An asymptotic solution of temperature field in a strip of a functionally graded material. *Int. Commun. Heat Mass Transf.* **2002**, *29*, 887–895. [CrossRef]
28. Lee, Y.-D.; Erdogan, F. Interface cracking of FGM coatings under steady-state heat flow. *Eng. Fract. Mech.* **1998**, *59*, 361–380. [CrossRef]
29. Wang, X.; Sudak, L.J. Three-Dimensional analysis of multi-layered functionally graded anisotropic cylindrical panel under thermomechanical loading. *Mech. Mater.* **2008**, *40*, 235–254. [CrossRef]
30. Yevtushenko, A.A.; Rozniakowska, M.; Kuciej, M. Transient temperature processes in composite strip and homogeneous foundation. *Int. Commun. Heat Mass Transf.* **2007**, *34*, 1108–1118. [CrossRef]
31. Yildirim, B.; Dag, S.; Erdogan, F. Three-Dimensional fracture analysis of FGM coatings under thermomechanical loading. *Int. J. Fract.* **2005**, *132*, 369–395. [CrossRef]
32. Erdogan, F.; Wu, B.H. Crack problem in FGM layers under thermal stresses. *J. Therm. Stresses* **1996**, *19*, 237–265. [CrossRef]
33. Guo, L.C.; Noda, N.; Ishihara, M. Thermal stress intensity factors for a normal surface crack in a functionally graded coating structure. *J. Therm. Stresses* **2007**, *31*, 149–164. [CrossRef]
34. Hsueh, C.H. Thermal stresses in elastic multilayer systems. *Thin Solid Film.* **2002**, *418*, 182–188. [CrossRef]
35. Zhuo, X.R.; Beom, H.G. Interface crack between a thin film and an orthotropic substrate under uniform heat flow. *Arch. Appl. Mech.* **2016**, *86*, 1019–1036. [CrossRef]
36. Bao, G.; Wang, L. Multiple cracking in functionally graded ceramic/metal coatings. *Int. J. Solids Struct.* **1995**, *32*, 2853–2871. [CrossRef]
37. Moya, J.S.; Sanchez-Herencia, A.J.; Requena, J.; Moreno, R. Functionally gradient ceramics by sequential slip casting. *Mater. Lett.* **1992**, *14*, 333–335. [CrossRef]

38. El-Borgi, S.; Erdogan, F.; Ben Hatira, F. Stress intensity factors for an interface crack between a functionally graded coating and a homogeneous substrate. *Int. J. Fract.* **2003**, *123*, 139–162. [CrossRef]
39. Wang, B.L.; Mai, Y.W.; Noda, N. Fracture mechanics analysis model for functionally graded materials with arbitrarily distributed properties. *Int. J. Fract.* **2002**, *116*, 161–177. [CrossRef]
40. Zhao, J.; Silberschmidt, V.V. Microstructure-Based damage and fracture modelling of alumina coatings. *Comp. Mat. Sci.* **2005**, *32*, 620–628. [CrossRef]
41. Benveniste, Y.; Baum, G. An interface model of a graded three-dimensional anisotropic curved interphase. *Proc. R. Soc. A.* **2007**, *463*, 419–434. [CrossRef]
42. Chen, Y.Z. Study of multiply-layered cylinders made of functionally graded materials using the transfer matrix method. *J. Mech. Mater. Struct.* **2011**, *6*, 641–657. [CrossRef]
43. Gurtin, M.E.; Murdoch, A.I. A continuum theory of elastic material surfaces. *Arch. Ration. Mech. Anal.* **1975**, *57*, 291–323. [CrossRef]
44. Hashin, Z. Thin interphase/imperfect interface in elasticity with application to coated fiber composites. *J. Mech. Phys. Solids* **2002**, *50*, 2509–2537. [CrossRef]
45. Kim, C.I.; Schiavone, P.; Ru, C.-Q. The effects of surface elasticity on an elastic solid with Mode-III crack: Complete solution. *Trans. ASME J. Appl. Mech.* **2010**, *77*, 021011. [CrossRef]
46. Kulchytsky-Zhyhailo, R.; Matysiak, S.J.; Bajkowski, A.S. Semi-Analytical solution of three-dimensional thermoelastic problem for half-space with gradient coating. *J. Therm. Stresses* **2018**, *41*, 1169–1181. [CrossRef]
47. Pasternak, I.M.; Pasternak, R.M.; Sulym, H.T. 2D boundary element analysis of defective thermoelectroelastic bimaterial with thermally imperfect but mechanically and electrically perfect interface. *Eng. Anal. Bound. Elem.* **2015**, *61*, 94–206. [CrossRef]
48. Pasternak, I.M.; Sulym, H.T.; Ilchuk, N.I. Interaction of physicomechanical fields in bodies with thin structural inhomogeneities: A survey. *J. Math. Sci.* **2021**, *253*, 63–83. [CrossRef]
49. Peng, X.-L.; Lee, X.-F. Thermoelastic analysis of functionally graded annulus with arbitrary gradient. *Appl. Math. Mech.* **2009**, *30*, 1211–1220. [CrossRef]
50. Piskozub, I.Z.; Sulym, H.T. Nonlinear deformation of a thin interface inclusion. *Mater. Sci.* **2018**, *53*, 600–608. [CrossRef]
51. Piskozub, Y.; Sulym, H. Effect of frictional slipping on the strength of ribbon-reinforced composite. *Materials* **2021**, *14*, 4928. [CrossRef]
52. Sulym, H.; Piskozub, Y.; Polanski, J. Antiplane deformation of a bimaterial with thin interfacial nonlinear elastic inclusion. *Acta Mech. Autom.* **2018**, *12*, 190–195. [CrossRef]
53. Sulym, H.T. *Bases of Mathematical Theory of Thermo-Elastic Equilibrium of Solids Containing Thin Inclusions*; Research and Publishing Center of NTSh: L'viv, Ukraine, 2007; 716p, (In Ukrainian). Available online: https://ua1lib.org/book/665574/5c937e (accessed on 19 January 2022).
54. Zhong, Z.; Yu, T. Analytical solution of cantilever functionally graded beam. *Compos. Sci. Technol.* **2007**, *67*, 481–488. [CrossRef]
55. Piskozub, I.Z.; Sulym, H.T. Asymptotics of stresses in the vicinity of a thin elastic interphase inclusion. *Mater. Sci.* **1996**, *32*, 421–432. [CrossRef]
56. Popina, S.Y.; Sulim, G.T. The limiting load for a brittle body with a thin-walled elastic inclusion. *Sov. Mater. Sci.* **1987**, *23*, 219–222. [CrossRef]
57. Sulym, H.T.; Popina, S.Y. Strength of a body with stochastic distribution of thin defects under the conditions of antiplane deformation. *Mater. Sci.* **1997**, *33*, 116–120. [CrossRef]
58. Kaczynski, A.; Matysiak, S.J. Stress singularities in a periodically layered composite with a transverse rigid line inclusion. *Arch. Appl. Mech.* **2010**, *80*, 271–283. [CrossRef]

materials

MDPI

*Article*

# Strain-Controlled Fatigue Behavior and Microevolution of 316L Stainless Steel under Cyclic Shear Path

**Xinna Liu [1], Shuai Zhang [2,*], Yanmei Bao [2], Zhongran Zhang [3,*] and Zhenming Yue [2]**

1 Rongcheng Campus, Harbin University of Science and Technology, Rongcheng 264300, China
2 School of Mechanical, Electrical and Information Engineering, Shandong University at Weihai, Weihai 264209, China
3 Naval Architecture and Port Engineering College, Shandong Jiaotong University, Weihai 264310, China
* Correspondence: 202037591@mail.sdu.edu.cn (S.Z.); zzhongran@163.com (Z.Z.)

**Abstract:** Based on the twin bridge shear specimen, the cyclic shear experiments were performed on 1.2 mm thin plates of 316L metastable austenitic stainless steel with different strain amplitudes from 1 to 5% at ambient temperature. The fatigue behavior of 316L stainless steel under the cyclic shear path was studied, and the microscopic evolution of the material was analyzed. The results show that the cyclic stress response of 316L stainless steel exhibited cyclic hardening, saturation and cyclic softening, and the fatigue life is negatively correlated with the strain amplitude. The microstructure was analyzed by using electron back-scattered diffraction (EBSD). It was found that grain refinement and martensitic transformation during the deformation process led to rapid crack expansion and reduced the fatigue life of 316L.

**Keywords:** cyclic shear; strain amplitude; cyclic response; martensitic transformation

**Citation:** Liu, X.; Zhang, S.; Bao, Y.; Zhang, Z.; Yue, Z. Strain-Controlled Fatigue Behavior and Microevolution of 316L Stainless Steel under Cyclic Shear Path. *Materials* **2022**, *15*, 5362. https://doi.org/10.3390/ma15155362

Academic Editors: Grzegorz Lesiuk and Dariusz Rozumek

Received: 9 July 2022
Accepted: 24 July 2022
Published: 4 August 2022



## 1. Introduction

With excellent strength, corrosion resistance, heat resistance, ductility and machinability, 316L stainless steel has been widely used in industrial applications [1–3]. The steel is usually made into parts that are subjected to certain cyclic loads at low, ambient or high temperatures, which will lead to fatigue failure of the parts [4]. Therefore, in order to improve the service life of materials under cyclic loading conditions, many scholars have continued to conduct in-depth studies on the microstructure and deformation mechanism of materials under cyclic loading. Pham et al. [5] conducted axial fatigue loading experiments on AISI 316L at room temperature, and found that the cyclic deformation response can be divided into three stages: cyclic hardening, followed by cyclic softening, and, finally, cyclic saturation before failure. In total, 316L stainless steel exhibits a cyclic hardening softening response in strain-controlled cyclic torsion [6]. The cyclic stress behavior of 316LN is determined mostly by the internal stress under non-proportional loading, and the initial cyclic hardening is greatly enhanced by the increase in loading non-proportionality [7]. Facheris et al. [8] carried out a set of uniaxial, torsional and multiaxial low-cycle fatigue and strain-controlled ratcheting tests. The effect of ratcheting on the mechanical response was found to be quantitatively stronger, causing a more pronounced drop of fatigue life. Zhou et al. [9] developed a cyclic constitutive model implemented to describe the cyclic behaviour of 316L including the hardening/softening and the strain range memory effect. During the uniaxial ratchetting deformation of 316L stainless steel, the dislocation density increases progressively, and twinning and strain-induced martensitic phases do not occur within the specified number of cycles [10]. Observations of the life-terminated dislocation arrangements by transmission electron microscopy showed that the dislocation microstructure depends essentially on the plastic strain amplitude, which in turn is strongly correlated to the stress amplitude and average stress [11]. It is believed that the austenite phase in some stainless steels is in a metastable state at ambient temperature due to the amount of Cr and Ni. In the process of monotonic or cyclic deformation, the deformation of the metastable austenite leads to the phase transformation from initial fcc austenite ($\gamma$) to final stable martensite($\alpha'$) [12–15]. Li et al. [16] performed a series of cyclic tests on 304L stainless steel with different loading paths, which showed that the growth rate of martensite content became higher and its distribution in the austenitic matrix was more uniform as the loading path increased non-proportionally. The sensitivity of the martensite induced by deformation depends on the chemical composition, temperature, degree of plastic deformation, and strain rate [17,18]. At the same time, strain-induced martensitic phase transformation can affect the ductility of austenitic stainless steels. It is believed that austenitic stainless steels exhibit significant work hardening, leading to the transformation from austenite to martensite [19]. Okayasu et al. [20] conducted an in-depth study of monotonic loading of austenitic steels at room and low temperatures and modeled the formation of martensite due to deformation. Das et al. [21] studied the effect of strain rate on strain induced martensite transformation characteristics. Some scholars also predicted the life of materials under fatigue loading. Branco et al. [22] developed the total strain energy density method to evaluate the fatigue life of notched specimens subjected to multiaxial loads, and the fatigue master curve can be efficiently generated from only two uniaxial strain-controlled tests, and a set of numerical simulations performed via single-element elastic-plastic models. Pelegatti et al. [23] established a robust procedure for durability design to estimate the strain–life curve, which was an effective way to deal with durability problems. Li et al. [24] proposed an energy-based model to predict the creep-fatigue, combined with high-low cyclic loading life. Because the fatigue fracture surface morphology can provide additional information for the analysis, many people have made a detailed analysis of the fracture surface. Fatigue fracture morphology is studied at different scales, and the most common fracture studies are observed using scanning electron microscopy (SEM) [25,26]. Post-failure applications of fracture surface measurements help to study the fatigue life and the fatigue damage accumulation in notched specimens subjected to severe stress gradient effects [27]. Macek et al. [28] studied the effect of asynchronous axial torsional strain-controlled on the fracture surface behavior of thin-walled tubular austenitic steel specimens. They also found that features of the post-failure fractures were related to the loading conditions and the fatigue life and the loading path significantly affects the surface topography. One of their conclusions is that the total volume fraction of martensite decreases when the sample temperature and the strain rate are increased. However, to date, although most researchers have studied the effects of stretching or combined torsion and tension on the microstructure and fracture behavior of austenitic stainless steel, few studies have studied the fatigue behavior of austenitic stainless steels under cyclic shear path.

Various forms of shear specimens have been used in the field of materials research, mainly applied to analyze forming properties and fracture behavior. Although the use of shear specimen's own geometry in the universal testing machine can achieve shear loading, it will produce a reaction torque in the process of loading, making the specimen deformation, resulting in shear path change [29,30]. Yin et al. [31] proposed the specimen for in-plane torsional experiments, which used the thinning method of machining grooves to form the experimental area. The shear loading is then completed by the relative rotation of the inner and outer rings. However, due to the need for mechanical machining of the specimen, there must be a hardening layer of interference, and the uniformity of the machined thickness cannot be guaranteed during the machining of the thin plate. Brosius et al. [32] proposed a new twin bridge shear specimen which can achieve large shear deformation by shearing the middle experimental region through the relative rotation of the inner and outer rings, and the stress-train change is measured by the torque sensor and the rotation angle. The experimental process does not have the high response movement of the unilateral shear specimen, no compensation is required, and the two experimental regions of the twin bridge shear specimen have the same deformation direction, which is suitable for studying the plate anisotropy.

Therefore, in this study, the principle of the twin bridge shear experiment is chosen to optimize the design of a shear fatigue test. Extensive strain-controlled experiments were conducted on 316L austenitic stainless steel by means of a designed cyclic shear fatigue testing machine. The main purpose is to investigate how deformation-induced martensite affects cyclic deformation and fatigue life, and to evaluate the effect of cyclic shear on the microstructural evolution of 316L austenitic stainless steel by microscopic analysis of test samples.

## 2. Experimental Procedure

### *2.1. Experimental Setup and Specimens*

To study the fatigue behavior of materials under the cyclic shear path, a cyclic shear fatigue testing machine based on the twin bridge shear specimen was designed. The physical drawing of the machine and the geometry of the test piece is shown in Figure 1. The reducer in the base provides power to the power shaft so that the power shaft can rotate. The torque sensor is fixedly connected with the reducer housing, so that the Torque sensor is relatively static with the base. The outer and inner rings of the specimen are in contact with the upper surface of the torque sensor and the upper surface of the power shaft, respectively. The middle pressing structure passes through the circular hole of the middle clamping plate and the central hole of the test piece in turn, and is connected with the power shaft through threads. The inner ring of the specimen is firmly fixed to the power shaft by rotating the bolt on the middle pressing structure. When the power shaft rotates, the inner ring of the specimen rotates with it. The ring-pressing structure is fixedly connected with the torque sensor by threads. The outer ring of the specimen is firmly fixed to the torque sensor by rotating the bolt on the ring-pressing structure. Therefore, when the reducer rotates, the power shaft and the inner ring of the specimen rotate together, but the outer ring of the specimen remains relatively stationary with the equipment, thus making the inner and outer rings of the specimen rotate relatively, and then completing the shear loading of the specimen. The connecting shaft is fixedly connected with the middle pressing structure through threads. The power shaft drives the connecting shaft to rotate, thus ensuring that the connecting shaft will not be deformed by the torsional force during the loading process. The amplifying disc is fixedly connected with the connecting shaft and rotates together with it, which can further improve the resolution of angle detection. One end of the encoder is fixed on the ring-pressing structure by magnetic suction, and the other end is in contact with the amplifying disc, so that the encoder can be driven to rotate when the amplifying disc rotates, thus recording the rotation angle. The specimen fixation method adopted by this device will not cause large deformation of the positioning piece due to the influence of external force, nor will it cause deformation of the key structure due to the action of torque, to better ensure the consistency of the experimental results [33]. At the same time, because one end of the encoder is fixed to the ring-pressing structure, and the other end is in contact with the amplification disc, this connection method effectively shields the impact of the transmission gap of the system on the experimental accuracy, so that the angular resolution can reach 0.002°.

### *2.2. Experimental Materials and Methods*

The commercial 316L metastable austenitic stainless steel sheet produced by the hot rolling process in this study has excellent heat resistance, ductility, corrosion resistance, and mechanical properties. The specimen used for the fatigue test is shown in Figure 1b. Figure 2 shows the metallographic image of the material with 98% austenite content. The chemical composition of the specimen measured by using EBSD is shown in Table 1.

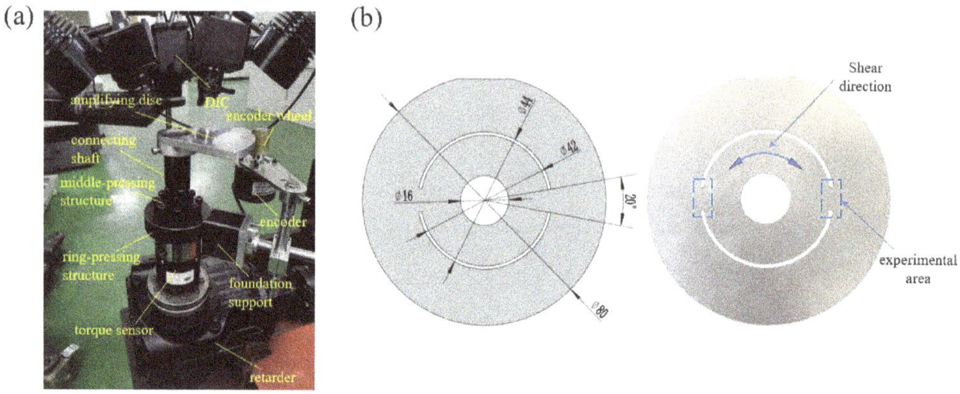


**Figure 1.** (**a**) Cyclic shear fatigue testing machine, (**b**) twin bridge shear specimen.

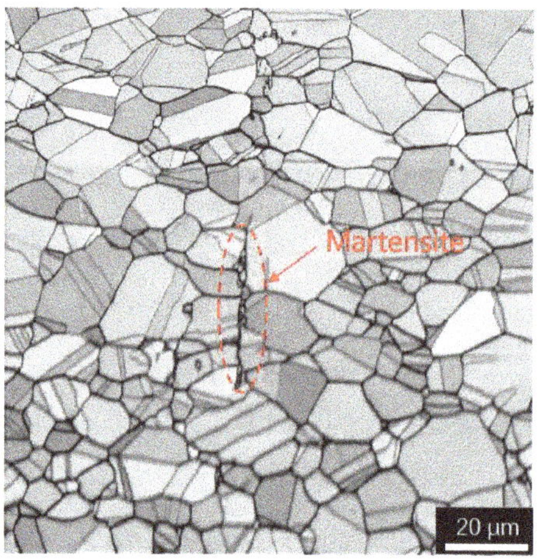


**Figure 2.** Microstructure of 316L before fatigue experiments.

**Table 1.** Chemical composition in weight percentage of the AISI 316L hot rolled plate.

| C | Si | Mn | P | S | Cr | Ni | Mo | N |
|---|---|---|---|---|---|---|---|---|
| 0.02 | 0.5 | 1.18 | 0.03 | 0.001 | 16.91 | 10.26 | 2.11 | 0.04 |

In order to verify the relationship between the rotation angle of the inner and outer rings and the strain, the monotonic shear loading experiment of 316L was carried out by using DIC. The angle acquisition of DIC and the encoder of the testing machine adopt the same time interval and start at the same time. The relationship between strain and angle is determined by interpolation [34]. The geometry of the 316L tensile specimen is shown in Figure 3a. The equivalent shear stress–strain and tensile stress–strain curves of 316L is shown in Figure 3b. It can be seen from the figure that the yield point and strengthening process of 316L are very similar under the tensile and shear paths, and stretching has better elongation. At the stage of failure, the curve decreases rapidly under the tensile state and slowly under the shear state. The von Mises stress and strain studied in this paper were regarded as equivalent stress and strain, and the von Mises equation was used to calculate the equivalent stress and strain.

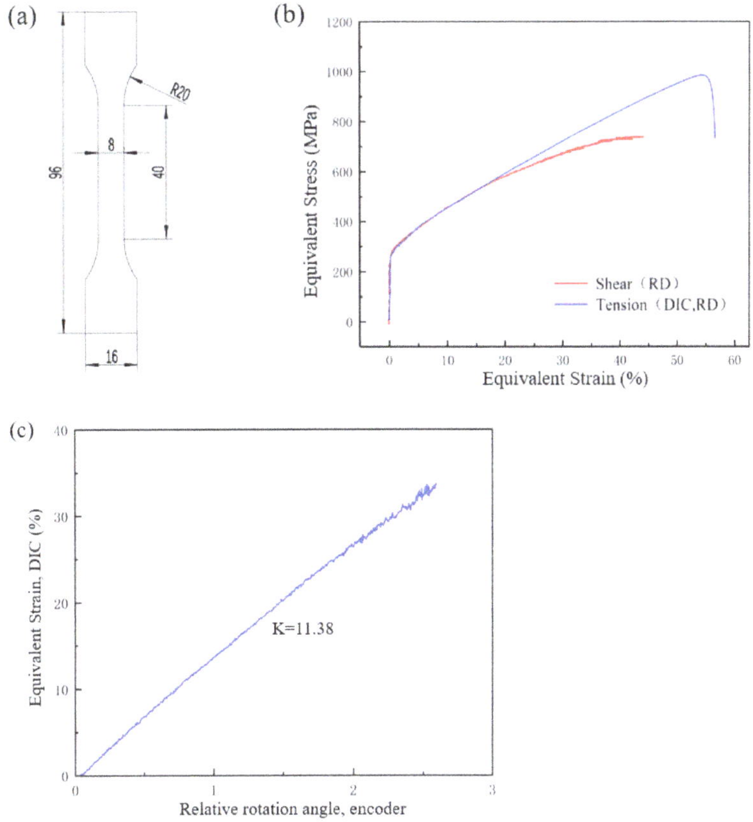


**Figure 3.** (**a**) Specimen geometry in monotonic tensile test, (**b**) comparison of tensile and shear, (**c**) relationship between equivalent strain and rotation angle.

From Figure 3c, a linear relationship between the rotation angle and the equivalent strain can be observed, which is consistent with the research conclusion of Yin et al. [34]. Therefore, the results of Figure 3c can be used to determine the relationship between the magnitude of strain amplitude and the loading angle in subsequent studies. At room temperature, the strain rate ($d\varepsilon/dt$) was constant at $2 \times 10^{-2}$ $S^{-1}$ and the strain-controlled symmetric shear loading tests were carried out under different total strain amplitudes ($\triangle\varepsilon/2$) ranging from 1 to 5%, the strain ratio was $-1$ ($R = \varepsilon_{max}/\varepsilon_{min} = -1$). Three tests were carried out under each strain condition, and a total of 15 groups of tests were carried out. After the fatigue test, the test section in the specimen was cut off by wire cutting. Then specimens were ground and polished with sandpaper of different specifications, and polished with 0.5 μm diamond polishing fluid and polishing cloth until the surface of the specimen had no scratches. Finally, these specimens were electropolished at room temperature. EBSD was used to analyze the original specimens and cracked specimen. The EBSD data were processed by ATEX software without data cleaning to obtain microstructure in different states.

## 3. Experimental Results and Analysis

Figure 4a,b show the stress response at strain amplitudes of 1% and 5%. The initial loading direction was defined as the positive direction of stress and strain. It can be seen from the figure that the distribution of the maximum positive stress and negative stress is approximately symmetrical under the same strain amplitude. The stress increases obviously at the initial stage of loading. With the increase in the number of cycles, the material was continuously hardened, then continuously softened until failure. Figure 4c shows the variation curves between angle and torque when the strain amplitude was 1%. The changes in shape are caused by the continuous hardening of the material, followed by continuous softening until failure. With the increase in the number of cycles, the strength of the material gradually increases, making the hysteretic curve gradually steeper. When the material hardens to a certain extent, the strength of the material begins to decrease gradually, and the hysteretic curve flattens gradually. The relationship between the stress amplitude and the number of cycles N under five different strain amplitudes is plotted in Figure 4d. It can be seen from the figure that the material hardens rapidly at the beginning of loading and the hardening amplitude increases with the increase in the strain amplitude, that is, the cyclic hardening of 316L is affected by the loading amplitude. With the increase in the strain amplitude, the stress amplitude increases, and the fatigue life decreases.

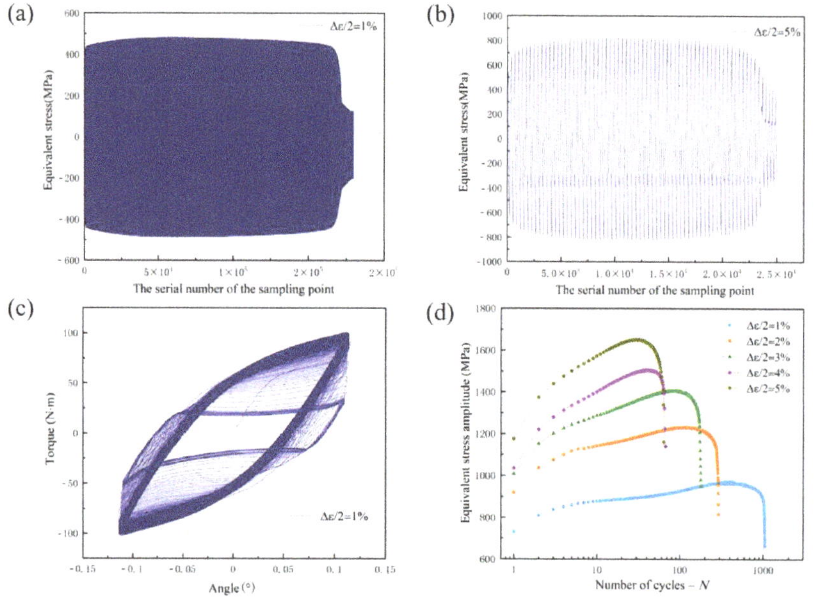


**Figure 4.** (**a**) Stress changes during 1% strain amplitude loading, (**b**) stress changes during 5% strain amplitude loading, (**c**) curves of relative rotation angle of specimen and torque under 1% strain amplitude, (**d**) cyclic stress response curves with different strain amplitudes.

The change rate of the cyclic stress response curve was used to analyze the stress change rate. The fatigue life of the specimen was defined as the number of cycles, and the first derivative of the stress amplitude in the fatigue life was carried out. The relationship between the stress change rate and the number of cycles under different strain amplitudes is shown in Figure 5a. As shown in Figure 5b, the fatigue life is divided into four stages by using 1% strain amplitude as an example: rapid change period, stabilization period, transition period, and failure period. Half the number of cycles at this failure point is defined as the half-life. In the rapid change period, the stress change rate decreases rapidly and tends to be stable gradually. The stabilization period includes the cyclic hardening and softening of the material. The hardening stage is above the zero axis and the softening stage is below the zero axis as shown in Figure 5b. With the increase in the number of cycles, the stress change in the hardening stage decreases linearly and the stress change rate in the softening stage increases linearly. The proportion of the stabilization period in the whole life cycle decreases with the increase in the strain amplitude. The stabilization period accounts for 76.5% of the whole life cycle during 1% strain amplitude. In other words, the length of the stabilization period determines the length of fatigue life. The longer the proportion of stabilization period, the longer the fatigue life. During the transition period, the stress change rate increases rapidly. At this time, obvious cracks have appeared in the specimen and the crack growth rate continues to increase. Finally, the material reaches the failure point, the specimen enters the failure period, and the stress change rate decreases rapidly, which indicate the fracture of the specimen.

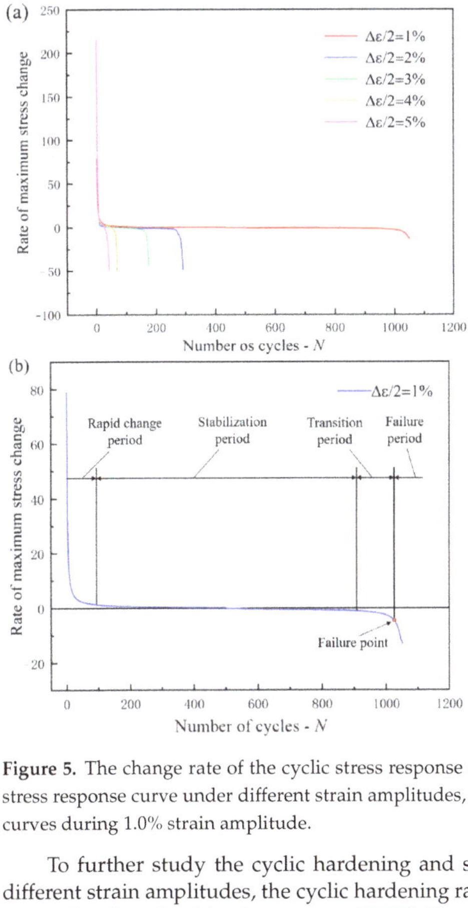


**Figure 5.** The change rate of the cyclic stress response curve of fatigue life: (**a**) the change rate of stress response curve under different strain amplitudes, (**b**) the change rate of cyclic stress response curves during 1.0% strain amplitude.

To further study the cyclic hardening and softening behavior of materials under different strain amplitudes, the cyclic hardening ratio (HR) and cyclic softening ratio (SR) are used to describe this phenomenon [33,35]. Figure 6 shows the relationship between the strain amplitude and cyclic hardening and softening. Under different strain amplitudes, the hardening ratio of the material increases with the increase in the strain amplitude and the softening ratio of the material almost remains unchanged. It is shown that the softening ability of 316L has little relation to the strain amplitude, and the strain amplitude affects the hardening ability of 316L.

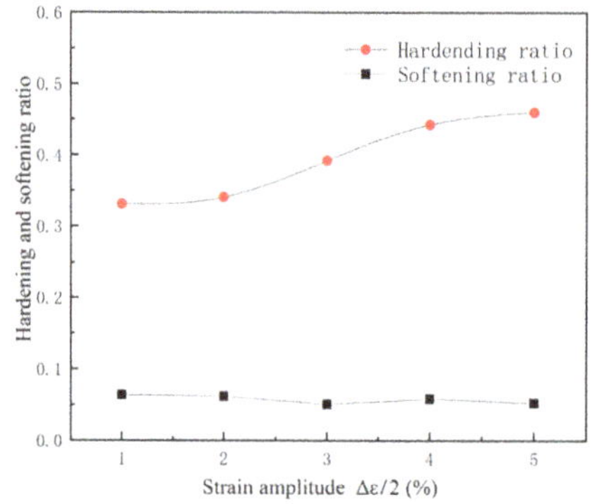


**Figure 6.** Cyclic hardening and softening ration under different strain amplitudes.

From the previous cyclic fatigue stress change diagram, it can be seen that the distribution of positive stress and negative stress is approximately symmetrical. In order to compare the stress change under cyclic fatigue with those under monotonic shear, the positive stress under cyclic fatigue was selected for analysis. The maximum positive stress at each strain amplitude was connected, which was defined as the maximum hardening curve (MH). The positive stress at the failure point at each strain amplitude was connected, which was defined as the failure strength curve (FS). According to the number of cycles at the failure point, the number of cycles corresponding to the half-life was calculated. Connecting the stress corresponding to the half-life of different strain amplitudes defined as the half-life stress curve (HS), MS, MH, HS, and FS were plotted in Figure 7. It can be seen from the figure that MH, HS, and FS are all approximately linearly related to the strain amplitude, and MH, HS, and FS are all higher than MS, which is caused by the hardening of the material under cyclic shear loading. MH and HS basically coincide at the same strain amplitude, which indicates that the hardening degree of 316L reaches the maximum at the half-life. To prove this point, the hysteresis curve at 1% strain amplitude is used as an example, as shown in Figure 8. It can be seen that the stress–strain curve at half-life is located at the maximum stress position of the entire hysteresis curve, which is the time of maximum hardening. In other words, the hardening and softening of 316L account for half of the fatigue life, respectively. Therefore, the strain amplitude under cyclic shear fatigue has a very important effect on the hardening and fatigue life of 316L. The fatigue life is approximately linear with the strain amplitude, and decreases linearly with the increase in the strain amplitude, as shown in Figure 9.

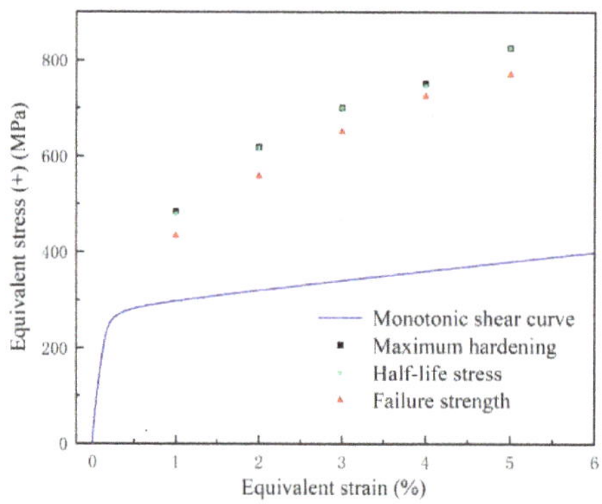


**Figure 7.** Monotonic shear curve, maximum hardening, half-stress, and failure strength relationship for 316L under cyclic shear path.

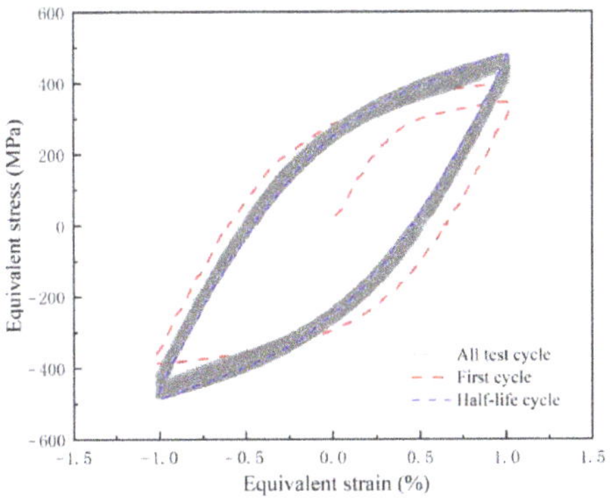


**Figure 8.** Hysteresis curve at 1% strain amplitude.

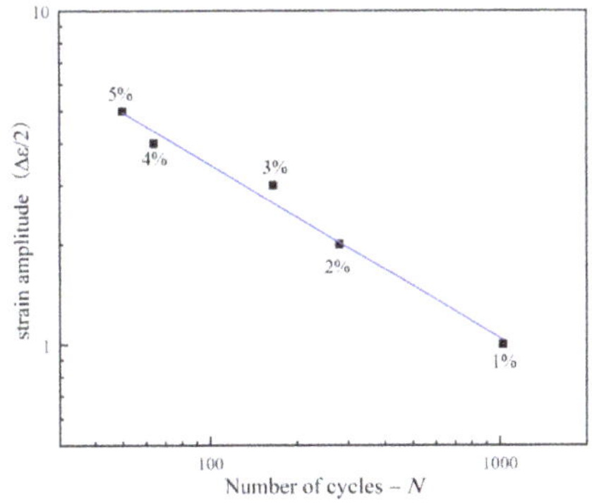


**Figure 9.** Fatigue life at different strain amplitude.

## 4. Microstructure Evolution during Cyclic Shear Loading

The EBSD technique was used to analyze the metallographic composition of the test specimens under cyclic fatigue shear path, and Figure 10 shows that the original material of 316L austenitic stainless steel was mainly austenite. The specimen produced a crack under the shear path, which was surrounded by martensite and the grain was refined. This means that the strain-induced austenite transforms into martensite with the crack initiation, and results in grain refinement around the crack. The tendency to the strain-induced martensite transformation has been explained in terms of the variation in the chemical free-energy difference between the austenite and martensite phases, referred to as the chemical driving force [18]. The hardening of austenitic stainless steels caused during fatigue cycle under a push-pull path is mainly due to deformation-induced martensite and dynamic strain aging (DAS). At room temperature, the decrease in fatigue life is due to the rapid crack growth induced by deformation-induced martensite hardening. The amount of martensite increases with the increase in strain amplitude, which is responsible for the rapid secondary hardening. Because no martensite was found at high temperature, the hardening was caused by dynamic strain aging [36]. Therefore, the hardening of 316L under cyclic shear path at room temperature is caused by martensite transformation. With the increase in cycle times, the number of martensite increases, which makes the crack propagate rapidly and leads to the decrease in fatigue life.

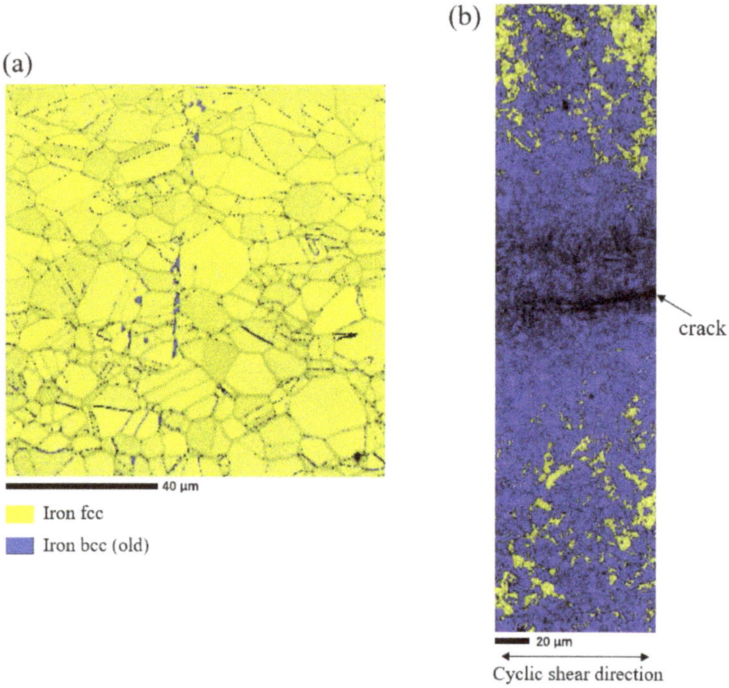


**Figure 10.** EBSD phase image of specimen fatigued cycle. (**a**) Phase image of the original specimen, (**b**) phase image of the specimen after experiment.

## 5. Conclusions

The strain-controlled fatigue properties of 316L austenitic stainless steel was investigated and the cyclic deformation microstructure was analyzed by EBSD in this work. The conclusions are as follows:

1. Under the cyclic shear path at room temperature, 316L exhibits cyclic hardening, saturation, and cyclic softening. The hardening rate is positively correlated with the strain amplitude.
2. The cyclic hardening and cyclic softening of 316L each account for approximately 50% of the life cycle, and the half-life coincides with the saturation period of cyclic hardening.
3. The deformation-induced martensitic transformation causes cyclic hardening of 316L austenitic stainless steel, whereas the fatigue life is reduced due to the rapid crack expansion caused by the deformation-induced martensitic transformation. The fatigue life is negatively correlated with the strain amplitude.

In conclusion, this study provided an in-depth investigation of the failure mechanism of 316L under shear cyclic path through experiments and microstructure observation, which provided a pre-theoretical guidance for the development of a life prediction model under cyclic shear path. The failure scenarios at different strain ratios and the development of life prediction models will be carried out in future work.

**Author Contributions:** Conceptualization, X.L.; Data curation, X.L. and S.Z.; Formal analysis, X.L.; Funding acquisition, Z.Y.; Investigation, X.L. and S.Z.; Methodology, Z.Z.; Project administration, Z.Z.; Software, Z.Z. and S.Z.; Validation, Y.B. and Z.Y.; Writing—original draft, X.L. and S.Z.; Writing—review and editing, Y.B. and Z.Y. All authors have read and agreed to the published version of the manuscript.

**Funding:** This research was funded by the financial support from National Natural Science Foundation of China (NO. 52175337 and 51975327).

**Institutional Review Board Statement:** Not applicable.

**Informed Consent Statement:** Not applicable.

**Data Availability Statement:** Not applicable.

**Conflicts of Interest:** The authors declare no conflict of interest.

## References

1. Facheris, G.; Janssens, K.G.F. Cyclic mechanical behavior of 316L: Uniaxial LCF and strain-controlled ratcheting tests. *Nucl. Eng. Des.* **2013**, *257*, 100–108. [CrossRef]
2. Hug, E.; Prasath Babu, R.; Monnet, I.; Etienne, A.; Moisy, F.; Pralong, V.; Enikeev, N.; Abramova, M.; Sauvage, X.; Radiguet, B. Impact of the nanostructuration on the corrosion resistance and hardness of irradiated 316 austenitic stainless steels. *Appl. Surf. Sci.* **2017**, *392*, 1026–1035. [CrossRef]
3. Fu, C.; Li, J.; Bai, J.; Li, Y.; Chen, Q.; Lei, G.; Lin, J.; Zhu, Z.; Meng, Y. Effect of helium bubbles on irradiation hardening of additive manufacturing 316L stainless steel under high temperature He ions irradiation. *J. Nucl. Mater.* **2021**, *550*, 152948. [CrossRef]
4. Mazánová, V.; Heczko, M.; Škorík, V.; Chlupová, A.; Polák, J.; Kruml, T. Microstructure and martensitic transformation in 316L austenitic steel during multiaxial low cycle fatigue at room temperature. *Mater. Sci. Eng. A* **2019**, *767*, 138407. [CrossRef]
5. Pham, M.S.; Holdsworth, S.R.; Janssens, K.G.F.; Mazza, E. Cyclic deformation response of AISI 316L at room temperature: Mechanical behaviour, microstructural evolution, physically-based evolutionary constitutive modelling. *Int. J. Plast.* **2013**, *47*, 143–164. [CrossRef]
6. Zhu, Y. Cyclic torsion behavior and the related thermal response of 316L stainless steel tube: Experiments and FE simulations. *Int. J. Mech. Sci.* **2017**, *128–129*, 17–22. [CrossRef]
7. Chang, B.; Zhang, Z. Cyclic deformation behavior of 316LN under non-proportional loading based on the analysis on the hysteresis loop and microstructure. *Mater. Sci. Eng. A* **2013**, *565*, 373–381. [CrossRef]
8. Facheris, G.; Janssens, K.G.F.; Foletti, S. Multiaxial fatigue behavior of AISI 316L subjected to strain-controlled and ratcheting paths. *Int. J. Fatigue* **2014**, *68*, 195–208. [CrossRef]
9. Zhou, J.; Sun, Z.; Kanouté, P.; Retraint, D. Experimental analysis and constitutive modelling of cyclic behaviour of 316L steels including hardening/softening and strain range memory effect in LCF regime. *Int. J. Plast.* **2018**, *107*, 54–78. [CrossRef]
10. Kang, G.; Dong, Y.; Wang, H.; Liu, Y.; Cheng, X. Dislocation evolution in 316L stainless steel subjected to uniaxial ratchetting deformation. *Mater. Sci. Eng. A* **2010**, *527*, 5952–5961. [CrossRef]
11. Spätig, P.; Heczko, M.; Kruml, T.; Seifert, H.P. Influence of mean stress and light water reactor environment on fatigue life and dislocation microstructures of 316L austenitic steel. *J. Nucl. Mater.* **2018**, *509*, 15–28. [CrossRef]
12. Lo, K.H.; Shek, C.H.; Lai, J.K.L. Recent developments in stainless steels. *Mater. Sci. Eng. R Rep.* **2009**, *65*, 39–104. [CrossRef]
13. Naghizadeh, M.; Mirzadeh, H. Modeling the kinetics of deformation-induced martensitic transformation in AISI 316 metastable austenitic stainless steel. *Vacuum* **2018**, *157*, 243–248. [CrossRef]
14. Li, Y.; Martín, D.S.; Wang, J.; Wang, C.; Xu, W. A review of the thermal stability of metastable austenite in steels: Martensite formation. *J. Mater. Sci. Technol.* **2021**, *91*, 200–214. [CrossRef]
15. Mao, W.; Gong, W.; Kawasaki, T.; Harjo, S. Effect of deformation-induced martensitic transformation on nonuniform deformation of metastable austenitic steel. *Mater. Sci. Eng. A* **2022**, *837*, 142758. [CrossRef]
16. Li, Y.; Yu, D.; Li, B.; Chen, X. Martensitic transformation of an austenitic stainless steel under non-proportional cyclic loading. *Int. J. Fatigue* **2019**, *124*, 338–347. [CrossRef]
17. Chen, X.; Wang, Y.; Gong, M.; Xia, Y. Dynamic behavior of SUS304 stainless steel at elevated temperatures. *J. Mater. Sci.* **2004**, *39*, 4869–4875. [CrossRef]
18. Talonen, J.; Hänninen, H. Formation of shear bands and strain-induced martensite during plastic deformation of metastable austenitic stainless steels. *Acta Mater.* **2007**, *55*, 6108–6118. [CrossRef]
19. Dan, W.J.; Zhang, W.G.; Li, S.H.; Lin, Z.Q. Finite element simulation on strain-induced martensitic transformation effects in TRIP steel sheet forming. *Comput. Mater. Sci.* **2007**, *39*, 593–599. [CrossRef]
20. Okayasu, M.; Fukui, H.; Ohfuji, H.; Shiraishi, T. Strain-induced martensite formation in austenitic stainless steel. *J. Mater. Sci.* **2013**, *48*, 6157–6166. [CrossRef]
21. Das, A.; Sivaprasad, S.; Chakraborti, P.C.; Tarafder, S. Correspondence of fracture surface features with mechanical properties in 304LN stainless steel. *Mater. Sci. Eng. A* **2008**, *496*, 98–105. [CrossRef]
22. Branco, R.; Prates, P.A.; Costa, J.D.; Berto, F.; Kotousov, A. New methodology of fatigue life evaluation for multiaxially loaded notched components based on two uniaxial strain-controlled tests. *Int. J. Fatigue* **2018**, *111*, 308–320. [CrossRef]
23. Pelegatti, M.; Lanzutti, A.; Salvati, E.; Srnec Novak, J.; De Bona, F.; Benasciutti, D. Cyclic Plasticity and Low Cycle Fatigue of an AISI 316L Stainless Steel: Experimental Evaluation of Material Parameters for Durability Design. *Materials* **2021**, *14*, 3588. [CrossRef]

24. Li, K.-S.; Wang, J.; Fan, Z.-C.; Cheng, L.-Y.; Yao, S.-L.; Wang, R.-Z.; Zhang, X.-C.; Tu, S.-T. A life prediction method and damage assessment for creep-fatigue combined with high-low cyclic loading. *Int. J. Fatigue* **2022**, *161*, 106923. [CrossRef]
25. Kowal, M.; Szala, M. Diagnosis of the microstructural and mechanical properties of over century-old steel railway bridge components. *Eng. Fail. Anal.* **2020**, *110*, 104447. [CrossRef]
26. Dutta, S.; Karmakar, A.; Roy, H.; Barat, K. Automatic estimation of mechanical properties from fractographs using optimal anisotropic diffusion and Voronoi tessellation. *Measurement* **2019**, *134*, 574–585. [CrossRef]
27. Macek, W.; Robak, G.; Żak, K.; Branco, R. Fracture surface topography investigation and fatigue life assessment of notched austenitic steel specimens. *Eng. Fail. Anal.* **2022**, *135*, 106121. [CrossRef]
28. Macek, W.; Pejkowski, Ł.; Branco, R.; Masoudi Nejad, R.; Żak, K. Fatigue fracture surface metrology of thin-walled tubular austenitic steel specimens after asynchronous loadings. *Eng. Fail. Anal.* **2022**, *138*, 106354. [CrossRef]
29. Bouvier, S.; Haddadi, H.; Levée, P.; Teodosiu, C. Simple shear tests: Experimental techniques and characterization of the plastic anisotropy of rolled sheets at large strains. *J. Mater. Process. Technol.* **2006**, *172*, 96–103. [CrossRef]
30. Yin, Q.; Zillmann, B.; Suttner, S.; Gerstein, G.; Biasutti, M.; Tekkaya, A.E.; Wagner, M.F.X.; Merklein, M.; Schaper, M.; Halle, T.; et al. An experimental and numerical investigation of different shear test configurations for sheet metal characterization. *Int. J. Solids Struct.* **2014**, *51*, 1066–1074. [CrossRef]
31. Yin, Q.; Soyarslan, C.; Isik, K.; Tekkaya, A.E. A grooved in-plane torsion test for the investigation of shear fracture in sheet materials. *Int. J. Solids Struct.* **2015**, *66*, 121–132. [CrossRef]
32. Brosius, A.; Yin, Q.; Güner, A.; Tekkaya, A.E. A New Shear Test for Sheet Metal Characterization. *Steel Res. Int.* **2011**, *82*, 323–328. [CrossRef]
33. Zhang, Z.; Yue, Z.; Qi, J.; Gao, J.; Qiu, Y. Low cycle fatigue performance of DP900 steel under cyclic shear paths. *Fatigue Fract. Eng. Mater. Struct.* **2021**, *45*, 22–39. [CrossRef]
34. Yin, Q.; Soyarslan, C.; Güner, A.; Brosius, A.; Tekkaya, A.E. A cyclic twin bridge shear test for the identification of kinematic hardening parameters. *Int. J. Mech. Sci.* **2012**, *59*, 31–43. [CrossRef]
35. Shao, C.W.; Zhang, P.; Liu, R.; Zhang, Z.J.; Pang, J.C.; Zhang, Z.F. Low-cycle and extremely-low-cycle fatigue behaviors of high-Mn austenitic TRIP/TWIP alloys: Property evaluation, damage mechanisms and life prediction. *Acta Mater.* **2016**, *103*, 781–795. [CrossRef]
36. She, M.; Liu, X.; He, G. The deformation-induced martensite and dynamic strain aging during cyclic deformation in AISI 321. *Mater. Res. Express* **2018**, *6*, 026530. [CrossRef]

*materials*

MDPI

*Article*

# Numerical Assessment of Damage Parameters for a Hard Interface Model

**Maria Letizia Raffa [1,*], Raffaella Rizzoni [2] and Frédéric Lebon [3]**

1 Laboratoire QUARTZ EA 7393, ISAE-Supméca, 93400 Saint-Ouen-sur-Seine, France
2 Department of Engineering, University of Ferrara, 44122 Ferrara, Italy
3 Aix Marseille Université, CNRS, Centrale Marseille, LMA, 13453 Marseille, France
* Correspondence: maria-letizia.raffa@isae-supmeca.fr

**Abstract:** Adhesive interfaces are suitable modelling tools to describe very thin elastic layers and the related occurring phenomena (such as damage, viscosity, friction, etc.), without using a volumetric description, which is often computationally prohibitive in a large-scale numerical simulation. A major drawback of these kinds of models is the identification of free parameters, because of the smallness of a direct observation scale. This paper proposes a numerical assessment of two model parameters, a damage energy threshold and a damage viscosity, of a hard interface model previously formulated by authors. The proposed assessment protocol uses macroscopic experimental data, available in the literature, on structural adhesives under standard characterization tests. The numerical results obtained give insights into the physical interpretation of these parameters.

**Keywords:** imperfect interface; adhesive; micro-cracking; analytical modelling; identification

**Citation:** Raffa, M.L.; Rizzoni, R.; Lebon, F. Numerical Assessment of Damage Parameters for a Hard Interface Model. *Materials* **2022**, *15*, 5370. https://doi.org/10.3390/ma15155370

Academic Editors: Grzegorz Lesiuk and Dariusz Rozumek

Received: 24 June 2022
Accepted: 2 August 2022
Published: 4 August 2022



## 1. Introduction

Adhesive interface modelling has been an expanding branch of solid mechanics research since the early 1940s. Goland and Reissner [1] were pioneers in modelling a thin adhesive as a weak interface and they were the first to define the spring-type interface, by assuming that the adherents were bonded by a continuous distribution of springs. They preconized that the thinness would involve a uniform distribution of the stresses field in the adhesive, and some years later, Gilibert and Rigolot [2] found a rational justification of this fact by means of the asymptotic expansion method, assuming that the thickness and the elastic properties of the adhesive had the same order of smallness $\varepsilon$.

During the eighties and nineties, the relaxation of the perfect interface approximation, i.e., continuity at the interfaces in displacements and stresses fields, was largely investigated, aiming to apply these theories to composite materials with coated fibres [3,4] or particles [5,6], or to the decohesion and nucleation problems in cohesive zones [7–9]. Many efforts were made to model damage [10,11], and other related physics such as friction [12,13] and viscosity [14], in adhesive interfaces. All these models have the advantage to allow a macroscopic description of a thin elastic adhesive and the occurring phenomena, without using a volumetric description, which is often computationally prohibitive in a large-scale numerical simulation.

A major drawback of these kinds of models is the identification of free parameters. Often, the identification protocols need direct experimental observations and this is not a trivial issue in very thin layers because of the small scale of the phenomena to be investigated. A fallback solution is to build on macroscale experimental tests on adhesive specimens and adhesive joint specimens and try to correlate microscopic model parameters to a macroscopic response [15].

The hard imperfect interface model with damage formulated by authors in [16] has been chosen to apply an estimation procedure of damage parameters. In particular, this interface model is suitable to describe structural adhesives as stiff as substrates [17]. In this case, the mechanical behaviour of the adhesive cannot be accurately described via a classic spring-like interface model (i.e., continuity in stresses field and discontinuity of displacements field [1]), but a hard condition considering also a jump in the stresses field is more indicated, as demonstrated by [17].

The imperfect interface model that is adopted for this numerical study has two free parameters, which is a great advantage of this kind of phenomenological models. These parameters represent respectively a damage viscosity ($\eta$) and a damage energy threshold ($\omega$) and they are related to the damage evolution law, as detailed in what follows. Note that the model sensitivity on these parameters has been already investigated by authors in [16], thus it is not the subject of this work.

This paper proposes a first attempt at a numerical assessment of the damage parameters $\eta$ and $\omega$. The proposed estimation procedure is based on experimental data available in the literature, which concerns macroscopic characterization tests on both bulk and joint structural adhesives commonly used in industry.

## 2. Materials and Methods

### *2.1. Overview of the Hard Imperfect Interface Model for Micro-Cracked Adhesive Joints*

In this section, a brief overview of the imperfect interface model is proposed. For an extensive description of the formulation, one refers to [16]. The model consists of a law of hard imperfect contact coupled with a damage evolution law. It is able to describe the mechanical behaviour of structural adhesives with micro-cracking damage. The transmission conditions reported below, prescribe jumps in the stresses $[[\sigma\ \mathbf{e}_3]]$ and displacements $[[\mathbf{u}]]$ fields across an interface of outward normal unit vector $\mathbf{e}_3$ between two adherents, thus describing the asymptotic behaviour of a very thin deformable adhesive made of a general anisotropic linear elastic material:

$$[[\mathbf{u}]] = \varepsilon\Big((\mathbf{K}^{33})^{-1}\Big(\langle\langle\sigma\,\mathbf{e}_3\rangle\rangle - \mathbf{K}^{\alpha 3}\langle\langle\mathbf{u}_{,\alpha}\rangle\rangle\Big) - \langle\langle\mathbf{u}_{,3}\rangle\rangle\Big), \tag{1}$$

$$[[\sigma\,\mathbf{e}_3]] = \varepsilon\Big(\Big(-\mathbf{K}^{\beta\alpha}\langle\langle\mathbf{u}_{,\beta}\rangle\rangle - \mathbf{K}^{3\alpha}(\mathbf{K}^{33})^{-1}\Big(\langle\langle\sigma\,\mathbf{e}_3\rangle\rangle - \mathbf{K}^{\beta 3}\langle\langle\mathbf{u}_{,\beta}\rangle\rangle\Big)\Big)_{,\alpha} - \langle\langle\sigma_{,3}\,\mathbf{e}_3\rangle\rangle\Big), \tag{2}$$

where $\varepsilon$ is the thickness of the adhesive, the symbols $[[(\cdot)]]$ and $\langle\langle(\cdot)\rangle\rangle$ are taken to denote the jump and the average of the quantity $(\cdot)$ across the interface separating the two adherents, respectively; Greek indexes ($\alpha, \beta = 1, 2$) are related to the in-plane ($x_1, x_2$) quantities; commas denote first derivatives and the summation convention is used. Matrices $\mathbf{K}^{ij}, i, j = 1, 2, 3$ are related to the elasticity coefficients $B_{ijkl}$ of the adhesive layer (from the constitutive equation in linear elasticity: $\sigma_{ij} = B_{ijkl}\,e_{kl}$). If the adhesive is modelled as isotropic, with Young's modulus $\bar{E}$ and Poisson's ratio $\bar{\nu}$, the matrices $\mathbf{K}^{ij}$ read as:

$$\mathbf{K}^{ii} = \frac{\bar{E}}{2(1+\bar{\nu})}\Big(\frac{2(1-\bar{\nu})}{(1-2\bar{\nu})}\,\mathbf{e}_i\otimes\mathbf{e}_i + \mathbf{e}_j\otimes\mathbf{e}_j + \mathbf{e}_k\otimes\mathbf{e}_k\Big),\quad i\neq j\neq k, \tag{3}$$

$$\mathbf{K}^{ij} = \frac{\bar{E}}{2(1+\bar{\nu})}\Big(\mathbf{e}_i\otimes\mathbf{e}_j + \frac{2\bar{\nu}}{(1-2\bar{\nu})}\,\mathbf{e}_j\otimes\mathbf{e}_i\Big),\quad j\neq i. \tag{4}$$

Engineering moduli $\bar{E}$ and $\bar{\nu}$ represent the effective mechanical properties of an isotropic microcracked material. Note that the assumption of a random distribution of microcracks is considered here, thus the resulting material remains isotropic. In the section below, three different micromechanical homogenization schemes are used to derive the elastic moduli of the microcracked adhesive material.

### *2.2. Micromechanical Homogenization Schemes for a Microcracked Adhesive*

Drawing on micromechanical homogenization theory [18–23] within the framework of effective field schemes, the effective elastic moduli of the microcracked adhesive are derived. Micro-cracking damage is represented here by a microcracks density parameter

Particularly, a generalized crack density [24] is adopted, allowing to by-pass the geometrical definition of the cracks, which is possible only for circular and regular cracks [25], and as a matter of fact, extending the generality of the interface model to any regular and irregular cracks shape. It should also be noted that the generalized microcracks density can be measured *postmortem* by X-ray micro-tomography [26,27].

#### 2.2.1. Kachanov-Sevostianov Scheme

The Kachanov-Sevostianov (KS) scheme [21,28] is a stress-based homogenization approach based on the non-interacting microcracks approximation (NIA) [25,29] and on Eshelby's theory [18]. In the 2D case, assuming the adhesive is an initial isotropic matrix embedding a random distribution of microcracks, the elastic potential in stresses (complementary energy density) of the effective medium yields the following structure for the effective Young's modulus $\bar{E}$ and Poisson's ratio $\bar{\nu}$:

$$\bar{E} = E_0 \left(1 + 2\, C_{ks}\, E_0\, \rho\right)^{-1}, \tag{5}$$

$$\bar{\nu} = \nu_0, \tag{6}$$

where $E_0$ and $\nu_0$ are the moduli of the undamaged matrix or the initial moduli of the adhesive before the damage, $\rho$ is the generalized microcracks density (i.e., damage parameter) and the constant $C_{ks}$ depends on the orientational distribution of defects. For a 2D random distribution of microcracks, $C_{ks} = \frac{\pi}{E_0}$ [21,22]. Note that Equation (6) is an assumption valid for all considered schemes.

#### 2.2.2. Welemane-Goidescu Scheme

The Welemeane-Goidescu (WG) scheme [30–33] is a strain-based homogenization approach based on the dilute limit approximation (for further details about dilute limit and NIA, one refers to [28]). Under the hypothesis of a 2D random distribution of microcracks embedded in an isotropic matrix, a linearized expression of the effective Young's modulus is found:

$$\bar{E} = E_0 \left(1 - 2\, C_{wg}\, E_0\, \rho\right), \tag{7}$$

with $C_{wg} = C_{ks} = \frac{\pi}{E_0}$ [31].

#### 2.2.3. Pan-Weng Scheme

Pan-Weng (PW) scheme [23] draws on Mori-Tanaka's theory [34] and it is a stress-based homogenization approach. In analogy with the previously cited schemes, the 2D case of a randomly distributed microcracks family is considered. Accordingly, the effective Young's modulus reads as:

$$\bar{E} = E_0 \left(1 - C_{pw}\, \rho\right)^{-1}, \tag{8}$$

where $C_{pw} = \frac{16}{3}(\nu_0^2 - 1)$ [23].

Note that by including Equations (5), (7) or (8) in the expression of the interface stiffness tensor (Equation (4)), a different dependency of the imperfect interface law on the crack density $\rho$, is obtained. Particularly, it has been previously established by authors that the KS scheme allows us to describe imperfect interfaces with ductile damaging behaviour, while the WG scheme is suitable to describe interfaces with brittle damaging behavior [16].

### 2.3. *A Description for the Micro-Cracking Damage Evolution*

In this section, a possible description of the evolving behaviour of the cracks density $\rho$ is given by drawing on further works by authors [11,16]. Damage in the adhesive joint is assumed to be caused by a microcracks accumulation and damage parameter $\rho$ is assumed to strictly increase in time. The evolution of the cracks density in the bulk adhesive of thickness $\varepsilon$ (i.e., cohesive damage) can thus be described by the following first-order ODE proposed in [11]:

$$\eta^{\varepsilon}\dot{\rho} = \left\{\omega^{\varepsilon} - \frac{1}{2}\,\mathbb{B}^{\varepsilon}_{,\rho}(\rho)(\mathbf{e}(\mathbf{u}^{\varepsilon}) : \mathbf{e}(\mathbf{u}^{\varepsilon}))\right\}_{+}, \tag{9}$$

where $\{\cdot\}_{+}$ denotes the positive part of the function, $\mathbb{B}^{\varepsilon}_{,\rho}(\rho)$ indicates the component-wise derivative of the effective stiffness tensor $\mathbb{B}^{\varepsilon}(\rho)$ of the micro-cracked adhesive with respect to the generalized cracks density $\rho$, and $\mathbf{e}(\mathbf{u})$ is the strain tensor under the small perturbation hypothesis.

By applying asymptotic expansions theory to Equation (9) under the assumption of *hard interface* [17], the expression of the evolution of cracks density in the interface (i.e., adhesive damage) is obtained, as in [16]:

$$\eta\dot{\rho} = \left\{\omega - \frac{1}{2}\,\mathbf{K}_{,\rho}(\rho)\begin{pmatrix}\langle\langle\mathbf{u}_{,1}\rangle\rangle\\ \langle\langle\mathbf{u}_{,2}\rangle\rangle\\ [[\mathbf{u}]] + \varepsilon\langle\langle\mathbf{u}_{,3}\rangle\rangle\end{pmatrix}\cdot\begin{pmatrix}\langle\langle\mathbf{u}_{,1}\rangle\rangle\\ \langle\langle\mathbf{u}_{,2}\rangle\rangle\\ [[\mathbf{u}]] + \varepsilon\langle\langle\mathbf{u}_{,3}\rangle\rangle\end{pmatrix}\right\}_{+}, \tag{10}$$

with initial condition $\rho(0) = \rho_0$.

In Equation (10), $\mathbf{u}$ is the displacement field, symbols $[[(\cdot)]]$ and $\langle\langle(\cdot)\rangle\rangle$ are taken to denote the jump and the average of the quantity $(\cdot)$ across the interface and $\mathbf{K}_{,\rho}(\rho)$ indicates the component-wise derivative of the stiffness tensor

$$\mathbf{K}(\rho) = \begin{pmatrix}\varepsilon\mathbf{K}^{11} & \varepsilon\mathbf{K}^{21} & \mathbf{K}^{31}\\ \varepsilon\mathbf{K}^{12} & \varepsilon\mathbf{K}^{22} & \mathbf{K}^{32}\\ \mathbf{K}^{13} & \mathbf{K}^{23} & \frac{1}{\varepsilon}\mathbf{K}^{33}\end{pmatrix} \tag{11}$$

with respect to the cracks density $\rho$.

### 2.4. Numerical Assessment Procedure

The adopted hard imperfect interface model (Equations (1) and (2)) integrating the damage evolution (Equation (10)) has two model parameters: $\eta$ and $\omega$ (respectively $\eta^{\varepsilon}$ and $\omega^{\varepsilon}$ in the bulk adhesive). In detail, $\eta$ (or $\eta^{\varepsilon}$) is a strictly positive constant parameter and represents a damage viscosity influencing the velocity of damage evolution; $\omega$ (or $\omega^{\varepsilon}$) is a strictly negative constant parameter and has the meaning of an energy threshold, after which damage evolution begins [16]. The goal of this work is to provide a first numerical assessment of these parameters.

The proposed assessment protocol uses macroscopic experimental data on structural adhesives under standard characterization tests. To this aim, experimental data from Murakami and coworkers [35] concerning tensile and torsional tests on the epoxy-based structural adhesive XA7416 (3M Japan Ltd., Tokyo, Japan) and from Kosmann and coworkers [36] concerning torsional test on epoxy adhesive Henkel Hysol EA9695 AERO 0.05 NW (Henkel AG and Co., Düsseldorf, Germany), were chosen. Experimental data were extracted from [35,36] by using the free online software *WebPlotDigitizer* [37]. Numerical simulations were carried out by numerically solving the differential problem Equations (1) and (2) with Equations (10) and (11), using the commercial software *Mathematica* [38]. The estimation procedure is schematized in the flow chart in Figure 1.

First, an estimation process in the bulk configuration is performed. To this aim, experimental results of a tensile test on a bulk adhesive specimen (Figure 7 in [35]) have been used to calculate the evolution of the experimental Young's modulus in time, shown in Figure 2.

Figure 2 shows that at the beginning of the test the Young's modulus is equal to its initially undamaged value $E_0$, then micro-cracking begins, which decreases the modulus. Three different numerical models of $E(\dot{\rho})$ have been obtained by integrating Equation (9) in the three schemes for damaged material: Equation (5) (for KS model), Equation (7) (for WG model) and Equation (8) (for PW model). Then, the numerical models were fitted to the experimental data by the least-square minimization method, to derive the damage parameters in the bulk adhesive configuration, $\eta^{\varepsilon}$ and $\omega^{\varepsilon}$ in Equation (9).

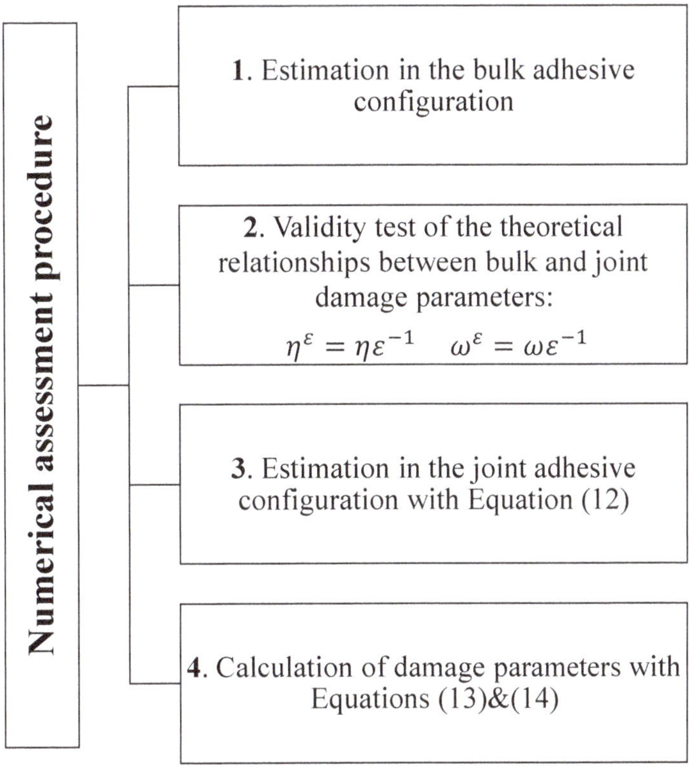


**Figure 1.** Flow chart detailing the numerical assessment procedure.

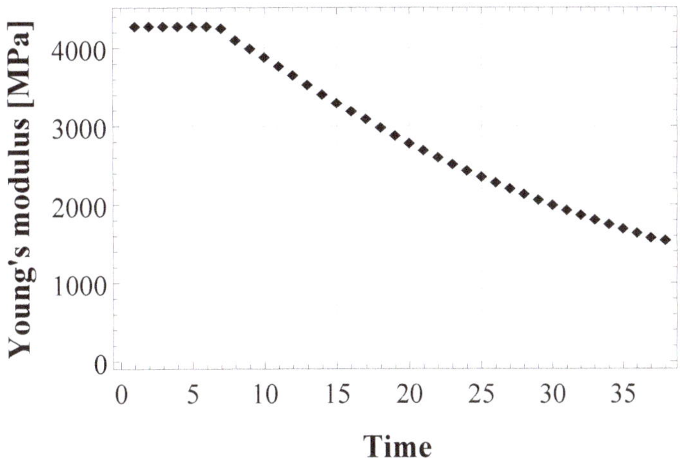


**Figure 2.** Evolution of the experimental Young's modulus of the epoxy-based adhesive XA7416, calculated from tensile test data by [35].

Then, the theoretical damage parameters in the joint (interface) configuration $\eta$ and $\omega$ have been calculated from relationships $\eta^{\varepsilon} = \eta\,\varepsilon^{-1}$ and $\omega^{\varepsilon} = \omega\,\varepsilon^{-1}$ and they have been used to calculate numerical curves and compared to experimental data. Since the theoretical parameters did not prove reliable for simulating experimental behaviour, a further step must be taken to identify damage parameters in the joint configuration.

An estimation process in the adhesive joint configuration was performed to identify damage parameters in the interface ($\eta$ and $\omega$). To this aim, shear strain-stress data of pure torsional tests on cylindrical butt-joint specimens were extracted from [35,36]. In this case, the analytical closed-form solution for the proposed hard interface model with damage (Equations (1) and (2) with (10)) has been formulated in [39] and represents the shear strain-stress ($\gamma - \tau$) response of the adhesive joint:

$$\gamma = \begin{cases} a\tau, & 0 \leq \tau \leq \tau_0 \\ a\tau + b\tau(\tau - \tau_0)^2(\tau + 2\tau_0), & \tau > \tau_0 \end{cases} \tag{12}$$

where $\tau_0$ is the damage initiation shear stress. Equation (12) was fitted to the experimental shear strain-stress curves (cf. Figure 10 by [35] and Figure 8 by [36]) to derive model parameters $a, b$ and $\tau_0$, then the damage parameters, $\eta$ and $\omega$, were estimated by using the following expressions:

$$\omega = -\frac{\tau_0^2 \varepsilon\, C_G}{2\, G_0}, \tag{13}$$

$$\eta = \frac{\varepsilon\, C_G^2}{6\, b\, \dot{\tau}\, G_0^2}. \tag{14}$$

where $G_0$ is the initially undamaged shear modulus of the adhesive joint and $C_G$ is the microstructural parameter depending on the adopted damaged material model. It can be equal to $C_{ks}, C_{wg}$ or $C_{pw}$ (see Equations (5), (7) and (8)).

For the numerical simulations of the cylindrical butt-joint specimens under torsion, the parameters related to the experimental configurations of the two different adhesive materials were extracted from [35,36] and reported in Table 1.

**Table 1.** Experimental parameters of the structural adhesives in joint configuration.

| Parameter | Epoxy-Based Adhesive XA7416 [35] | Epoxy-Based Adhesive Henkel Hysol EA9695 [36] |
|---|---|---|
| Adhesive thickness $\varepsilon$ [mm] | 0.3 | 0.05 |
| Loading rate $\dot{\tau}$ [MPa/s] | $6.67 \times 10^{-2}$ | $9.14 \times 10^{-1}$ |
| Shear modulus $G_0$ [MPa] | 671.88 | 672.91 |

## 3. Results

### *3.1. Numerical Assessment in the Bulk Configuration*

The behaviour of the apparent Young's modulus of the epoxy-based adhesive XA7416 is represented in Figure 2. At the beginning of the tensile test (elastic domain), it is constant and equal to $E_0 = 4282$ MPa, then it degrades in time as an effect of the micro-cracks accumulation. This experimental finding was used to assess damage parameters in the bulk adhesive. Figure 3 shows the three numerical models obtained by implementing the evolution of the cracks density (Equation (9)) in the three schemes for damaged material, i.e., Equation (5), (7) and (8), respectively for KS, WG, and PW scheme.

Numerical results obtained with KS and PW model are quite similar and very close to the experimental behaviour, showing a gradual degradation of the Young's modulus from its initial value $E_0$ until reaching 36% of $E_0$ at the end of the simulated tensile test. On the contrary, the WG model immediately after the onset of damage accumulation, corresponding at a strain of 0.4%, goes to zero abruptly. Note that time in this study is only indicative, so no units are necessary.

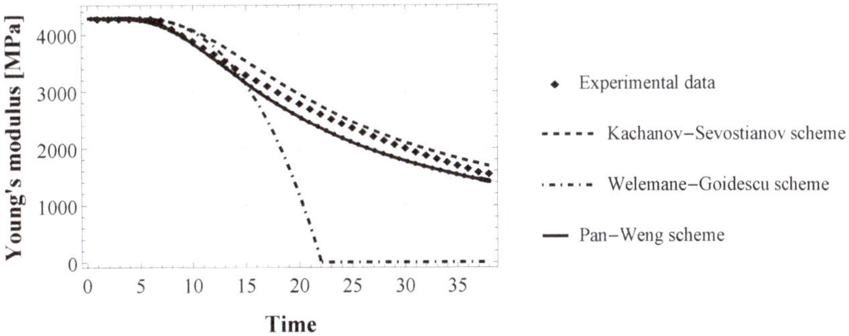


**Figure 3.** Evolution of the Young's modulus of the epoxy-based adhesive XA7416: fitting of the proposed numerical model (with Kachanov-Sevostianov, Welemane-Goidescu and Pan-Weng scheme) to the experimental data extracted by authors from experiments by [35].

Damage parameters estimated in the bulk adhesive configuration are the same for all numerical models as reported in Table 2.

### *3.2. Numerical Assessment in the Joint Configuration*

Figure 4 shows the experimental behaviour under pure torsion of two S45C carbon-steel cylinders joined by an epoxy adhesive XA7416 [35] (see Table 1 for the characteristics of the experimental configuration). The elastic part of the experimental $(\gamma - \tau)$ curve has been used to estimate the initially undamaged shear modulus of the adhesive joint, which results in $G_0 = 672$ MPa. Moreover, in Figure 4 are represented the three best-fitted numerical curves for KS, WG and PW model, obtained from Equation (12) by using the same estimates for all three models, namely $a = 1.5829 \times 10^{-3}$, $b = 1.0541 \times 10^{-6}$ and the damage initiation stress $\tau_0 = 50.25$ MPa. Then, the damage parameters $\omega$ and $\eta$ were calculated from Equations (13) and (14), respectively, by using the proper microstructural parameter $C_G$, and reported in Table 2.

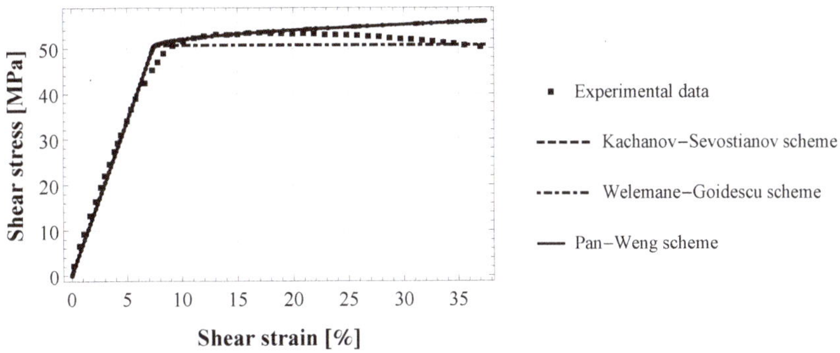


**Figure 4.** Torsional shear strain-stress $(\gamma - \tau)$ response of cylindrical butt-joint specimens of epoxy-based adhesive XA7416: fitting of the proposed numerical model (with Kachanov-Sevostianov, Welemane-Goidescu and Pan-Weng scheme) to the experimental data by [35].

#### Influence of the Adhesive Type

The experimental data by [36] were chosen to investigate the influence of the adhesive material type on the estimated parameters. Figure 5 shows the experimental behaviour under torsion of two steel cylinders joined by an epoxy adhesive Henkel Hysol EA9695 AERO 0.05 NW [36] (see Table 1 for the characteristics of the experimental configuration). The elastic part of the experimental $(\gamma - \tau)$ curve has been used to estimate the initially undamaged shear modulus of the adhesive joint, which results in $G_0 = 673$ MPa. Moreover, in Figure 5 is represented the best-fitted numerical curve for KS scheme, obtained from Equation (12) by using the following estimates: $a = 2.2224 \times 10^{-3}$, $b = 9.9261 \times 10^{-7}$ and the damage initiation stress $\tau_0 = 50.77$ MPa. Then, the damage parameters $\omega$ and $\eta$ were calculated from Equations (13) and (14), respectively, by using the proper microstructural parameter $C_G$ for the KS scheme, and the estimated parameters are reported in Table 2.

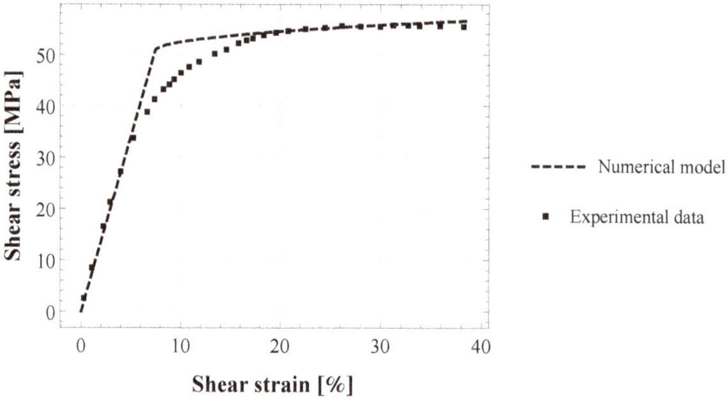


**Figure 5.** Torsional shear strain-stress $(\gamma - \tau)$ response of cylindrical butt-joint specimens of epoxy-based adhesive Henkel Hysol EA9695: fitting of the proposed numerical model (with Kachanov-Sevostianov scheme) to the experimental data by [36].

**Table 2.** Numerically estimated damage parameters.

| Configuration/Homog. Scheme/Adhesive Type | Damage Threshold $\omega$ [MJ/mm$^2$] | Damage Viscosity $\eta$ [MJ$\cdot$ s/mm$^2$] |
|---|---|---|
| Bulk/KS, WG and PW/XA7416 | −0.3 | 230 |
| Joint/KS and WG/XA7416 | −3.54 | 62.19 |
| Joint/PW/XA7416 | −2.62 | 33.95 |
| Joint/KS/EA9695 | −0.60 | 0.80 |

## 4. Discussion

The present work is a first attempt at coping with the identification of damage parameters for interface models with damage. One of the main questions that arises is whether these parameters represent an intrinsic property of the adhesive material, or they depend on the model adopted to describe the damage. For this reason, two numerical assessment procedures in both bulk and joint adhesive configuration are carried out by using three different homogenization schemes integrated into our hard interface model with damage evolution. Moreover, two different adhesive materials are studied.

Concerning the numerical assessment in the bulk configuration, the three damaged material models give the same damage threshold $\omega$ and damage viscosity $\eta$, as reported in Table 2. This result suggests that these parameters are related to the adhesive material nature as intrinsic properties, regardless of the model adopted. Moreover, the KS and PW model give very similar results, as shown in Figure 3, and are able to reproduce the asymptotic degradation of the experimental elastic modulus. This is related to the fact that both KS and PW micromechanical homogenization schemes are drawn on the approximation of non-interacting cracks. This assumption extends the accuracy of these schemes to high values of crack density (for $\rho > 60\%$ as established in [28]). On the contrary, the WG scheme remains accurate to a lower density value than the first two models, and this is because it is based on the dilute limit assumption. These results are thus consistent with the theory [28,40].

As well-established in literature, the mechanical behaviour of the bulk adhesive can be very different from that of the adhesive joint, and this can explain why the damage parameters assessed in the joint configuration are different from that of the bulk adhesive configuration. Note that in [16] it is assumed that $\eta^{\varepsilon}$ and $\omega^{\varepsilon}$ are volumetric densities and thus inversely proportional to the non-dimensional interphase thickness $\varepsilon$: $\eta^{\varepsilon} = \eta\,\varepsilon^{-1}$ and $\omega^{\varepsilon} = \omega\,\varepsilon^{-1}$. Actually, the results obtained by numerical estimation disprove this hypothesis, as shown in Table 2.

As in the bulk configuration, also in the joint case, KS and PW give very similar results, the two curves are superposed as shown in Figure 4. It is worth highlighting that the numerical simulations of the pure torsional test on epoxy adhesive XA7416 are carried out in a force-controlled mode to be consistent with the experimental configuration, by imposing a shear stress rate $\dot{\tau}$. For this reason, numerical results, regardless of the considered homogenization scheme, cannot reproduce the experimental softening behaviour occurring for a shear strain $\gamma > 23\%$.

Concerning the numerical assessment in the joint configuration, the resulting damage parameters depend on the considered damage scheme, unlike what is found in the bulk configuration assessment. Particularly, damage parameters depend on the microstructural parameter $C_G$ (see Equations (13) and (14)). KS and WG schemes have the same microstructural parameter $C_G = C_{ks} = C_{wg} = \frac{\pi}{E_0}$ and this explains why the same damage parameters are found. Damage parameters estimated for the PW scheme are slightly different even if they have the same order of magnitude (see Table 2). Moreover, all the numerical curves, regardless of the scheme, are in good agreement with the experimental pattern. This result is consistent with previous numerical insights found in [39].

Figure 5 shows the comparison between the numerical model and the experimental data on cylindrical butt-joint with epoxy-based adhesive Henkel Hysol EA9695 by [36]. In this case, only the KS scheme was used in order to compare the resulting parameters with them estimated for the other adhesive in a similar joint configuration. It results that the estimated parameters $\eta$ and $\omega$ are quite smaller than that obtained for the epoxy adhesive XA7416, as highlighted in Table 2. This fact can be due to the influence of the material's elastic properties, although no precise information is available from [36] concerning bulk properties of adhesive Henkel Hysol EA9695. Another possible explanation is the influence of the experimental configuration. In fact, although the test configurations of Murakami [35] and Kosmann [36] are quite similar (steel cylindrical adhesively bonded butt-joint under torsion), some main differences still remain, such as the adhesive thickness and the loading rate. In detail, the film thickness in Kosmann's experiments is thinner and the loading rate is higher than that in Murakami's experiments (see Table 1). Additionally, the geometrical parameters of the joint specimens (not reported here) are also different. However, further investigations on different adhesive materials are necessary to derive a correlation between model damage parameters and the above properties. Moreover, it is crucial to have the same test configuration in order to provide reliable comparisons.

This study has some main limitations. First, the proposed estimation protocol is based on macroscopic experimental data, such as a tensile test on bulk adhesive specimens and a torsional test on bonded tubular butt-joints, even if the model parameters to be estimated are related to the material constitutive behaviour (microscale). However, this is a very usual choice in experimental mechanics because of the difficulty to carry out direct mensuration of damage at the interface level. Second, the considered damage evolution law is a unique function of mechanical loadings (see Equation (10)), nevertheless it is established that other physics, such as environmental conditions, can contribute to the damage evolution. However, the interface model formulated by [16] can be readily generalized in order to account for multi-physics couplings as done for example by [41]. Third, the viscoplasticity and viscoelasticity typical of structural adhesives were not considered in the proposed model. Multi-physics coupling and viscosity aspects could be the object of further works. Lastly, because of a lack of experimental data available in the literature, this study uses only two kinds of adhesive materials for the estimation of model parameters. As a perspective, more adhesives should be compared in order to establish reliable correlations between adhesive material properties and damage parameters $\eta$ and $\omega$.

## 5. Conclusions

This paper proposes a first attempt at a numerical assessment of the damage parameters of viscosity $\eta$ and energy threshold $\omega$ of the imperfect interface model previously formulated by authors in [16]. The proposed estimation procedure draws on experimental data available in the literature, concerning macroscopic characterization tests on a structural adhesive commonly used in industry, in both bulk (tensile test) and joint (torsional test) configuration.

Several points arise from this preliminary numerical assessment and main are related to the physical meaning of model parameters $\eta$ and $\omega$:

- **They could represent an intrinsic material property** as the performed numerical estimation on the bulk configuration gives the same values of parameters regardless of the homogenization scheme used.
- **They could depend on the adhesive configuration** in agreement with the fact that the mechanical behaviour of an adhesive in bulk configuration is different from that of the same adhesive in joint form. In fact, the estimated parameters in the joint configuration are different from that found in the bulk one.
- In the joint configuration, parameters **could depend on the adopted homogenization scheme.**
- **They could depend on the adhesive material properties and on the test configuration**, however further investigation is needed to elucidate this point.
- The theoretical relationships between parameters in the bulk and in the joint configuration, assumed to be inversely proportional to the adhesive thickness ($\eta^{\varepsilon} = \eta\, \varepsilon^{-1}$ and $\omega^{\varepsilon} = \omega\, \varepsilon^{-1}$) [11,16,17] is not fulfilled.

Moreover, unanswered issues still remain, such as:

- the dependency of the model parameters on the considered microstructure (i.e., the shape of the porosity), particularly the influence of the microstructural parameter $C_G$ could be investigated;
- the dependency of the model parameters on the type of structural adhesive, to this aim several adhesive materials could be compared.
- In the above case, the possible dependency on the test configuration must be eliminated by using the same configuration for all different adhesives.

To corroborate these preliminary numerical insights and to answer the above-cited open questions, it will be crucial to expand this study by setting up a hybrid numerical/experimental protocol of identification. Within future research, several adhesive materials should be compared and *ad hoc* test configurations should be studied to investigate the correlation between parameters and physical properties.

**Author Contributions:** Conceptualization, M.L.R., F.L. and R.R.; methodology, M.L.R.; software, M.L.R.; validation, M.L.R., F.L. and R.R.; formal analysis, M.L.R.; investigation, M.L.R.; resources, M.L.R.; data curation, M.L.R.; writing—original draft preparation, M.L.R.; writing—review and editing, M.L.R., F.L. and R.R.; visualization, M.L.R.; supervision, F.L.; project administration, M.L.R., F.L. and R.R.; funding acquisition, R.R. All authors have read and agreed to the published version of the manuscript.

**Funding:** This work has been supported by the University of Ferrara via FAR grants 2020 and 2021.

**Institutional Review Board Statement:** Not applicable.

**Informed Consent Statement:** Not applicable.

**Data Availability Statement:** No new data were created or analyzed in this study. Data sharing is not applicable to this article.

**Conflicts of Interest:** The authors declare no conflict of interest.

## References

1. Goland, M.; Reissner, E. The stresses in cemented joints. *J. Appl. Mech.* **1944**, *11*, A17–A27. [CrossRef]
2. Gilibert, Y.; Rigolot, A. Asymptotic analysis of double adhesive bonded joints loaded in shear tension. *J. Méc. Appl.* **1979**, *3*, 341–372.
3. Benveniste, Y. The effective mechanical behaviour of composite materials with imperfect contact between the constituents. *Mech. Mater.* **1985**, *4*, 197–208. [CrossRef]
4. Hashin, Z. Thermoelastic properties of fiber composites with imperfect interface. *Mech. Mater.* **1990**, *8*, 333–348. [CrossRef]
5. Hashin, Z. Thermoelastic properties of particulate composites with imperfect interface. *J. Mech. Phys. Solid* **1991**, *39*, 745–762. [CrossRef]
6. Hashin, Z. Extremum principles for elastic heterogenous media with imperfect interfaces and their application to bounding of effective moduli. *J. Mech. Phys. Solid* **1992**, *40*, 767–781. [CrossRef]
7. Needleman, A. Micromechanical modelling of interfacial decohesion. *Ultramicroscopy* **1992**, *40*, 203–214. [CrossRef]
8. Bövik, P. On the modelling of thin interface layers in elastic and acoustic scattering problems. *Quart. J. Mech. Appl. Math.* **1994**, *47*, 17–42. [CrossRef]
9. Benveniste, Y.; Miloh, T. Imperfect soft and stiff interfaces in two-dimensional elasticity. *Mech. Mater.* **2001**, *33*, 309–323. [CrossRef]
10. Frémond, M. Adhérence des solides. *J. Méc. Théorique Appl.* **1987**, *6*, 383–407.
11. Bonetti, E.; Bonfanti, G.; Lebon, F.; Rizzoni, R. A model of imperfect interface with damage. *Meccanica* **2017**, *52*, 1911–1922. [CrossRef]
12. Raous, M.; Cangémi, L.; Cocou, M. A consistent model coupling adhesion, friction and unilateral contact. *Comput. Methods Appl. Mech. Eng.* **1999**, *177*, 383—399. [CrossRef]
13. Del Piero, G.; Raous, M. A unified model for adhesive interfaces with damage, viscosity, and friction. *Eur. J. Mech. A Solid* **2010**, *29*, 496–507. [CrossRef]
14. Chaboche, J.-L.; Feyel, F.; Monerie, Y. Interface debonding model: A viscous regularization with a limited rate dependency. *Int. J. Solid Struct.* **2001**, *38*, 3127–3160. [CrossRef]
15. Corigliano, A.; Mariani, S. Simulation of damage in composites by means of interface models: Parameter identification. *Comp. Sci. Technol.* **2001**, *61*, 2299–2315. [CrossRef]
16. Raffa, M.L.; Lebon, F.; Rizzoni, R. A micromechanical model of a hard interface with micro-cracking damage. *Int. J. Mech. Sci.* **2022**, *216*, 106974. [CrossRef]
17. Rizzoni, R.; Dumont, S.; Lebon, F.; Sacco, S. Higher order model for soft and hard interfaces. *Int. J. Solid Struct.* **2014**, *51*, 4137–4148. [CrossRef]
18. Eshelby, J.D. Book Review: Progress in solid mechanics. Vol. 1. Edited by IN SNEDDON and R. HILL, North-Holland Publishing Company, Amsterdam, 1960, 448 pp., 100s. *J. Mech. Phys. Solid* **1961**, *9*, 67. [CrossRef]
19. Budiansky, B.; O'Connell, R.J. Elastic moduli of a cracked solid. *Int. J. Solid Struct.* **1976**, *12*, 81–97. [CrossRef]
20. Andrieux, S.; Bamberger, Y.; Marigo, J.J. Un modèle de matériau microfissuré pour les bétons et les roches. *J. Méc. Théor. Appl.* **1986**, *5*, 471–513.
21. Kachanov, M. Elastic solids with many cracks and related problems. *Adv. Appl. Mech.* **1994**, *30*, 259–445.
22. Mauge C.; Kachanov M. Effective elastic properties of an anisotropic material with arbitrarily oriented interacting cracks. *J. Mech. Phys. Solid* **1994**, *42*, 561–584. [CrossRef]
23. Pan, H.H.; Weng, G.J. Elastic moduli of heterogeneous solids with ellipsoidal inclusions and elliptic cracks. *Acta Mech.* **1995**, *110*, 73–94. [CrossRef]
24. Bruno, G.; Kachanov, M.; Sevostianov, I.; Shyam, A. Micromechanical modelling of non-linear stress-strain behavior of polycrystalline microcracked materials under tension. *Acta Mater.* **2019**, *164*, 50–59. [CrossRef]
25. Sevostianov, I.; Kachanov, M. On some controversial issues in effective field approaches to the problem of the overall elastic properties. *Mech. Mat.* **2014**, *69*, 93–105. [CrossRef]
26. Maurel-Pantel, A.; Lamberti, M.; Raffa, M.L.; Suarez, C.; Ascione, F.; Lebon, F. Modelling of a GFRP adhesive connection by an imperfect soft interface model with initial damage. *Comp. Struct.* **2020**, *239*, 112034. [CrossRef]
27. Dumont, V.; Badulescu, C.; Adrien, J.; Carrere, N.; Thévenet, D.; Maire, E. Experimental investigation of porosities evolution in a bonded assembly by means of X-ray tomography. *J. Adh.* **2021**, *97*, 528–552. [CrossRef]
28. Kachanov, M.; Sevostianov, I. *Micromechanics of Materials, with Applications*; Springer: Cham, Switzerland, 2018; Volume 249.
29. Kachanov, M.; Sevostianov, I. On quantitative characterization of microstructures and effective properties. *Int. J. Solid Struct.* **2005**, *42*, 309–336. [CrossRef]
30. Welemane, H.; Cormery, F. Some remarks on the damage unilateral effect modelling for microcracked materials. *Int. J. Damage Mech.* **2002**, *11*, 65–86. [CrossRef]
31. Welemane, H.; Goidescu, C. Isotropic brittle damage and unilateral effect. *Compte Rendu Méc.* **2010**, *338*, 271–276. [CrossRef]

32. Goidescu, C.; Welemane, H.; Kondo, D.; Gruescu, C. Microcracks closure effects in initially orthotropic materials. *Eur. J. Mech. A Solid* **2013**, *37*, 172–184. [CrossRef]
33. Goidescu, C.; Welemane, H.; Pantalé, O.; Karama, M.; Kondo, D. Anisotropic unilateral damage with initial orthotropy: A micromechanics-based approach. *Int. J. Damage Mech.* **2015**, *24*, 313–337. [CrossRef]
34. Mori, T.; Tanaka, K. Average stress in matrix and average elastic energy of materials with misfitting inclusions. *Acta Metal.* **1973**, *21*, 571–574. [CrossRef]
35. Murakami, S.; Sekiguchi, Y.; Sato, C.; Yokoi, E.; Furusawa, T. Strength of cylindrical butt joints bonded with epoxy adhesives under combined static or high-rate loading. *Int. J. Adh. Adhes.* **2016**, *67*, 86–93. [CrossRef]
36. Kosmann, J.; Klapp, O.; Holzhüter, D.; Schollerer, M.J.; Fiedler, A.; Nagel, C.; Hühne, C. Measurement of epoxy film adhesive properties in torsion and tension using tubular butt joints. *Int. J. Adh. Adhes.* **2018**, *83*, 50–58. [CrossRef]
37. Rohatgi, A. WebPlotDigitizer. Version: 4.4. Available online: https://automeris.io/WebPlotDigitizer (accessed on 4 July 2022).
38. Wolfram Research, Inc. *Mathematica*; Version 12.1; Wolfram Research, Inc.: Champaign, IL, USA, 2020.
39. Raffa, M.L.; Rizzoni, R.; Lebon, F. A model of damage for brittle and ductile adhesives in glued butt joints. *Technologies* **2021**, *9*, 19. [CrossRef]
40. Sevostianov, I.; Kachanov, M. Effective properties of heterogeneous materials: Proper application of the non-interaction and the "dilute limit" approximations. *Int. J. Eng. Sci.* **2012**, *58*, 124–128. [CrossRef]
41. Serpilli, M.; Rizzoni, R.; Lebon, F.; Dumont, S. An asymptotic derivation of a general imperfect interface law for linear multiphysics composites. *Int. J. Solids Struct.* **2019**, *180–181*, 97–107. [CrossRef]

materials

Article

# Constitutive Law Identification and Fatigue Characterization of Rigid PUR Elastomers 80 ShA and 90 ShA

Krzysztof Junik [1], Grzegorz Lesiuk [1,*], Szymon Duda [1], Krzysztof Jamroziak [1], Wojciech Błażejewski [1], Paweł Zielonka [1], Tomasz Socha [2], Arkadiusz Denisiewicz [2], Krzysztof Kula [2] and Anna Szczurek [1]

1 Department of Mechanics, Materials Science and Biomedical Engineering, Faculty of Mechanical Engineering, Wroclaw University of Science and Technology, Smoluchowskiego 25, 50370 Wrocław, Poland
2 The Faculty of Civil Engineering, Architecture and Environmental Engineering, University of Zielona Góra, ul. prof. Z. Szafrana 1, 65516 Zielona Góra, Poland
* Correspondence: grzegorz.lesiuk@pwr.edu.pl; Tel.: +48-713-204-216

**Abstract:** This paper presents the results of a study of polyurethane rigid (PUR) elastomers in terms of the constitutive law identification, and analyses the effect of polyurethane elastomers' hardness on fatigue properties. The research objects were PUR materials based on 4,4′-diphenylmethane diisocyanate (MDI) with the hardness of 80 ShA and 90 ShA, typically used in various industrial applications. Based on the performed experimental campaign under static and cyclic loading, the constitutive model proposed by Ogden is most appropriate. In addition, a hybrid numerical–experimental analysis (using FEM-DIC) of diabolo specimens' behaviour is carried out in fatigue tests. Based on the performed fatigue test, it is worth noting that the energy approach describes the fatigue process synonymously compared to the displacement or strain approach. Finally, simple fatigue characteristics were analyzed and statistically validated for both PUR material configurations.

**Keywords:** hardness; polyurethane; fatigue; numerical analysis

**Citation:** Junik, K.; Lesiuk, G.; Duda, S.; Jamroziak, K.; Błażejewski, W.; Zielonka, P.; Socha, T.; Denisiewicz, A.; Kula, K.; Szczurek, A. Constitutive Law Identification and Fatigue Characterization of Rigid PUR Elastomers 80 ShA and 90 ShA. *Materials* **2022**, *15*, 6745. https://doi.org/10.3390/ma15196745

Academic Editors: Francesco Iacoviello and Andrea Spagnoli

Received: 25 July 2022
Accepted: 22 September 2022
Published: 28 September 2022



## 1. Introduction

A material called "polyurethane" is widely used with various properties and functionalities. The number of varieties of polyurethane allows this material to be adapted to the needs and desired properties, such as stiffness and flexibility. Among the wide range of purposes, the following can be distinguished: insulation (e.g., of buildings, pipelines, household appliances), depreciation (e.g., in the furniture industry), adhesives and coatings, material for mattresses, clothing, shoe soles, rollers, tires, and auto parts. Rubber and polymeric materials are also commonly used in vehicle suspension systems mainly due to their hyperelastic characteristics, which include low weight, corrosion resistance, and a high capacity for vibration damping and energy absorption. By shaping various mechanical parts and material modifications (composites, layered structures, hybrid joints), it is possible to achieve the appropriate stiffness characteristics. Structural components made of elastomers such as polyurethane or rubber (mainly) work excellently in compression and shear stress states, because they can be easily damaged during tensile stress [1]. One of the excellent examples of structural components used in suspension systems is metal–elastomer (such as the considered PUR material) bushings, which are indirectly related to the research topic undertaken in recent papers [2–4]. As already mentioned, they are used to connect individual elements of the suspension system. Suspension bushings are one of the smaller components of the chassis, but they cause many problems during vehicle operation due to fatigue loading [5–10]. The contribution of the material selection for the chassis system is substantial, and the knowledge about material properties is essential in fatigue lifetime prediction. It allows for properly designing the control arms and links, and the seating of all components in the vehicle's structure. In comparison with rubber, PUR, with its properties, seems to be ideal for use in suspension systems [2]. It retains its elastic properties at temperatures as low as −40 °C. The structure of this polymer resembles a tangled line, which becomes intertwined when stretched, making it very difficult to break. It should be noted that common polyurethane configurations found in the marketplace come in two grades [2–4,11]: soft rubber replacement (70–80 ShA) and hard rubber replacement (90–95 ShA).

Therefore, an essential cognitive objective of this paper is the comparative analysis of the behaviour of this material in different varieties of hardness.

As the fatigue nature of elastomers, rubber-like materials [12], are different from the microscopic perspective, an excellent review of fatigue and fatigue crack growth rate analysis was performed in the review papers [8,13]. Based on this, it can be concluded that there exists in the literature three major competitive approaches in fatigue curves description, similar to metal's description of fatigue curves: strain-based models (strain predictor [5,14]), stress-based models (stress predictor [15,16]), and energy-based models (energy predictor [17,18]). However, based on the literature review, it can be concluded that several attempts to describe the fatigue phenomenon under displacement control mode were performed in the paper [10]. Recently, [19] summarized the experimental fatigue campaign for rubber elastomers testing strategy. Excluding the ASTM D4482 standard, volumetric geometries of specimens were successfully tested in fatigue testing: diabolo specimens/hourglass shape/3D dumbbell specimens: [20–28]; dumbbell, dogbone specimens: [20,21,29–33]; cylindrical: [34–38]; ring: [39]; disc-shaped: [40,41]; and thin-film: [7,42].

In general, the fatigue lifetime can be expressed as:

$$N_f = \alpha(D)^{\kappa} \tag{1}$$

where:

$\alpha$ and $\kappa$ are experimentally determined constants;
$D$—specific damage parameter (predictor).

As reported in various experimental papers, the displacement-controlled experiment provides reliable fatigue data values [10,43,44]. An analysis of the literature reveals a significant lack of evaluation of the effect of PUR elastomeric material hardness on fatigue properties. Therefore, the main objective is to evaluate the fatigue properties of this material in the range of Shore 80 and 90 hardness. The selection of these levels of hardness is motivated by practical purposes. The PUR-type materials of 80 ShA hardness are an interesting alternative to rubber in automotive applications, assuming the conventional operation of automotive vehicles. On the other hand, materials of 90 ShA hardness are offered to the automotive market as stiffer and intended for racing applications. From an application perspective, the issue of the impact of hardness is becoming of interest to the automotive industry and other industries.

Considering the above, this article fills a gap in the results of fatigue studies of polyurethanes with two different hardness levels of 80 and 90 ShA.

## 2. Materials and Methods

The objects of interest were the two groups of polyurethane materials with levels of hardness of 80 ShA and 90 ShA. Duroplastic polyurethane (based on MDI for methylene diphenyl 4,4′-diisocyanate) was obtained by casting in an automated molding unit—a compound with proprieties fulfilling the designer's requirements is formed from several mixed compounds. The resulting mixture is then cast into a preheated mold, and left with the mix in an oven to cure. The curing time depends on the degree of hardness required. After that, the component is demolded and the product is removed from the mold, before being returned to the oven for about 12 h for annealing. This stage is followed by mechanical processing of the over-molded parts, i.e.: turning and grinding until the final product is obtained. The study identified the target structure of the polyurethane chain and individual bonds by Raman spectroscopy.

To reveal the differences in the chemical composition of the two types of polyurethane samples, the Raman spectroscopy measurements were carried out using a Raman spectrometer LabRAM HR800 Horiba Jobin Yvon (Kyoto, Japan). The sample excitation was provided with a He–Ne laser source working at 632.8 nm. The measurements were performed in a 4000–10 $cm^{-1}$ region with a spectral resolution of 2.5 $cm^{-1}$. The received spectra were processed with the extraction of the background.

### 2.1. Static Tensile Tests and Numerical Identification of Constitutive Law

Static tensile testing is one of the primary testing methods for determining the basic mechanical properties of structural materials. During the test, the material's response, in the form of elongation, to a given tensile load propagating at a constant rate is recorded. Specimens used for this type of test, as well as the test process itself for elastomers, are standardized and described in ASTM D412. The test was prepared following the mentioned standard.

The finite element method is applied to deliver strain–displacement curves and model parameters. Several material models were assessed, such as reduced polynomial, Arruda–Boyce, and Ogden.

The Arruda–Boyce model is expressed as:

$$W = Nk_Bt\sqrt{n}\left[\beta\lambda_{chain} - \sqrt{n}\ln\left(\frac{\sinh\beta}{\beta}\right)\right] \tag{2}$$

where:

$n$ is the number of chain segments;
$k_B$—Boltzmann constant;
$t$—temperature expressed in Kelvins;
$N$—the number of chains in the network of a cross-linked structure.

The Ogden [45] model is represented as:

$$W = \sum_{i=1}^{N} \frac{2\mu_i}{\alpha_i^2}\left(\overline{\lambda}_1^{\alpha_i} + \overline{\lambda}_2^{\alpha_i} + \overline{\lambda}_3^{\alpha_i} - 3\right) \tag{3}$$

where $\mu_i$ and $\alpha_i$ material constants.

For compressible materials, Bergstrom [46] added an additional component:

$$W = \sum_{i=1}^{N} \frac{2\mu_i}{\alpha_i^2}\left(\overline{\lambda}_1^{\alpha_i} + \overline{\lambda}_2^{\alpha_i} + \overline{\lambda}_3^{\alpha_i} - 3\right) + \sum_{i=1}^{N} \frac{1}{D_i}\left(J^{el} - 1\right)^{2i} \tag{4}$$

where:

$D_i$ denotes volumetric change parameter.

The experimental data from uniaxial and planar tests were uploaded to determine model parameters. The evaluation shows that Ogden 3 parametric fits the experiment data accurately. The numerical model shows acceptable conformity with provided experiment data. A numerical model is represented by a solid geometry of a quarter of the diabolo-shaped specimen according to Figure 1. Finite element analysis was performed using Abaqus software in a static term.

This modelling aims to obtain the relationship between strain and displacement for this specimen geometry. This approach allows the representation of the fatigue data in the ε–N relationship. Boundary conditions need to be applied to the geometry to provide adequate loading conditions. To keep the symmetry of the specimen, the boundaries along the x- and z-axis were provided. Transfer of the load to the sample is pursued by coupling the reference points to the inner surfaces (Figure 2a). This specific connection results in a design of the real specimen used for the fatigue test. Two reference points were used to provide fixed support (U1 = U2 = U3 = 0; translation in all three directions is pinned), and the load is supplied by a displacement boundary condition applied to the second reference point.

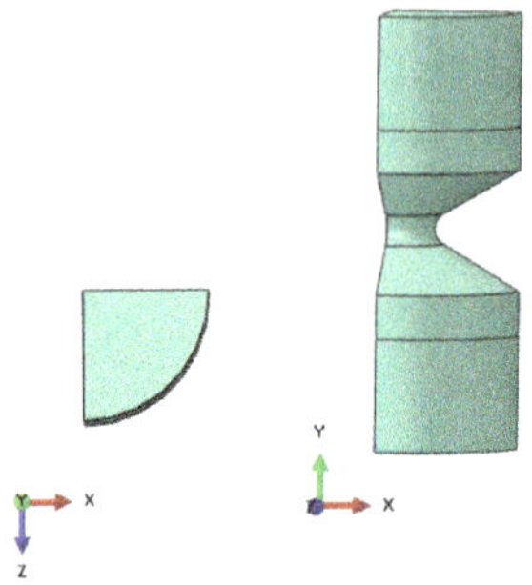


**Figure 1.** The geometry of a diabolo-shaped specimen.

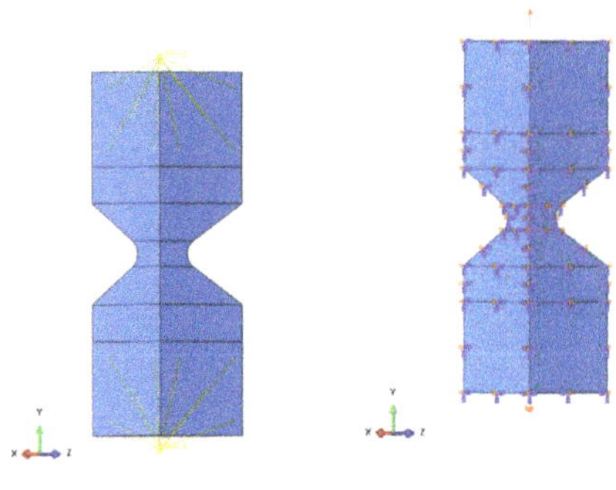

(**a**) (**b**)

**Figure 2.** Representation of boundary conditions applied to the specimen, (**a**) coupling connection, (**b**) symmetry boundary conditions.

The presented geometry with the applied loading conditions meshed was into a finite continuum object. The mesh applied to the object consists of 39,402 quadratic hexahedral elements of type C3D20RH. The element size set for this model is 2 mm. However, some regions were enriched with additional nodes. Figure 3 shows the final meshed geometry; additionally, areas with the most significant number of nodes are noticeable. This enriched region allowed for obtaining more adequate results.

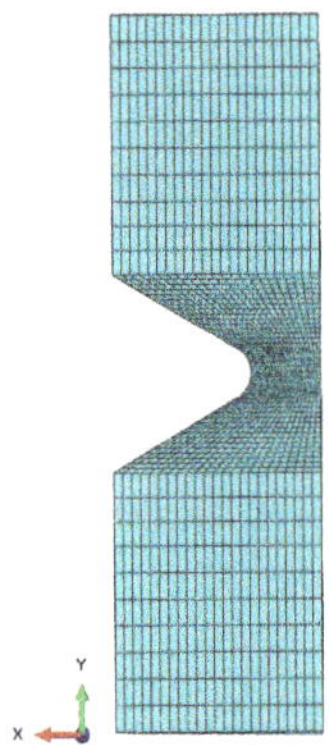


**Figure 3.** The meshed object used in FEA.

This finite element model was applied to investigate two material models, polyurethane 80 and 90 ShA. The simulation was run to reflect the static tensile test of this diabolo-shaped specimen and provide strain values concerning the displacement.

Additionally, the results from the FEM were assessed by applying the digital image correlation method. The Dantec Q-400 system with 2 cameras (4.8 megapixels) (Skovlunde, Denmark) was used to validate the strain pattern and value. The DIC was correlated with the FEM and the results were obtained from the Istra4D software.

### 2.2. Fatigue Tests

Fatigue tests were performed using the displacement control mode method for two displacement ratios $R_d = 0$. For the experimental campaign, a special type of specimen, diabolo, was designed, as shown in Figure 4. An essential part of the griping system was to develop a proper connection between the polyurethane specimens and metal insert.

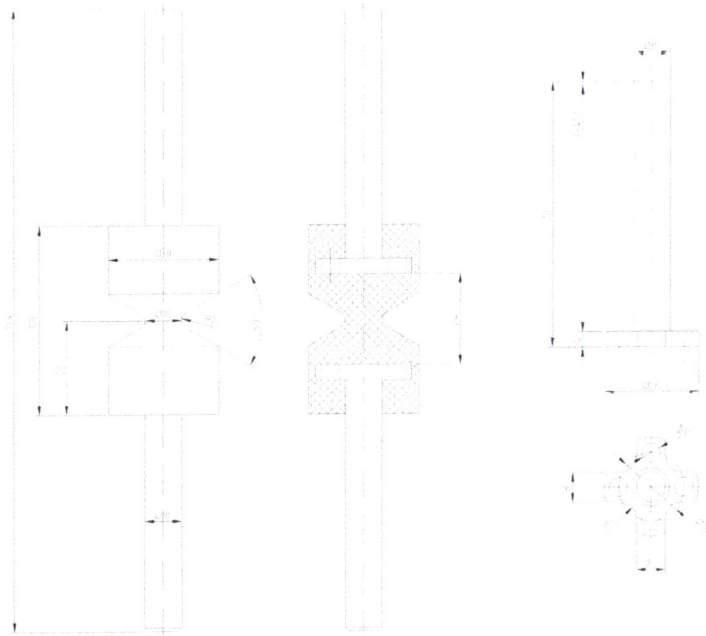

**Figure 4.** Designed diabolo specimen with metallic fixture for fatigue machine (in mm).

Specimens designed in this way were manufactured by casting. Numerical analyses made it possible to determine a wide range of fatigue loads to calculate all fatigue parameters, such as stress, strain, or strain energy density, based on which fatigue diagrams were constructed afterwards.

All experiments were performed on an MTS858 Bionix (Chesterfield, MI 48051, USA) testing machine (Figure 5) with constant amplitude loads. The specimens were loaded cyclically at a frequency of 2–3 Hz. The loading frequency was chosen to avoid temperature rise during the experiment. The temperature was monitored periodically using a pyrometer. During the fatigue phase, it did not exceed a difference of 3–4 degrees Celsius. Experiments were carried out in an air-conditioned laboratory under RT conditions.

The control signal was displacement. Further, fatigue diagrams of d–N, ε–N, and W–N were constructed based on the computations, enabling a comparison of fatigue life of PUR materials with a hardness of 80 ShA and 90 ShA. A 50% decrease in specimen stiffness caused by the development of fatigue damage was used as the failure criterion.

**Figure 5.** MTS858 Bionix fatigue test stand used for PUR fatigue testing.

## 3. Results and Discussion

Figure 6 presents Raman spectra of polyurethane samples. In spectra of both types of material, bands characteristic of polyurethane structure are observed. Bands characteristic for aromatic ring vibrations are present at 639 cm$^{-1}$, 865 cm$^{-1}$, and 1615 cm$^{-1}$ [47–49]. Bands confirming amide groups are observed at 1255 cm$^{-1}$ and 1540 cm$^{-1}$, characteristic of C-N and N-H stretching of amide II [48–50]. Bands characteristic for N=C=O stretching and $CH_2$ bending vibrations are observed at 1438 cm$^{-1}$ [48,50]. The band observed at 1185 cm$^{-1}$ confirms C-O-C links [51]. The band at 1700 cm$^{-1}$ confirms the presence of hydrogen-bonded carbonyl groups (C=O), and the band at 1735 cm$^{-1}$, free carbonyl groups [52]. In the presented spectra, the ratio of hydrogen bonded to free C=O groups is higher for the 90 ShA material. The strong bands at 2869 cm$^{-1}$, and 2923 cm$^{-1}$ are characteristic of $CH_2$ stretching [49]. Despite the convergent chemical structure of both types of studied polyurethanes, the Raman spectra reveal chemical differences between them with bands marked with "*", "**", "***". In the spectrum of the 80 ShA sample, additional bands characteristic for $CH_2$ rocking at 752 cm$^{-1}$, for $CH_3$ bending as four bands in the range of 1364–1416 cm$^{-1}$, and for $CH_2$ deformation vibrations at 1490 cm$^{-1}$ are observed. This is connected with the higher content of politetrahydrofuran, containing unbranded hydrocarbon chains used for the synthesis of 80 ShA material in comparison to the 90 ShA material. The higher amount of hydrocarbon chains constituting the soft segments [53] in the polyurethane structure in the red material, and the higher amount of hydrogen-bonded carbonyl groups constituting hard segments in the polyurethane structure in the yellow material affect the mechanical properties of polyurethanes, causing higher hardness of the yellow material.

### *3.1. Static Tensile Test Results and Constitutive Law Identification*

In total, ten dumbbell specimens (type S1) were cut from molded plates. Before testing, all specimens were conditioned 24 h/23 ± 2 °C, 50 ± 10% RH. A tensile test was performed under displacement control mode with a rate 500 mm/min. During the test, force, displacement, and strain were measured using an extensometer for elastomeric materials. Tensile stress–strain curves are shown in Figures 7 and 8 for 80 ShA and 90 ShA, respectively.

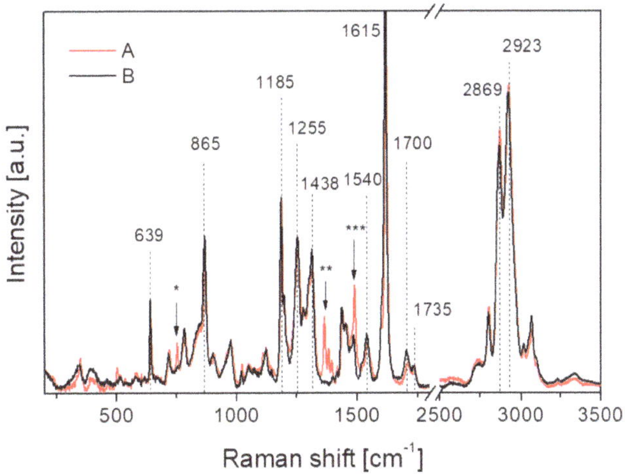


**Figure 6.** Raman spectra of polyurethanes: A—80 ShA sample, B—90 ShA sample.

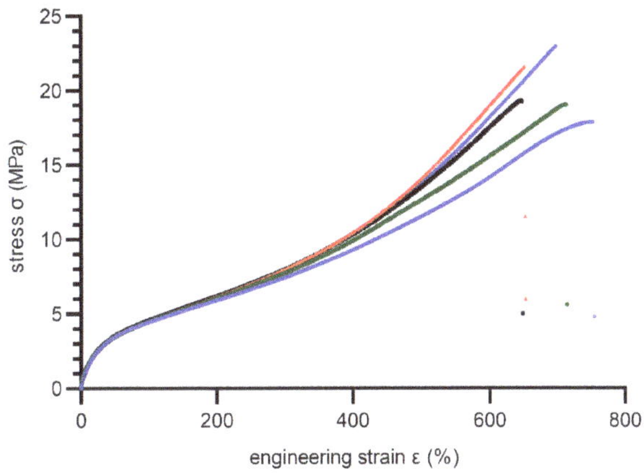


**Figure 7.** Stress–strain curves obtained during the tensile test for 80 ShA PUR material (solid lines represents different specimens, dots–measurement points after break).

As is noticeable, the 80 ShA material exhibits larger elongation at break compared to the 90 ShA material. Summarized results of the tensile test are included in Table 1.

Abaqus CEA software allows for assessing the hyperelastic models by checking the stability of the strain in a certain range. For this purpose, two experimental data sets were provided to choose the best-fitted hyperelastic material model. The uniaxial and planar test data were counted and evaluated for choosing the more accurate model. Analysis was performed for models such as reduced polynomial, Ogden, and Arruda–Boyce, and are presented in Figures 9 and 10.

Based on the performed numerical procedure, the best-fitting test was performed for the Ogden model. Obtained parameters are listed in Tables 2 and 3.

To compare the results from the FEM, 3 values of strain were chosen and correlated with DIC images in terms of the reaction force acting on the geometry. The principal strain 1, which shows maximum strain for every data point mapped as a color plot on the image, was taken for steps 60, 30, and 15, and presented in Figures 11–13. The comparison is based on the strain values, which is identical for DIC and FEM. Finally, the reaction force is a value that exhibits discrepancy.

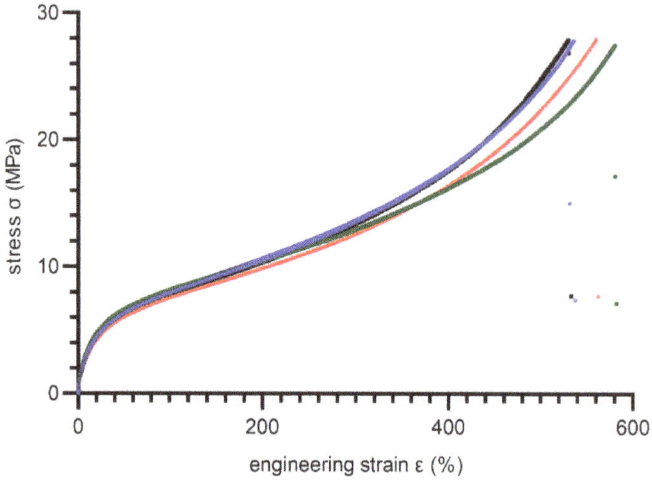


**Figure 8.** Stress–strain curves obtained during the tensile test for 90 ShA PUR material (solid lines represents different specimens, dots–measurement points after break).

**Table 1.** Tensile test results analysis for 80 ShA and 90 ShA material configuration.

| Specimen ID | UTS—Ultimate Tensile Strength in MPa | A—Elongation at Break In % |
|---|---|---|
| PU80_#1 | 17.9 | 749.2 |
| PU80_#2 | 19.4 | 646.4 |
| PU80_#3 | 21.6 | 651.0 |
| PU80_#4 | 19.0 | 710.4 |
| PU80_#5 | 23.7 | 711.0 |
| PU80 (median ± std.dev) | 19.4 ± 2.3 | 710.4 ± 43.9 |
| PU90_#1 | 28.0 | 530.9 |
| PU90_#2 | 27.9 | 529.5 |
| PU90_#3 | 28.1 | 559.6 |
| PU90_#4 | 27.5 | 579.5 |
| PU90_#5 | 27.9 | 535.3 |
| PU90 (median ± std.dev) | 27.9 ± 0.2 | 535.3 ± 21.9 |

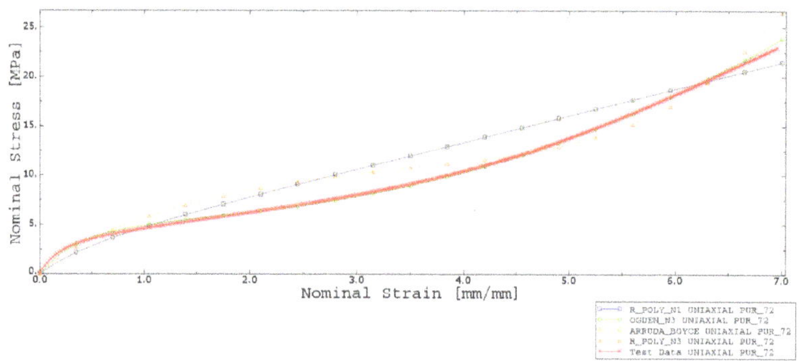


**Figure 9.** Stress–strain curves—experimental (red curve—test data) and numerical data fitting (Ogden, Arruda–Boyce, and reduced polynomial with 1 and 3 model parameters) for 80 ShA.

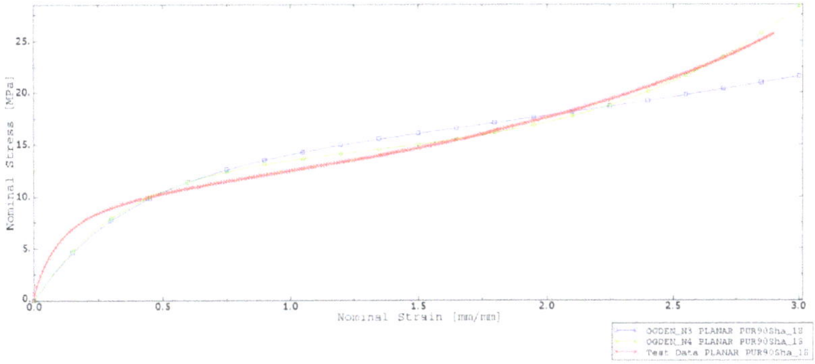


**Figure 10.** Stress–strain curves—experimental (red curve—test data) and numerical data fitting (Ogden, Arruda–Boyce, and reduced polynomial with 1 and 3 model parameters) for 90 ShA.

**Table 2.** Ogden model parameters for 80 ShA.

| N | $\mu_i$ | $\alpha_i$ |
|---|---|---|
| 1 | 4.34400372 | −0.380731347 |
| 2 | 0.210081339 | 3.47276528 |
| 3 | $4.921467147 \times 10^{-3}$ | 6.93803394 |

**Table 3.** Ogden model parameters for 90 ShA.

| N | $\mu_i$ | $\alpha_i$ |
|---|---|---|
| 1 | −707.693690 | 1.13390568 |
| 2 | 316.165559 | 1.34789067 |
| 3 | 401.397194 | 0.904727974 |
| 4 | $8.765965905 \times 10^{-3}$ | −7.00870716 |

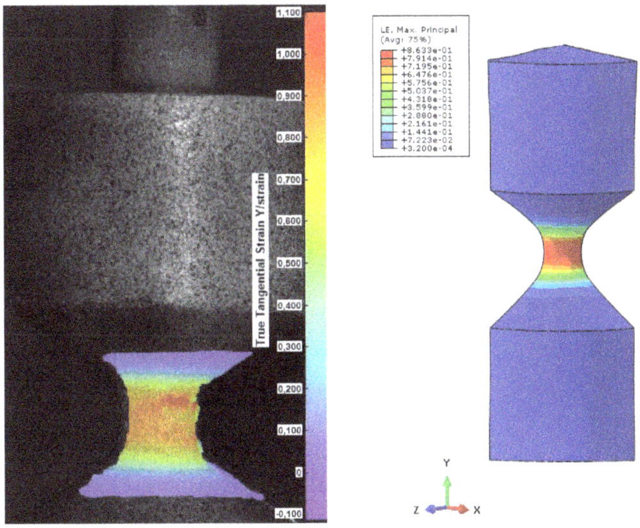


**Figure 11.** Comparison of DIC and FEM in terms of the obtained value of strain for step 60 (force DIC = 591 N, FEM force = 828 N).

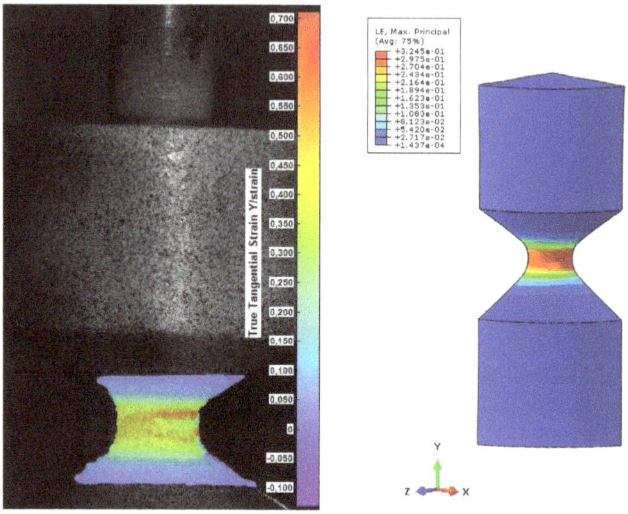


**Figure 12.** Comparison of DIC and FEM in terms of the obtained value of strain for step 30 (force DIC = 465 N, FEM force = 468 N).

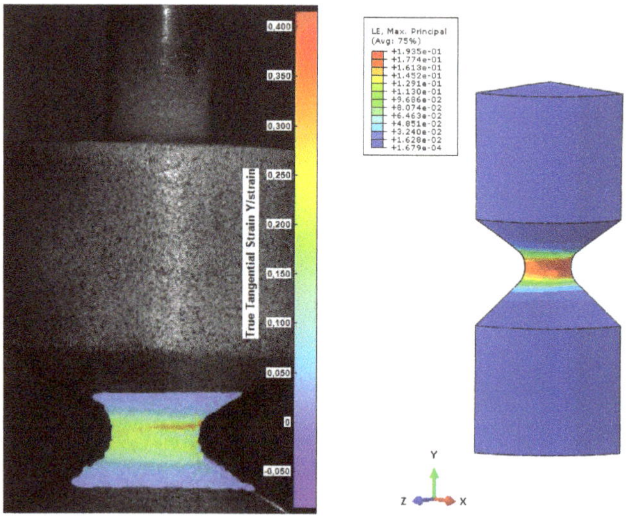


**Figure 13.** Comparison of DIC and FEM in terms of the obtained value of strain for step 15 (force DIC = 280 N, FEM force = 302 N).

According to the validation, after a certain strain level, the values (reaction force) diverge. In Figure 11, the reaction force for DIC is lower than 200 N. It is caused by incorrect adhesion along PUR and the provided pattern. Due to the high elongation of the material, the thin stochastic pattern on the surface is deboned, influencing the results of strain.

### 3.2. Fatigue Results

Fatigue experimental tests were conducted on a Bionix MTS858 machine by controlling the displacement signal. The tests were conducted on specially designed volumetric diabolo-type specimens. All results reported below are related to two hardness states, 80 ShA and 90 ShA, with a displacement ratio R = 0. The tests were conducted until the initial stiffness of the specimen drops by 50%. Typical power-law models (for displacement d, and strain ε fatigue parameters) were used in the description of the fatigue curves for better comparison with other materials available in the literature:

$$d_{max} = \alpha\left(N_f\right)^n \quad (5)$$

$$\varepsilon_{max} = A\left(N_f\right)^m \quad (6)$$

$$W_s = W_0\left(N_f\right)^\gamma \quad (7)$$

where:

$\alpha$ $A$, $m$, $n$, $W_0$, $\gamma$—experimentally determined constants.

Due to the complexity of the diabolo-type specimen shape, and the difficulty of directly measuring strains with the typical strain gauge, extensometer-type sensors are used. The relationship between displacement and maximum principal strain in the specimen was calibrated based on the DIC-FEM analyses performed in the previous chapters. In this way, it was possible to proceed with the construction of fatigue diagrams. Figure 14 shows the calibration curve of the strain–displacement relationship that was controlled during the experiment.

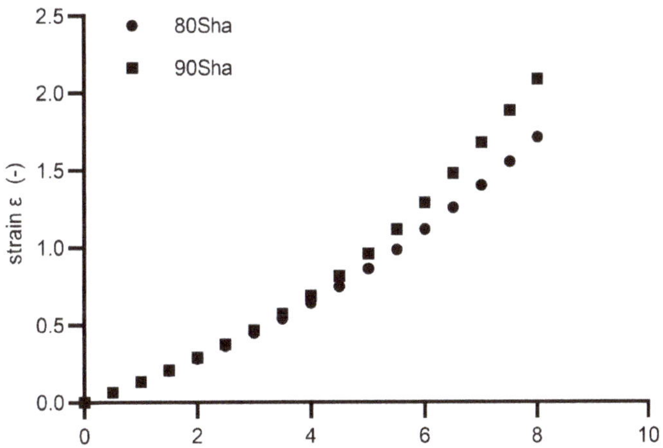


**Figure 14.** Nominal strain vs. displacement for diabolo specimens used in the experimental campaign.

Similarly, a calibration between the energy parameter U was performed based on numerical calculations. The calibration results are shown in Figure 15. Fatigue curves for the 80 ShA and 90 ShA materials for the initial control signal, displacement, are shown in Figure 16. The 95% confidence intervals are also marked on this diagram. The results show that the differences in fatigue life levels are statistically significant. A similar phenomenon is observed when the quantity describing the fatigue process is strain, shown in Figure 17. On the other hand, the question arises whether, in describing elastomeric materials, assuming such quantities as strain or stress, by directly reproducing the description of the fatigue phenomenon observed in metals, is appropriate. In this case, these characteristics are significantly different in the tensile test. Therefore, a universal quantity such as strain energy was used—Figure 18. Energy as a criterion quantity dimensionally combines force and displacement, which, in the engineering sense, allows us to consider both together in describing the phenomenon.

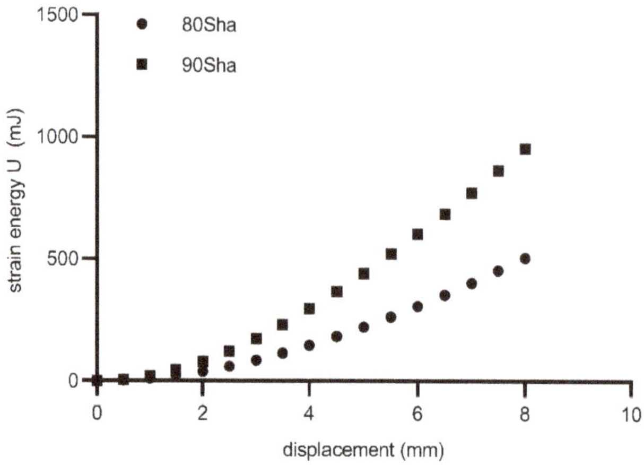


**Figure 15.** Energy vs. displacement for diabolo specimens used in the experimental campaign.

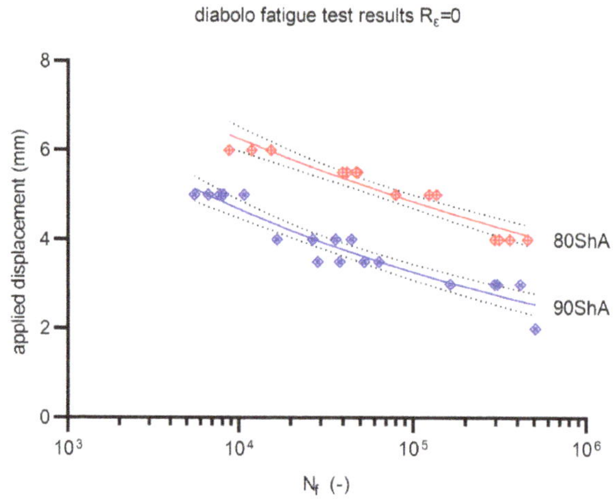


**Figure 16.** Displacement control mode fatigue data for R = 0 (80 ShA and 90 ShA).

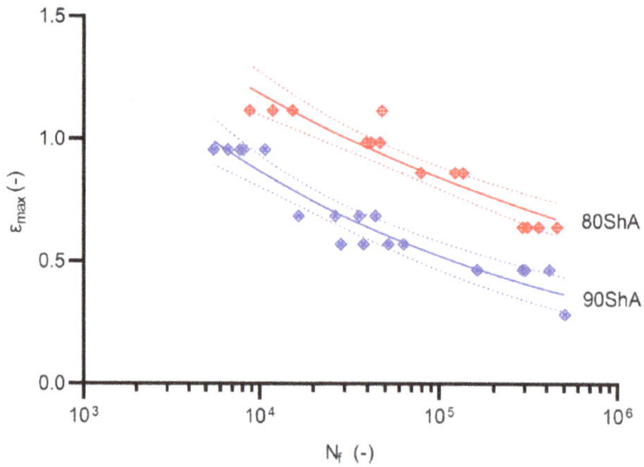


**Figure 17.** Fatigue data for R = 0 represented by $\varepsilon_{max}$-N approach (80 ShA and 90 ShA).

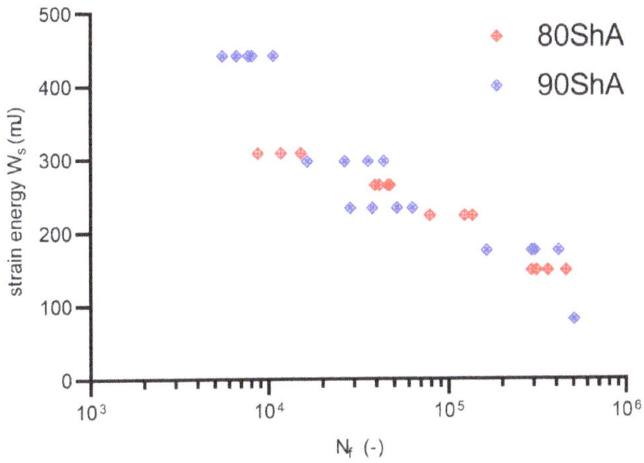


**Figure 18.** Fatigue data for R = 0 represented by $W_s$–N energy approach (80 ShA and 90 ShA).

Statistical fitting and analyses were performed in Graphpad PRISM environment. All results are collected in Table 4.

**Table 4.** Fatigue models data fitting for 80 ShA and 90 ShA.

| | $\alpha$ (mm) | $n$ (−) | $R^2$ | $A$ (−) | M (−) | $R^2$ | $W_0$ (mJ) | $\gamma$ (−) | $R^2$ |
|---|---|---|---|---|---|---|---|---|---|
| 80 ShA | 17.32 | −0.1106 | 0.91 | 4.627 | −0.148 | 0.87 | 1870 | −0.189 | 0.91 |
| 90 ShA | 19.62 | −0.1556 | 0.90 | 6.592 | −0.2201 | 0.91 | 5362 | −0.283 | 0.92 |

The energy description also results in a better fit of the curves, as evidenced by the greater or equal $R^2$ values obtained for the underlying energy model compared to the displacement or strain models. Energy as a criterion quantity unambiguously describes the fatigue process, and further observations in the field of PUR materials modeling should be performed from the energy perspective.

## 4. Conclusions

All the research and analysis conducted on PUR materials in the hardness range from 80 ShA and 90 ShA allows us to draw the following conclusions:

- Ogden's model provides the best description of the behavior of polyurethane material based on MDI for methylene diphenyl 4,4′-diisocyanate for both hardness types in the range of 80 ShA and 90 ShA;
- The higher content of politetrahydrofuran containing unbranded hydrocarbon chains is characteristic for 80 ShA material compared to the 90 ShA material—this allows for the mechanical properties of both materials to change in the static range in the sense of stress–strain curves in uniaxial tensile testing. The 90 ShA material is characterized by higher stiffness and stress levels characteristic of 100%, 200%, and 300% strain levels;
- Regarding strain approach, the fatigue process of PUR elastomers for 80 ShA and 90 ShA hardness show statistically significant differences. Under displacement control mode, the 80 ShA material exhibits an apparently higher fatigue strength;
- The fatigue process analysis from the strain energy point of view makes it possible to describe it unambiguously. In the case analyzed, there are no significant differences in the *W*–*N* fatigue curves for the 80 ShA and 90 ShA hardness levels;
- The energy approach is also characterized by a better statistical fitting of the measurement data to the energy model. It enables the accurate prediction of the fatigue life of

components, regardless of the ranges and types of loads during fatigue tests based on the force or displacement control mode of fatigue experiments.

**Author Contributions:** Conceptualization, K.J. (Krzysztof Junik) and G.L.; methodology, K.J. (Krzysztof Junik); software, S.D.; validation, T.S., A.D., and K.K.; formal analysis, W.B. and G.L.; investigation, K.J. (Krzysztof Junik), A.S., and P.Z.; resources, K.J. (Krzysztof Junik); data curation, S.D. and K.J. (Krzysztof Jamroziak); writing—original draft preparation, G.L. and K.J. (Krzysztof Junik); writing—review and editing, K.J. (Krzysztof Junik); visualization, S.D.; supervision, G.L. and K.J. (Krzysztof Jamroziak); project administration, K.J. (Krzysztof Jamroziak); funding acquisition, G.L. All authors have read and agreed to the published version of the manuscript.

**Funding:** This research received no external funding. The APC was funded by Wroclaw University of Science and Technology.

**Informed Consent Statement:** Not applicable.

**Acknowledgments:** The publication was prepared as a part of the Support Programme of the Partnership between Higher Education and Science and Business Activity Sector financed by the City of Wroclaw.

**Conflicts of Interest:** The authors declare no conflict of interest.

## References

1. Gabryelewicz, M. *Motor Vehicle Chassis and Bodies: Braking and Steering, Suspension, and Body Systems;* Wydawnictwa Komunikacji i Łaczności: Warszawa, Poland, 2014. (In Polish)
2. Junik, K.; Snowacki, K.; Duda, S.; Towarnicki, K.; Correia, J.A. Impact of hardness on the fracture and tear characterization of rigid pur materials used in suspension systems of vehicles. *Eng. Fail. Anal.* **2021**, *127*, 105510. [CrossRef]
3. Lesiuk, G.; Junik, K.; Smolnicki, M.; Correia, J.A.F.O.; De Jesus, A.M.P.; Babiarczuk, B.; Otczyk, K. Structural integrity assessment of rigid polyurethane components using energy methods. *Procedia Struct. Integr.* **2018**, *13*, 1595–1599.
4. Cerit, M.; Nart, E.; Genel, K. Investigation into effect of rubber bushing on stress distribution and fatigue behaviour of anti-roll bar. *Eng. Fail. Anal.* **2010**, *17*, 1019–1027. [CrossRef]
5. Kim, W.D.; Lee, H.J.; Kim, J.Y.; Koh, S.K. Fatigue life estimation of an engine rubber mount. *Int. J. Fatigue* **2004**, *26*, 553–560. [CrossRef]
6. Lee, H.J.; Kim, W.D.; Choi, B.I.; Woo, C.S.; Kim, J.Y.; Koh, S.K. Effect of Mean Stress on the Fatigue Life of Engine Mount and Life Prediction. In Proceedings of the KSME Conference the Korean Society of Mechanical Engineers, Jeonju, Korea, 1–3 November 2001; pp. 99–104.
7. Wiggins, M.J.; Anderson, J.M.; Hiltner, A. *Biodegradation of Polyurethane under Fatigue Loading;* Elsevier: Amsterdam, The Netherlands, 2003.
8. Mars, W.V.; Fatemi, A. A literature survey on fatigue analysis approaches for rubber. *Int. J. Fatigue* **2002**, *24*, 949–961. [CrossRef]
9. Mars, W.V.; Fatemi, A. Factors that affect the fatigue life of rubber: A literature survey. *Rubber Chem. Technol.* **2004**, *77*, 391–412. [CrossRef]
10. Holzweber, J.; Major, Z. Characterization of the fatigue behavior of TPU's. In *XIIth Youth Symp Exp Solid Mech;* 2014; pp. 44–47. Available online: https://www.imeko.org/publications/ysesm-2014/IMEKO-YSESM-2014-010.pdf (accessed on 30 May 2022).
11. Lesiuk, G.; Myszka, W.; Snowacki, K.; Junik, K. Strength testing of polyurethane elastomers used in the automotive industry using fracture mechanics methods. *Autobusy Tech. Eksploat. Syst. Transp.* **2016**, *17*, 1106–1112. (In Polish)
12. Zhao, J.; Li, Q.; Shen, X. Finite element analysis and structure optimization for improving the fatigue life of rubber mounts. *J. Macromol. Sci. Part A Pure Appl. Chem.* **2008**, *45*, 479–484. [CrossRef]
13. Tee, Y.L. Dynamic Characteristics of Filled Natural Rubber for Earthquake Energy Applications. Ph.D. Thesis, University of Malaya, Kuala Lumpur, Malaysia, 2018.
14. Li, Q.; Zhao, J.; Zhao, B. Fatigue life prediction of a rubber mount based on test of material properties and finite element analysis. *Eng. Fail. Anal.* **2009**, *16*, 2304–2310. [CrossRef]
15. Brunac, J.B.; Gérardin, O.; Leblond, J.B. On the heuristic extension of Haigh's diagram for the fatigue of elastomers to arbitrary loadings. *Int. J. Fatigue* **2009**, *31*, 859–867. [CrossRef]
16. Ayoub, G.; Naït-Abdelaziz, M.; Zaïri, F.; Gloaguen, J.; Charrier, P. Fatigue life prediction of rubber-like materials under multiaxial loading using a continuum damage mechanics approach: Effects of two-blocks loading and R ratio. *Mech. Mater.* **2012**, *52*, 87–102. [CrossRef]
17. Mars, W.V. Cracking energy density as a predictor of fatigue life under multiaxial conditions. *Rubber Chem. Technol.* **2002**, *75*, 1–17. [CrossRef]
18. Verron, E.; Andriyana, A. Definition of a new predictor for multiaxial fatigue crack nucleation in rubber. *J. Mech. Phys. Solids* **2008**, *56*, 417–443. [CrossRef]

19. Tee, Y.L.; Loo, M.S.; Andriyana, A. Recent advances on fatigue of rubber after the literature survey by Mars and Fatemi in 2002 and 2004. *Int. J. Fatigue* **2018**, *110*, 115–129. [CrossRef]
20. Ayoub, G.; Naït-Abdelaziz, M.; Zaïri, F. Multiaxial fatigue life predictors for rubbers: Application of recent developments to a carbon-filled SBR. *Int. J. Fatigue* **2014**, *66*, 168–176. [CrossRef]
21. Wang, Y.; Yu, W.; Chen, X.; Yan, L. Fatigue life prediction of vulcanized natural rubber under proportional and non-proportional loading. *Fatigue Fract. Eng. Mater. Struct.* **2008**, *31*, 38–48. [CrossRef]
22. Huneau, B.; Verron, E. Description of fatigue damage in carbon black filled natural rubber. *Fatigue Fract. Eng. Mater. Struct.* **2008**, *31*, 1031–1038. [CrossRef]
23. Flamm, M.; Spreckels, J.; Steinweger, T.; Weltin, U. Effects of very high loads on fatigue life of NR elastomer materials. *Int. J. Fatigue* **2011**, *33*, 1189–1198. [CrossRef]
24. Le Saux, V.; Marco, Y.; Calloch, S.; Doudard, C.; Charrier, P. Fast evaluation of the fatigue lifetime of rubber-like materials based on a heat build-up protocol and micro-tomography measurements. *Int. J. Fatigue* **2010**, *32*, 1582–1590. [CrossRef]
25. Shangguan, W.-B.; Wang, X.-L.; Deng, J.-X.; Rakheja, S.; Pan, X.-Y.; Yu, B. Experiment and modeling of uniaxial tension fatigue performances for filled natural rubbers. *Mater. Des.* **2014**, *58*, 65–73. [CrossRef]
26. Ayoub, G.; Zaïri, F.; Naït-Abdelaziz, M.; Gloaguen, J.M. Modeling the low-cycle fatigue behavior of visco-hyperelastic elastomeric materials using a new network alteration theory: Application to styrene-butadiene rubber. *J. Mech. Phys. Solids* **2011**, *59*, 473–495. [CrossRef]
27. Marco, Y.; le Saux, V.; Calloch, S.; Charrier, P. X-ray computed μ-tomography: A tool for the characterization of fatigue defect population in a polychloroprene rubber. *Procedia Eng.* **2010**, *2*, 2131–2140. [CrossRef]
28. Zine, A.; Benseddiq, N.; Naït Abdelaziz, M. Rubber fatigue life under multiaxial loading: Numerical and experimental investigations. *Int. J. Fatigue* **2011**, *33*, 1360–1368. [CrossRef]
29. Narynbek Ulu, K.; Huneau, B.; le Gac, P.Y.; Verron, E. Fatigue resistance of natural rubber in seawater with comparison to air. *Int. J. Fatigue* **2016**, *88*, 247–256. [CrossRef]
30. el Fray, M.; Altstädt, V. Fatigue behaviour of multiblock thermoplastic elastomers. 1. Stepwise increasing load testing of poly(aliphatic/aromatic-ester) copolymers. *Polymer* **2003**, *44*, 4635–4642. [CrossRef]
31. Hainsworth, S.V. An environmental scanning electron microscopy investigation of fatigue crack initiation and propagation in elastomers. *Polym. Test.* **2007**, *26*, 60–70. [CrossRef]
32. Engineering, M.; Baumard, T.L.M.; Thomas, A.G.; Busfield, J.J.C. Plastics, Rubber and Composites Fatigue peeling at rubber interfaces Fatigue peeling at rubber interfaces. *Plast. Rubber Compos.* **2012**, *41*, 296–300. [CrossRef]
33. Wu, Y.-P.; Zhao, W.; Zhang, L.-Q. Improvement of Flex-Fatigue Life of Carbon-Black-Filled Styrene-Butadiene Rubber by Addition of Nanodispersed Clay. *Macromol. Mater. Eng.* **2006**, *291*, 944–949. [CrossRef]
34. Fukahori, Y.; Sakulkaew, K.; Busfield, J.J.C. Elastic–viscous transition in tear fracture of rubbers. *Polymer* **2013**, *54*, 1905–1915. [CrossRef]
35. Harbour, R.J.; Ali, A.E.; Ae, F.; Mars, W.V. Fatigue crack orientation in NR and SBR under variable amplitude and multiaxial loading conditions. *J. Mater. Sci.* **2008**, *43*, 1783–1794. [CrossRef]
36. Berton, G.; Cruanes, C.; Lacroix, F.; Méo, S.; Ranganathan, N. Study of the Fatigue Behavior of the Polychloroprene Rubber with Stress Variation Tests. *Procedia Eng.* **2015**, *101*, 413–420. [CrossRef]
37. Mostafa, A.; Abouel-Kasem, A.; Bayoumi, M.R.; El-Sebaie, M.G. Effect of carbon black loading on the swelling and compression set behavior of SBR and NBR rubber compounds. *Mater. Des.* **2009**, *30*, 1561–1568. [CrossRef]
38. Andriyana, A.; Chai, A.B.; Verron, E.; Johan, M.R. Interaction between diffusion of palm biodiesel and large strain in rubber: Effect on stress-softening during cyclic loading. *Mech. Res. Commun.* **2012**, *43*, 80–86. [CrossRef]
39. Toki, S.; Fujimaki, T.; Okuyama, M. Strain-induced crystallization of natural rubber as detected real-time by wide-angle X-ray diffraction technique. *Polymer* **2000**, *41*, 5423–5429. [CrossRef]
40. Kaang, S.; Jin, Y.W.; Huh, Y.-I.; Lee, W.-J.; Bin Im, W. A test method to measure fatigue crack growth rate of rubbery materials. *Polym. Test.* **2006**, *25*, 347–352. [CrossRef]
41. Zhou, Y.; Jerrams, S.; Betts, A.; Farrell, G.; Chen, L. The influence of particle content on the equi-biaxial fatigue behaviour of magnetorheological elastomers. *Mater. Des.* **2015**, *67*, 398–404. [CrossRef]
42. Weng, G.; Huang, G.; Lei, H.; Qu, L.; Nie, Y.; Wu, J. Crack initiation and evolution in vulcanized natural rubber under high temperature fatigue. *Polym. Degrad. Stab.* **2011**, *96*, 2221–2228. [CrossRef]
43. Belkhiria, S.; Hamdi, A.; Fathallah, R. Strain-based criterion for uniaxial fatigue life prediction for an SBR rubber: Comparative study and development. *Proc. Inst. Mech. Eng. Part L J. Mater. Des. Appl.* **2020**, *234*, 897–909. [CrossRef]
44. Jerabek, M.; Major, Z.; Lang, R.W. Strain determination of polymeric materials using digital image correlation. *Polym. Test.* **2010**, *29*, 407–416. [CrossRef]
45. Ogden, R.W. Large deformation isotropic elasticity–on the correlation of theory and experiment for incompressible rubberlike solids. *Proc. R. Soc. Lond. A Math. Phys. Sci.* **1972**, *326*, 565–584. [CrossRef]
46. Bergstrom, J.S. *Mechanics of Solid Polymers: Theory and Computational Modeling*; Elsevier: Amsterdam, The Netherlands, 2015.
47. Diniz, F.B.; De Andrade, G.F.; Martins, C.R.; De Azevedo, W.M. A comparative study of epoxy and polyurethane based coatings containing polyaniline-DBSA pigments for corrosion protection on mild steel. *Prog. Org. Coat.* **2013**, *76*, 912–916. [CrossRef]

48. Parnell, S.; Min, K.; Cakmak, M. Kinetic studies of polyurethane polymerization with Raman spectroscopy. *Polymer* **2003**, *44*, 5137–5144. [CrossRef]
49. de Sá, S.F.; Ferreira, J.L.; Matos, A.S.; Macedo, R.; Ramos, A.M. A new insight into polyurethane foam deterioration—The use of Raman microscopy for the evaluation of long-term storage conditions. *J. Raman Spectrosc.* **2016**, *47*, 1494–1504. [CrossRef]
50. Xu, L.; Li, C.; Ng, K.Y.S. In-Situ Monitoring of Urethane Formation by FTIR and Raman Spectroscopy. *J. Phys. Chem. A* **2000**, *104*, 3952–3957. [CrossRef]
51. Ghobashy, M.M.; Abdeen, Z.I. Radiation Crosslinking of Polyurethanes: Characterization by FTIR, TGA, SEM, XRD, and Raman Spectroscopy. *J. Polym.* **2016**, *2016*, 1–9. [CrossRef]
52. Weakley, A.T.; Warwick, T.; Bitterwolf, T.E.; Aston, D.E. Multivariate analysis of micro-Raman spectra of thermoplastic polyurethane blends using principal component analysis and principal component regression. *Appl. Spectrosc.* **2012**, *66*, 1269–1278. [CrossRef] [PubMed]
53. Kumagai, S.; Motokucho, S.; Yabuki, R.; Anzai, A.; Kameda, T.; Watanabe, A.; Nakatani, H.; Yoshioka, T. Effects of hard- and soft-segment composition on pyrolysis characteristics of MDI, BD, and PTMG-based polyurethane elastomers. *J. Anal. Appl. Pyrolysis* **2017**, *126*, 337–345. [CrossRef]

materials

*Article*

# Low Cycle Fatigue of G20Mn5 Cast Steel Relation between Microstructure and Fatigue Life

**Antonin Bermond [1,2,*], Claire Roume [1], Jacques Stolarz [1], Matthieu Lenci [1], Jean-François Carton [2] and Helmut Klocker [1]**

1 Mines Saint-Etienne, University Lyon, CNRS, UMR 5307 LGF, SMS Centre, F-42023 Saint Etienne, France
2 Safe Metal, 1 Boulevard de la Boissonnette, F-42110 Feurs, France
* Correspondence: antonin.bermond@safe-metal.com

**Abstract:** Cast steel is commonly used to produce structural and safety parts. Foundry processes allow producing parts from scrap steel directly to the required dimensions without any forming operation. Cast components may, however, exhibit macro- and micro-shrinkage porosities. The combined effect of macro- and micro-shrinkages on the fatigue behavior of cast steel has been characterized in the literature. Macro-shrinkages may nowadays be eliminated by adequate positioning of risers. However, micro-shrinkages will always be present in cast steel components. Present work addresses the influence of micro-shrinkage porosity on a G20Mn5 cast steel. G20Mn5 (normalized) ingots have been cast under industrial conditions, but ensuring the absence of macro-porosities. Solidification leads to two very different microstructures prior to the normalization treatment: columnar dendrites beneath the surface (**S**kin) and equiaxed microstructures close to the center (**C**ore). First, metallographic observations of the whole ingot revealed the same grain size in both areas. Fatigue samples were extracted, by differentiating two sampling volumes corresponding to columnar (S) and equiaxed solidification (C), respectively. The distribution of micro-porosities was determined on all samples by Micro-CT-scans. Core samples exhibit micro-porosities with volumes 1.7 larger than Skin samples. Low cycle fatigue tests (3 levels of fixed plastic strain) were run on both sample series (C, S). Results follow a Manson–Coffin law. Core specimens exhibit lower fatigue life than Skin specimens. The differences in fatigue life have been related successfully to the differences in micro-porosities sizes.

**Keywords:** G20Mn5 cast steel; columnar and equiaxed dendrites; micro-shrinkages; low-cycle fatigue; CT-scan; damage evolution

**Citation:** Bermond, A.; Roume, C.; Stolarz, J.; Lenci, M.; Carton, J.-F.; Klocker, H. Low Cycle Fatigue of G20Mn5 Cast Steel Relation between Microstructure and Fatigue Life. *Materials* **2022**, *15*, 7072. https://doi.org/10.3390/ma15207072

Academic Editors: Grzegorz Lesiuk and Dariusz Rozumek

Received: 13 September 2022
Accepted: 5 October 2022
Published: 11 October 2022



## 1. Introduction

Cast steel is commonly used to produce structural and safety parts for trucks, railways and construction equipment. Steel foundry processes allow producing parts directly to the required dimensions without any forming operation. Moreover, it uses only scrap steel, so it is an ecological process that preserves Earth's resources. A historical overview of casting may be found in [1]. The cast components of interest in the present work are submitted to fatigue loading.

Fatigue is damage resulting from cyclic loading. The fatigue of rolled and forged metallic components has been studied since the end of the nineteenth century. The main mechanisms of fatigue failure are described extensively in [2–4]. Plastic strain leads to clusters with higher dislocation densities than average [4]. Cyclic straining increases this heterogeneity [5]. After a sufficient number of cycles, cells or planar arrangements of dislocations are created. The slip is concentrated in thin persistent slip bands forming extrusions and intrusions on the surface [4]. Multiplication of extrusions and intrusions initiates a crack, which begins to propagate in shear mode (stage I of fatigue crack initiation). In stage II of fatigue crack initiation, the crack progressively aligns perpendicular to the load axis. The fatigue life corresponds to the number of cycles to initiate a dominant crack and to propagate the latter until catastrophic failure occurs. Crack initiation can be as high as some 90% of the total fatigue life [5]. Due to the variability of the initial flaw populations, fatigue results exhibit considerable scatter in the number of cycles to failure [2–5].

Three main loading domains may be distinguished in fatigue analyses [4]. Under giga-cycle (more than $10^6$ cycles) and high-cycle fatigue, the material deforms primarily elastically. The number of cycles to failure under such high-cycle fatigue has traditionally been characterized in terms of the stress range. The stresses associated with low-cycle fatigue cause appreciable plastic deformation prior to failure, and the fatigue life is characterized in terms of the strain range. A major application of low-cycle fatigue is the prediction of fatigue life in regions ahead of a stress concentration. The low-cycle approach to fatigue has found particularly extensive use in ground-vehicle industries [2].

Under low-cycle (high stress fatigue) the effect of small defects on the fatigue life is increased. Murakami [6] gives an extensive overview of the effects of small defects, nonmetallic inclusions, notches and cracks on the fatigue life of forged or rolled steels. Zerbst and Klinger [7] analyzed fatigue crack initiation by non-metallic inclusions, pores, micro-shrinkages and un-welded regions.

*Compared to rolled or forged work pieces, cast ones may contain shrinkage porosities* [1,8–10]. These porosities control to a large extend crack initiation (stages I, II). The formation of porosities during the solidification of steel is linked to two major phenomena [1,8–10]: the volume shrinkage of the steel and the release of gaseous elements. The first phenomenon gives rise to the formation of porosities called *micro-shrinkage*. A typical example of a micro-shrinkage porosity observed in G20Mn5 steel is shown in Figure 1. These porosities exhibit a large number of branches. Shrinkages are internal or external *macroscopic defects* resulting from the same phenomenon. Niyama presented one of the first criteria for shrinkage prediction [11]. Historically, acceptable classes of defects were defined. No distinction has been made between micro-shrinkages and sink marks [12–15]. Beckermann et al. [16] analyzed the effect of porosity size distribution on the fatigue behavior of 8630 cast steel. Serrano-Munoz analyzed the effect of close surface defects on the fatigue life [17]. Carlson and Beckermann suggested a dimensionless Niyama criterion [18], and Shouzhu et al. [19] suggested new metal feeding rules to prevent sink mark formation.

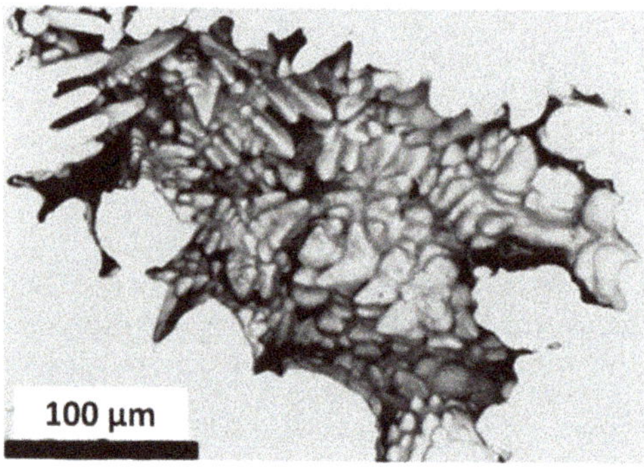


**Figure 1.** Micro-shrinkage porosity.

To the best of our knowledge, no detailed analyses of the influence of micro-shrinkages without the presence of macro porosities on low cycle fatigue damage of cast steels are available. Nevertheless, Han [20] highlighted that cracks initiate around those porosities in G20Mn5QT grade during high cycle fatigue (HCF) tests; as well as Wu in a G20Mn5N grade during low cycle fatigue (LCF) tests [21]. Moreover, Nagel studied the impact of internal and surface porosities inside cast specimens in G20Mn5Q grade during HCF tests [22]. Various types of porosities were produced thanks to two geometries of cast specimens, with or without a thicker central cross-section. It appeared that specimens cast with this larger shoulder contain a bigger central shrinkage, which reduces fatigue life. However, these kinds of porosities are rather macro-shrinkages linked to the hot spot in the middle of the specimen, which is not fed with liquid steel when it solidifies, than micro-shrinkages, which are formed according the same phenomenon between dendrites.

The present paper addresses the fatigue life of normalized G20Mn5 cast steel and the influence of the solidification conditions (columnar or equiaxed dendrites) on its microstructure, especially the size and the shape of micro-shrinkage porosities; then the influence of these micro-porosities on fatigue damage. This steel grade is the most produced by Safe Metal and commonly used for structural parts. Special ingots were produced with oversized risers and the absence of macro-porosity was controlled. The paper is organized as follows. The details of materials elaboration and experimental methods are given in Section 2. Section 3 is dedicated to the microstructural and monotonic mechanical characterizations. In Section 4, low-cycle fatigue test results are presented and correlated to the microstructure of the G20Mn5 steel.

## 2. Materials and Methods

### 2.1. Ingots Casting

Several ingots of G20Mn5 steel were cast specifically for this work, following the usual process of Castmetal Feurs foundry. The spray of the casting design (Figure 2) and the casting parameters were defined in order to avoid macro-shrinkages. Special care was taken to optimize feeding, ingots geometry and the use of the pencil core. The ingots composition (Table 1) was determined by spark emission spectrometry (Thermo Fisher ARL 4462, Walthamm, MA, USA). The analysis was conducted on a sample taken in the molted bath.

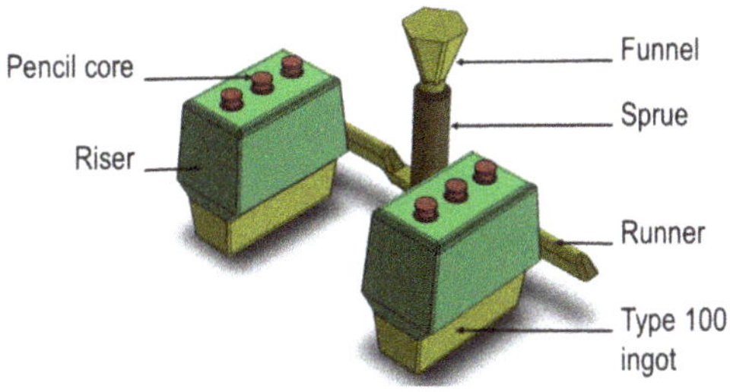


**Figure 2.** Casting device.

**Table 1.** G20Mn5 chemical composition (%wt).

| C | Mn | Si | S | P | Ni | Cr | Mo | V | Al | Fe |
|---|---|---|---|---|---|---|---|---|---|---|
| 0.18 | 1.12 | 0.36 | 0.009 | 0.011 | 0.07 | 0.26 | 0.05 | 0.005 | 0.05 | 97.8 |

The ingots were separated from the riser by oxygen cutting. Ultrasonic inspection was conducted on each ingot. A normalizing heat treatment at 880°C followed by air cooling was applied.

### 2.2. Microstructure

On a first ingot, macroscopic etching was done on a full cross section prior to the normalization treatment. A 22% HCl solution at 70 °C was used during 20 min. As expected, the delimitation between columnar and equiaxed dendrites areas has been revealed (Figure 3a). Figure 3b,c show zooms of regions, respectively, in the columnar and equiaxed zone. On another ingot, after normalization, microstructural characterizations were conducted at different locations. Samples with a size of 20 × 10 × 10 mm were cut following the blueprint given in Figure 3d. Each sample was coated using compression mounting method in a Simplimet 1000 Buehler device. The samples were polished on an automatic Automet 250 Buehler device (Buehler, Leinfelden-Echterdingen, Germany) with

SiC disks or diamond past, until mirror finish step. After polishing the samples, they were rinsed in ethanol, dried with compressed air and stored in a desiccator. The reactant used to reveal grain boundaries was Nital (5 $HNO_3$/95 ethanol, vol.). Images were acquired with an optical microscope (Olympus GX 51, Evident Europe GmbH-French Branch, Rungis, French) at a magnification of ×200, then analyzed with a method defined in Olympus Stream software (https://www.olympus-ims.com/fr/microscope/stream2/ (accessed on 12 September 2022)).

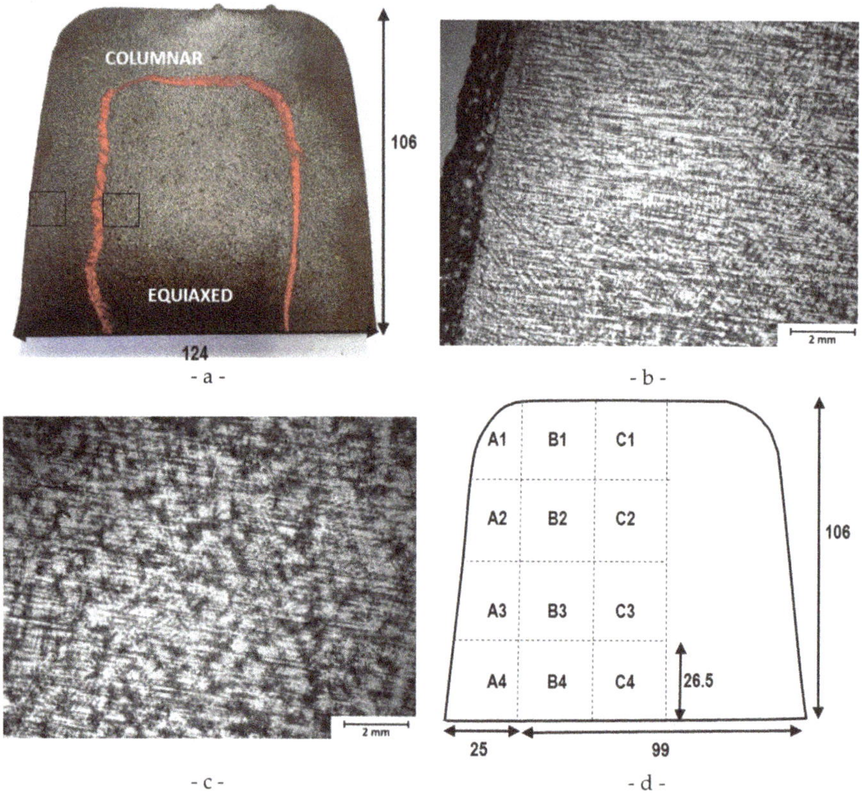


**Figure 3.** (**a**) Ingot cross section prior to normalization treatment. Red: boundary between columnar and equiaxed the black squares correspond to the zooms presented in figures (**b**,**c**). (**b**) Columnar zone, (**c**) equiaxed zone. (**d**) Sample cutting localization and labels for microstructure analyzes on an ingot after normalizing heat treatment.

A specific characterization procedure was developed to study porosities size, shape and localization in fatigue test pieces. X-ray Computed Tomography data were acquired with a General Electric Nanotom 180 micro-tomograph device (General Electric, Boston, MA, USA). The analyzed volume was a cylinder of 5 mm in diameter and 10 mm in length. The size of the voxel is 6.25 μm. The data were processed with Phoenix DatosX software (https://www.bakerhughes.com/waygate-technologies/ndt-software/phoenix-datosx-industrial-ct-scanning-software (accessed on 12 September 2022)). Porosity characteristics were then studied with Avizo software (version 2020). Considering the voxel size, only porosities containing at least 10 voxels, equivalent to a sphere with a diameter higher than 14 μm, are kept for further analyses.

## 2.3. Mechanical Tests

### 2.3.1. Sample Preparation

Mechanical test specimens were milled from bars cut into two ingots by wire electro discharge machining. The positions of the bars in the ingots are given in Figure 4a. It is worth noting that bars are cut either on the border (red) or in the center (blue) of the ingots. Additionally, some bars have their axis along ingot length and other have their axis along ingot height. The same specimen geometry (Figure 4b) was used for monotonic tensile and low cycle fatigue tests.

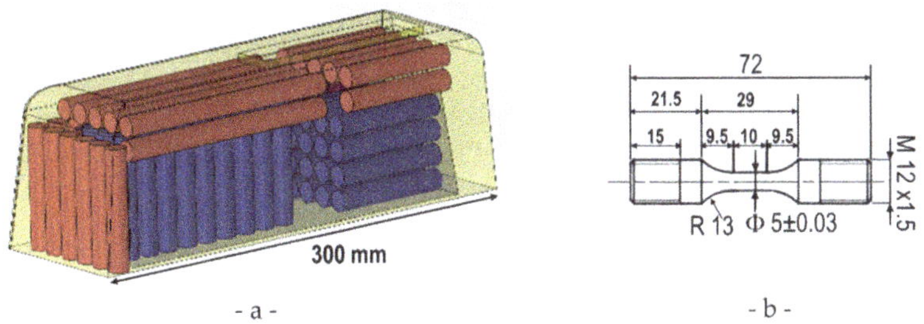


**Figure 4.** (**a**) Type 100 ingot with indication of wire cutting positions (skin in red, core in blue) for cylindrical bars. (**b**) Tensile and fatigue sample geometry.

### 2.3.2. Monotonic Mechanical Tests

Tensile tests and compression tests were conducted on samples taken either close to the surface (red) or from the center of ingots (blue), in horizontal or in vertical direction (Figure 4a). The tensile test pieces geometry is given on Figure 4b. The initial gage diameter and gage length are, respectively, 5 mm and 10 mm. The tests were conducted on a Schenck RMC100 device controlled with Instron Console and Wavematrix. Compression test pieces were cylinders with a diameter of 10 mm and a height of 15 mm. The tests were conducted on a servohydraulic Schenck compression machine. The mechanical tests were strain rate or velocity controlled. The compression tests were done at a constant strain rate of 0.01 $s^{-1}$. The monotonic tensile tests were conducted at constant cross-head velocity of 1 mm/s. Vickers hardness tests were also carried out on a Wolpert Testwell macrohardness under a load of 10 kgf and on a Matra MVK-1H microhardness testing machine under loads of 10 gf inside the ferrite and 25 gf inside ferrite and pearlite.

### 2.3.3. Low Cycle Fatigue Tests

Low-cycle fatigue tests were conducted with an electro-mechanical Schenck RMC100 testing machine (Carl Schenck AG, Darmstadt, Germany). The test conditions are summarized in Table 2. Three levels of plastic strain range were defined in order to cover a wide range of low cycle fatigue. At each strain level, respectively, ten samples from the ingot skin and ten samples from the core were tested. In order to avoid damaging the fracture surface, a shutoff parameter was defined to stop every fatigue test when the maximal (tensile) stress ($\sigma_{max}$) dropped more than 20% compared to a control stress ($\sigma_{cont.}$). The values of the control stress were measured during each test. For the lowest loading ($\Delta\varepsilon_p/2 = 0.02\%$), this value has been determined at cycle n° 500. For the two other loadings ($\Delta\varepsilon_p/2 = 0.1\%$, 0.4%), the value of the control stress ($\sigma_{cont.}$) was determined at cycle n° 100.

**Table 2.** Low-cycle fatigue test parameters. Three different levels of plastic deformation. The plastic load ratio ($\epsilon_p^{min}/\epsilon_p^{max}$) is equal to $-1$.

| Strain Level | $\frac{\Delta\varepsilon_p}{2}$ (%) | Signal (Strain) | Censoring (Cycles) | Shutoff |
|---|---|---|---|---|
| 1 | 0.02 | Triangular | $10^6$ | $\sigma_{max} \leq 0.8\,\sigma_{cont.}$ |
| 2 | 0.1 | | $10^5$ | |
| 3 | 0.4 | | $10^4$ | |

## 3. Microstructure and Mechanical Characterization

### 3.1. Microstructure

After the normalizing treatment, a ferritic–pearlitic microstructure was expected. Optical micrographs were acquired at different locations in the ingot, either in the center or in the skin part of the cross section. Figure 5 displays two typical examples of the ferritic–pearlitic microstructure observed, respectively, at the ingot skin an at the ingot center. The average grain size was measured on six samples taken from the ingot skin and on six samples taken from the core of the slab. The average equivalent diameter (Table 3) is almost the same from one sample to another. No significant difference was observed between samples taken from the core or the skin.

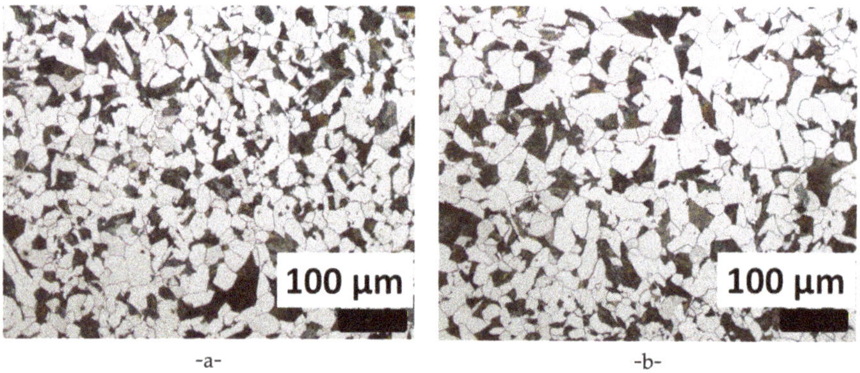


**Figure 5.** Metallographic observations after Nital etching in: (**a**) the skin (A4) and (**b**) core (C4) parts.

**Table 3.** Mean grain size in the skin and in the core of the ingot. $\bar{d}$ designates the average equivalent grain diameter.

| Location | $\bar{d}$ (µm) | G Index |
|---|---|---|
| Skin | 16.4 | 8.8 |
| Core | 16.9 | 8.9 |

The populations of inclusions were characterized prior to etching. In accordance with the chemical composition of the G20Mn5 steel grade, mainly manganese sulfides (MnS) especially MnS type III (Figure 6) were observed. The composition was checked by EDS analysis. Based on [23] indexes of inclusion cleanliness were determined (Table 4). For both fine and thick series (corresponding, respectively, to inclusions with a thickness below or above 8 µm), indexes are very close between the skin and the core of the ingot.

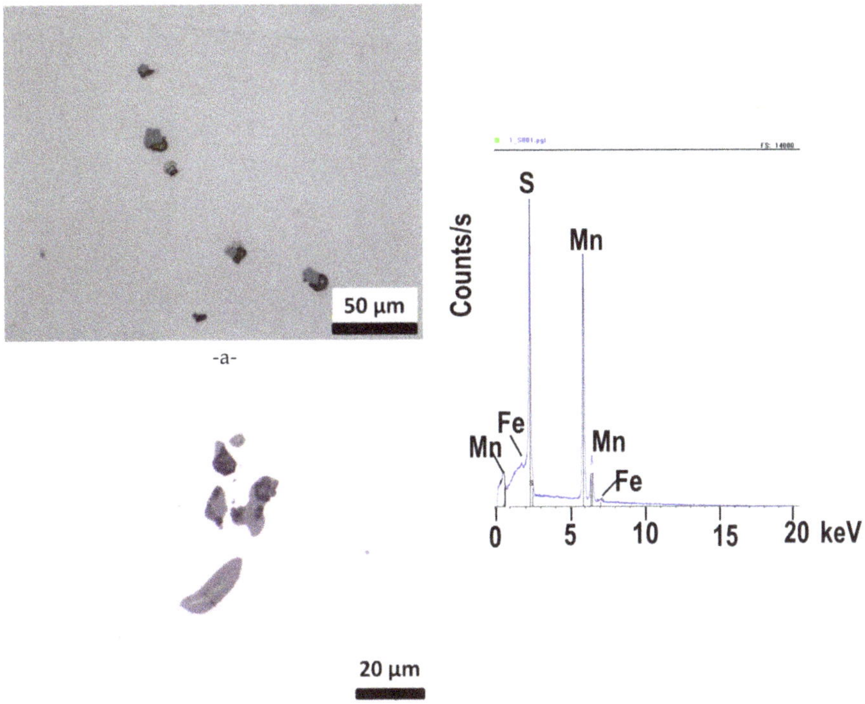


-b- -c-

**Figure 6.** MnS type III (**a**,**b**), typical EDS observation of one of the particles (**c**).

**Table 4.** Indexes of inclusion cleanliness.

| Location | MnS Type III (Fine) | MnS Type III (Thick) |
|---|---|---|
| Skin | 1.13 | 0.60 |
| Core | 1.08 | 0.68 |

Special attention was given to porosities characterization. Three-dimensional X-ray tomography observations were made on several fatigue samples prior to mechanical testing. A significant difference between samples taken from the core and those taken from the skin of the ingot appears. Table 5 summarizes the mean values of the porosity distributions. The mean void volume fraction is slightly larger in the skin of the ingot than in the core. However, the main difference concerns the void volume distribution. The number of voids per unit volume is much smaller in the samples taken from the core of the ingot. The mean micro-shrinkage volume is much larger in the core than in the skin of the ingot. Figure 7 depicts the probability density as function of the void (micro-shrinkage) volume for the core and skin of the ingot. The skin of the ingot is characterized by a rather sharp peak, whereas in the core of the ingot very large voids (0.003 $mm^3$) are observed. The core contains thus a smaller amount of micro-shrinkages than the skin; but much higher volumes are observed.

**Table 5.** Mean values of the porosity distribution in the skin and the core of the ingot. $\overline{f}$ and $\overline{N}$ correspond, respectively, to the mean void volume fraction and number of voids per unit volume. $\overline{v}$ and $\overline{d}$ designate, respectively, the mean void (micro-shrinkage) volume and diameter.

| Location | $\overline{f}$ (%) | $\overline{N}$ (mm$^{-3}$) | $\overline{v}$ (μm$^3$) | $\overline{d}$ (μm) |
|---|---|---|---|---|
| skin | 0.0104 | 403 | 5.3 $10^4$ | 46.6 |
| core | 0.00947 | 234 | 8.8 $10^4$ | 55.2 |

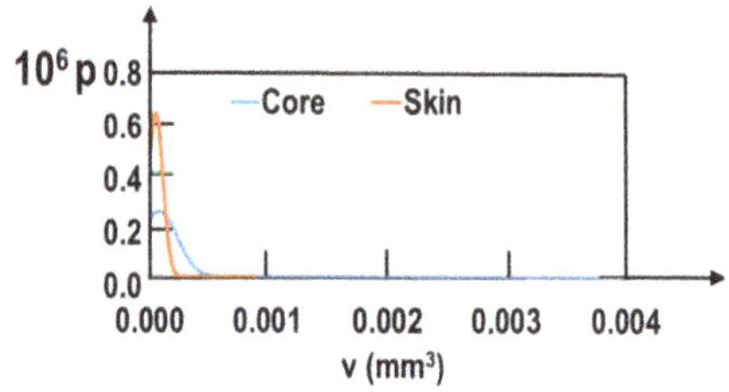


**Figure 7.** Probability density distribution p in function of the void volume v.

The probability distributions of voids in the core and skin are different. The mechanical effect of voids depends on the size, but also on the shape and location in the sample. The shape of voids is often characterized by their sphericity. Figure 8a depicts the sphericity as a function of the void volume. Obviously, the sphericity decreases strongly when the void volume is increased. Thus, only small voids exhibit a quasi-spherical shape. For voids with a volume larger than $v_c$ = 0.00025 mm$^3$, the sphericity is smaller than 0.4 (red dashed lines in Figure 8a). For both locations (ingot skin or ingot core), voids with volumes larger $v_c$ are present. Among all platonic solids, the tetrahedron exhibits the smallest sphericity (0.671). As, in present work, the voids are associated with micro-shrinkages, the description depicted in Figure 8b seems more adequate. The total surface of the object (sum all external surfaces) was determined. Figure 8b shows the total void surface as a function of the void volume. For both locations (skin or core), a perfectly linear relationship between the void surface and the void volume is observed. This linear relation implies quasi-two-dimensional void shapes as represented schematically in Figure 8b. If the "thickness" h is always small compared to the in plane dimensions ($d_1$,$d_2$) a linear relationship between the void surface and void volume is satisfied, even if the in-plane shape is much more complex in reality. The shapes shown in Figure 8b satisfy the condition of a constant ratio volume/total surface and allow small values of the sphericity parameter. Indeed, as an example, we assume $d_1$ = $d_2$. Then, a sphericity of 0.4 leads to h/$d_1$ = 0.085.

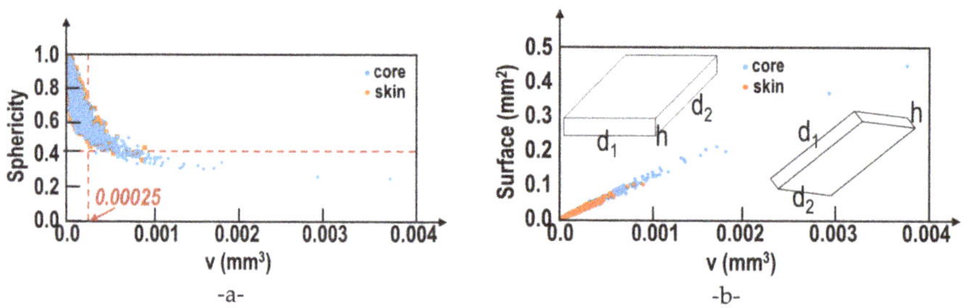


**Figure 8.** Micro-shrinkage characterization. (**a**) Sphericity in function of the volume v, and (**b**) total surface in function of the volume v.

The microstructural observations may be summarized as follows. After the normalizing treatment, a ferritic–pearlitic microstructure was observed at the ingot skin and in the ingot center. The average grain size is almost the same in the skin and the core of the ingot. The inclusion distribution in the skin and the core of the ingot are very close. The main difference between the ingot skin and core corresponds to the presence of larger micro-shrinkages in the core than in the skin of the ingot.

### 3.2. Monotonic Mechanical Properties

Vickers macro-hardness and micro-hardness tests were conducted at different locations in the skin and core of the ingot. For macro-hardness, 12 indents were done on each part. Microhardness indents were located either in ferrite or in pearlite. For each phase, 30 indents and 40 indents were done in the skin parts and core part, respectively. The results are given in Table 6. The difference in hardness between the ingot skin and core is contained in the error bars. The hardness of the skin and core are thus considered identical. The same conclusion holds for the micro-hardnesses in both phases.

**Table 6.** Monotonic test results. Average values from three tests for tensile tests and tens of tests for hardness in each batch (core or skin).

| | Core | Skin |
|---|---|---|
| UTS (Mpa) | 518 | 528 |
| $Rp_{0.2}$ (Mpa) | 293 | 309 |
| A% (%) | 37% | 37% |
| Ferrite micro-hardness (HV) | 113 | 121 |
| Pearlite micro-hardness (HV) | 303 | 291 |
| Macro-hardness (HV) | 140 | 142 |

Tensile and compression tests were conducted on samples from the ingot skin and core. Monotonic test results are presented in Figure 9. Monotonic tensile and compression behavior for the ingot skin and core are identical. The tensile curves exhibit the Portevin-Le Chatelier effect and Lüders bands typical of low-carbon steels. Very high strain hardening at small strains (<5%) is observed. Comparison of tensile and compression curves hints that significant damage in tenson occurs after 5% (point A). However, macroscopic necking occurs much later at ~18% (point B). The significant effect of damage at low stress triaxiality (prior to necking) corresponds to the large initial void volume fractions. However, the large strain hardening (compression curves) leads to a significant ultimate strength of about 520 MPa and a large ductility of 37%. The monotonic test results are summarized in Table 6.

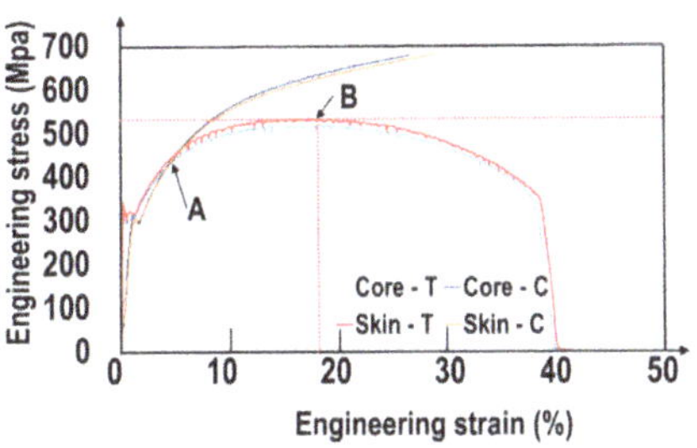


**Figure 9.** Tensile and compressive curve for skin and core specimens.

## 4. Low Cycle Fatigue Tests

Low cycle fatigue tests, with $R_{\varepsilon}^{p} = -1$, were conducted for three levels of plastic strain ($\Delta\varepsilon^{P}/2$ = 0.02%, 0.1% and 0.4%, respectively) on 10 samples from the ingot core and

10 samples of the ingot skin. First, a control value of the stress ($\sigma_{cont.}$) was *determined during* the test. The tests were stopped when the tensile stress became smaller than $0.8\sigma_{cont.}$. From now on, the number of cycles achieved precisely at $\sigma_{max} = 0.8\sigma_{cont.}$ will be called $N_f$. The final rupture was obtained by simple tension. This is the classical way to preserve a fracture surface with $R_\varepsilon^p = -1$.

### *4.1. Low Cycle Fatigue S-N-Curves*

Figure 10 shows the corresponding results. In each column, the plastic strain increases from top to bottom. In each row samples from the ingot core and skin are shown. Comparing the results corresponding to different plastic strain levels (column) shows the sharp decrease in the number of cycles for increasing plastic strain for samples from the ingot core as well as samples from the skin.

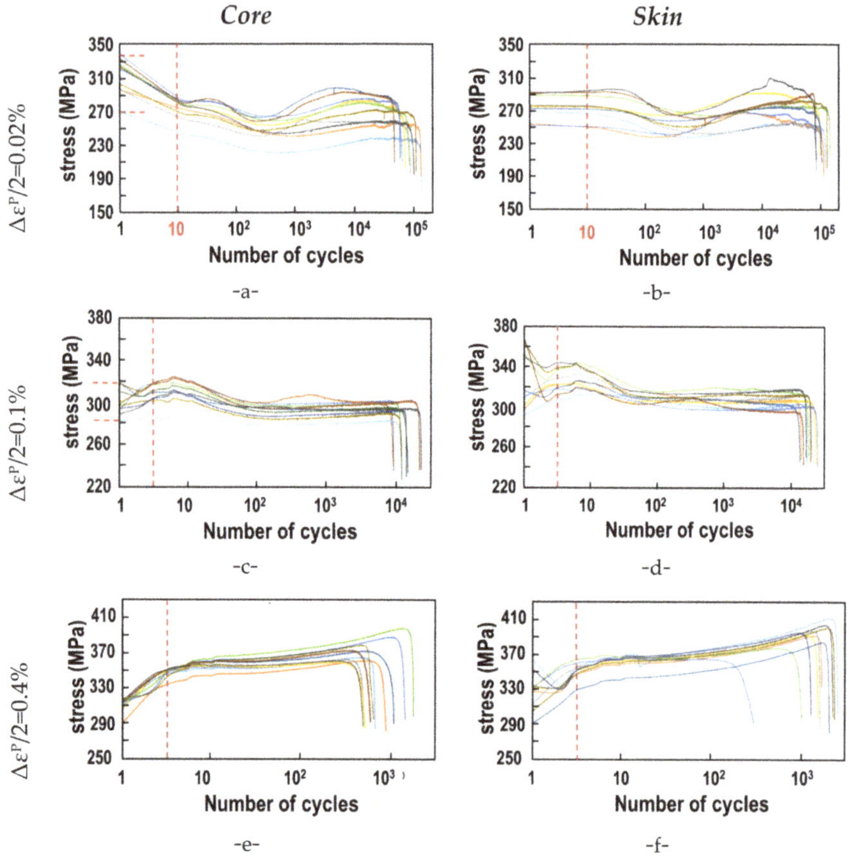


**Figure 10.** Fatigue test results (stress vs. number of cycles) for different values of plastic strain $\Delta\varepsilon^P/2$ = 0.1% (**a**,**b**); 0.2% (**c**,**d**) and 0.4% (**e**,**f**). Samples for the ingot core are represented in (**a**,**c**,**e**) and from the ingot skin in (**b**,**d**,**f**).

A significant dispersion of the stress has to be noted. At the very beginning of the tests, dispersion may be attributed to differences in the microstructure, but also to small misorientations of the samples. Hence, vertical dashed lines delimit on each graph a very small portion of the S-N-curves, which will not be considered in the interpretation. The following discussion concerns only the parts to the right of the red dashed lines. Even to the right of the dashed lines, significant dispersion is observed. This dispersion decreases if the applied plastic strain and hence the applied stress increase. At the lowest level of plastic strain (0.02%) during accommodation, a decrease of the stress is observed. At the intermediate strain increment, no significant softening nor hardening occurs. Finally, at the highest plastic strain level (0.4%), significant strain hardening occurs. In Section 4.2, the number of cycles will be analyzed by Manson Coffin's approach. In Section 4.3, a damage analysis will allow to distinguish between subcritical and final crack growth.

### 4.2. Number of Cycles at Failure

#### 4.2.1. Manson–Coffin Parameters

The results presented in Figure 10 allow determining the mean number of cycles at failure for all the considered samples (Table 7). For all the considered plastic strains, samples from the ingot skin exhibit an average number of cycles at failure ($N_f$) larger than specimens from the core.

**Table 7.** Number of cycles at failure (Nf) as function of the plastic strain.

| $\Delta\varepsilon^P/2$ | | | | | |
|---|---|---|---|---|---|
| 0.02% | | 0.1% | | 0.4% | |
| Core | Skin | Core | Skin | Core | Skin |
| 92339 | 108684 | 13796 | 17940 | 835 | 1697 |

Under low cycle fatigue conditions, Manson and Coffin [24,25] postulated a simple power–law relation between the numbers of cycles at failure ($N_f$) and the plastic strain level ($\Delta\varepsilon_P$).

$$\frac{\Delta\varepsilon_P}{2} = \varepsilon'_f\, N_f^c \tag{1}$$

$\varepsilon'_f$ and c are material parameters. Based on the results of Table 7, the Manson–Coffin parameters for skin and core samples taken separately and for all samples were determined (Table 8).

**Table 8.** Manson–Coffin coefficients.

| | $\varepsilon'_f$ | c |
|---|---|---|
| Skin | 0.2996 | 0.714 |
| Core | 0.8822 | 0.626 |
| Global | 0.5536 | 0.677 |

Figure 11 shows a graphical representation of the plastic strain increment as a function of the number of cycles at failure. This representation leads to a small difference between the behaviors of samples from the ingot skin and ingot core. The mean behavior (skin+ core samples) is represented in solid green (MC). The global behavior is very close to the one determined by Han [20] for G20Mn5QT cast steel at R = −1 (green dashed line). However, a significant difference with the curve from Wu's study [21] who characterized the material at R = 0.1 (dotted green line). This difference hints at a significant influence of the average stress on the fatigue life. Finally, looking at the total strain by adding Basquin's law (yellow curve) allows us to match HFC results obtained in a previous study.

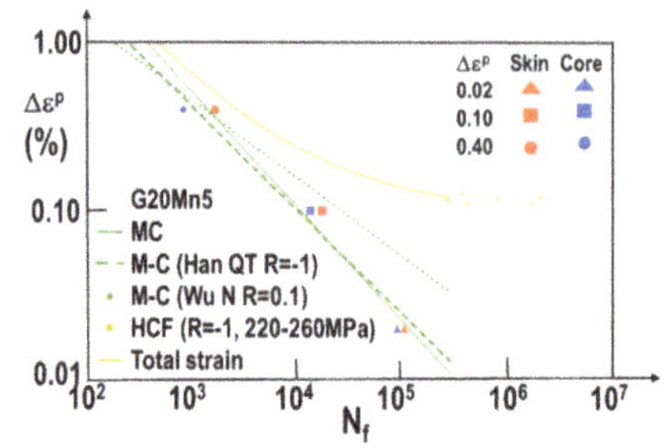


**Figure 11.** Manson–Coffin plot based on experimental values obtained in this work and compared to literature data. The latter are Han et al. work on quenched G20Mn5 and Wu et al. work on normalized G20Mn5. HCF data (yellow points) were obtained with applied stress tests. Corresponding elastic strain values are used to plot the results on this strain vs. number of cycles to rupture graph.

#### 4.2.2. Difference between Ingot Core and Skin

The number of cycles at failure based on the combination of skin and core samples gives only a first indication of the material behavior under cyclic loading and allows to show good correspondence with Han's results. A more detailed analysis of the number of cycles at failure is given in Table 9. The mean values correspond to Table 7. The main focus is now on the dispersion of the results (standard deviation/mean value). Results corresponding to the ingot core exhibit always a larger dispersion than samples from the ingot skin. For both series the dispersion increases with the applied plastic strain. For each applied plastic strain, the number of cycles at failure ($N_f$) between samples from the skin and the core of the ingot differs significantly (15% to 51%). However, the stabilized stress exhibits almost the same value between both series.

**Table 9.** Detailed results from low cycle fatigue tests.

| | | $\Delta\varepsilon^P/2$ | | | | | |
|---|---|---|---|---|---|---|---|
| | | 0.02% | | 0.1% | | 0.4% | |
| | | Core | Skin | Core | Skin | Core | Skin |
| Number of cycles at failure $N_f$ | mean | 92339 | 108684 | 13796 | 17940 | 835 | 1 697 |
| | stand. dev. | 29511 | 20524 | 4703 | 3776 | 433 | 704 |
| | stand. dev./mean | 34% | 20% | 34% | 21% | 52% | 41% |
| | max. | 139875 | 148455 | 22677 | 24428 | 1741 | 2617 |
| | Min. | 48754 | 83775 | 8937 | 13195 | 473 | 298 |
| $\frac{(N_f^{skin}-N_f^{core})}{N_f^{skin}}$ | | 15% | | 30% | | 51% | |
| $\sigma_{stab}$. (MPa) | | 251.5 | 251.7 | 292.2 | 308.6 | 362.2 | 369.4 |

### *4.3. Remaining Load Carrying Capacity during Cycling*

The number of cycles at failure gives a first precious indication of the fatigue life expectancy for a fixed load level ($\Delta\varepsilon^P/2$). However, for different successive loadings, the remaining load carrying capacity of the material after a given number of cycles is significant. Figure 12 depicts the stress as a function of the normalized number of cycles ($N/N_f$) for all considered loadings.

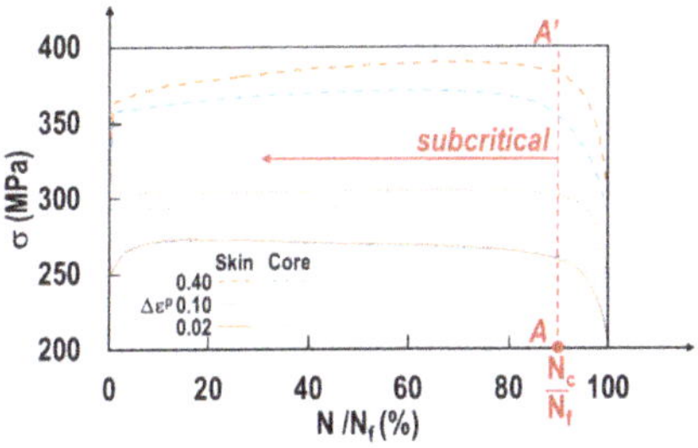


**Figure 12.** Stress as function of the normalized number of cycles for all considered loadings.

Remarkably, for all considered strain levels and both sample locations, the sharp stress decrease occurs after about 90% of the fatigue life time (red dashed line AA'). This corresponds to the transition from subcritical to critical crack growth at a number of cycles $N_c$. In the following paragraph, a method to exactly determine $N_c$ is elaborated. The number of cycles at transition $N_c$ can be determined. The crack length at $N_c$ is not determined. This method is based on damage analysis. A damage variable *D* may be defined based on the stress response [26]. The most simple definition of the damage variable assuming isotropic material behavior is as follows:

$$\tilde{\sigma} = \sigma(1 - D) \tag{2}$$

$\tilde{\sigma}$ and $\sigma$ designate, respectively, the apparent stress in the damaged material and the corresponding stress in the sound material. Under monotonic loading, the apparent stress $\tilde{\sigma}$ varies from $\sigma$ (sound material) to zero ($D = 1$) at failure. Under fatigue loading, the cycling is stopped prior to the final failure and the samples are broken under simple tension. Nevertheless, the evolution of the damage parameter in the function of the normalized number of cycles N/Nf gives precious indications on the remaining load carrying capacity. Lemaitre and Chaboche [26] suggested different damage variables *D*.

Three different damage variables were determined in present work. The first $D(\sigma)$ compares the stress at a given cycle, to the stress after stabilization $\sigma_{stab}$.

$$D(\sigma) = 1 - \frac{\sigma(N)}{\sigma_{stab}} \tag{3}$$

where $\sigma(N)$ and $\sigma_{stab}$ correspond, respectively, to the reached tensile stress at cycle *N* and at the stabilized cycle. The second damage measure is based on the ratio between the apparent Young's modulus and the value of Young's modulus at stabilization:

$$D\,(E) = 1 - \frac{E(N)}{E_{stab}} \tag{4}$$

where $E(N)$ and $E_{stab}$ correspond, respectively, to the Yound modulus measured at a given cycle and at the stabilized cycle. The last damage measure considered in present work is based on the difference between the tensile and the compressive stresses at each cycle normalized by the stress after stabilization.

$$D(TC) = 1 - \left| \frac{\sigma_T(N) - \sigma_C(N)}{\sigma_{stab}} \right| \tag{5}$$

where $\sigma_T(\mathrm{N})$ and $\sigma_C(\mathrm{N})$ correspond, respectively, to the compressive stresses reached at a given cycle and at the stabilized cycle.

These most common damage variables are depicted on Figure 13 as a function of the normalized number of cycles $N/N_f$ for the three considered plastic strain levels. Negative values correspond to initial softening. The damage varies slowly up to a value $N/N_f$ ~0.9. Then, a very steep increase follows.

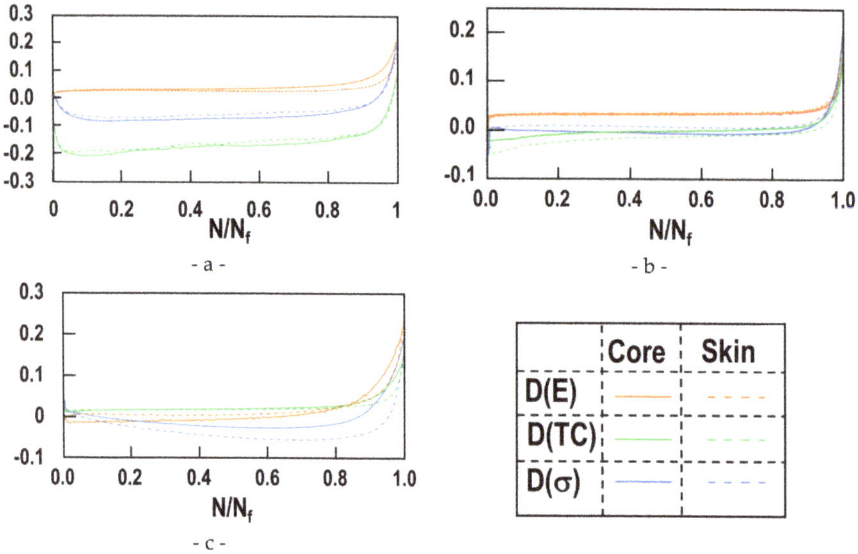


**Figure 13.** Comparison of three most common damage variables for all considered strain levels $\Delta\varepsilon^P/2$: 0.01% (**a**), 0.2% (**b**), and 0.4%(**c**).

At a given plastic strain, the three damage variables indicate the same transition point from slow variation to steep increase. Any of the three common damage variables may thus be used to determine transition from damage initiation (sub-critical growth) to macroscopic growth. The combined knowledge of the numbers of cycles at failure ($N_f$) and the transition from the initiation stage to accelerated growth gives precious results for determining the remaining load carrying capacity at a given number of cycles. As the three common damage variables D(E), D(TC) and D(σ) lead to similar transition points, we suggest working with D(σ). As the current stress at each cycle is measured for all common fatigue experiments, D(σ) is known.

To determine the transition from initiation to rapid growth, both regimes have been represented by straight lines as suggested in [26]. The intercept between both lines determines the critical number of cycles $N_c$ at the transition. Figure 14 depicts the corresponding results for all considered plastic strains and both sample locations. For all considered load levels and locations, the transition between both regimes occurs at $N_c/N_f > 92\%$. The initiation regime is thus always larger than 90% of the sample lifetime. Remarkably, transition occurs at the smallest ratio $N_c/N_f$ for the core sample loaded at $\Delta\epsilon^p/2 = 0.4\%$. The combination of large applied remote stress and the of large micro-shrinkages leads to a reduced lifetime ($N_f$) and a reduced initiation state ($N_c/N_f$).

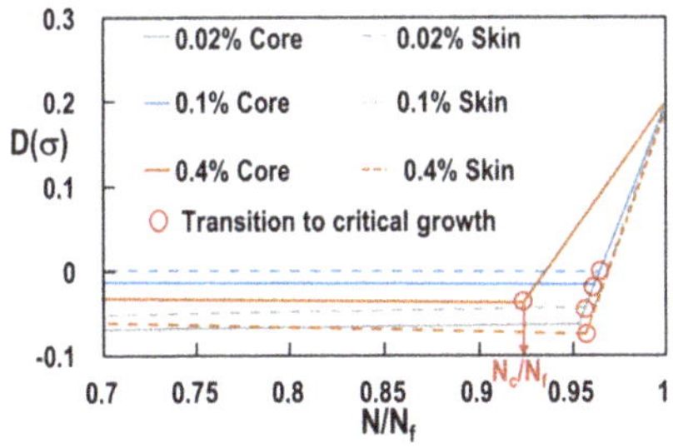


**Figure 14.** Damage D(σ) as a function of the normalized number of cycles. Determination of the critical number of cycles NC corresponding to the transition from sub-critical to critical growth.

## *4.4. Microscopic Analysis of Damage and Murakami Type Law*

### 4.4.1. Different Damage Parameters

Let us briefly summarize the previous results. The ferritic–pearlitic microstructure observed after normalizing treatment exhibits the same grain size and same inclusion distribution in the skin and the core of the ingot. The monotonic material behavior (stress–strain curve, hardness and micro-hardness in both phases) of samples from the skin and core of the ingot are very close. The sole significant differences between skin and core samples are the presence of larger micro-shrinkages in the core than in the skin, and a reduced number of cycles at failure ($N_f$) as well as a reduced initiation state ($N_c/N_f$) for samples from the core compared to samples from ingot skin. This hints to a close relation between the initial porosity and the number of cycles at failure ($N_f$). Moreover, the micro-shrinkages exhibit a strictly linear relationship between the volume and the surface. Defects exhibiting a large volume under tomographic observation should thus give rise on the sample fracture surface to defects with large areas.

The fracture surfaces of all samples were characterized by SEM-observations. These observations confirmed that initiation occurs always at micro-shrinkage porosities. In samples from the ingot core, cracking starts in most cases (80%) at a single large micro-shrinkage porosity. In samples from the ingot skin, initiation at single and at multiple shrinkage porosities were observed. In the present work, only examples with single crack initiation are shown. We found that multiple crack initiation was favored if the plastic strain was increased ($\Delta\varepsilon_P$ = 0.4%,). If the plastic strain is increased, the remote stress increases as well, and thus more initial defects will correspond to a critical size.

Murakami et al. [27] closely analyzed the relationship between fatigue life time and initial defect area. The relevance of several damage variables based on this concept is analyzed. Figure 15 depicts a typical fracture surface and explains the measurements made to determine different damage variables.

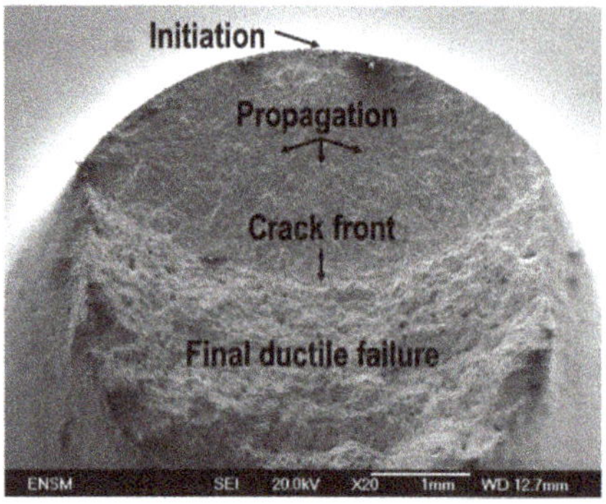


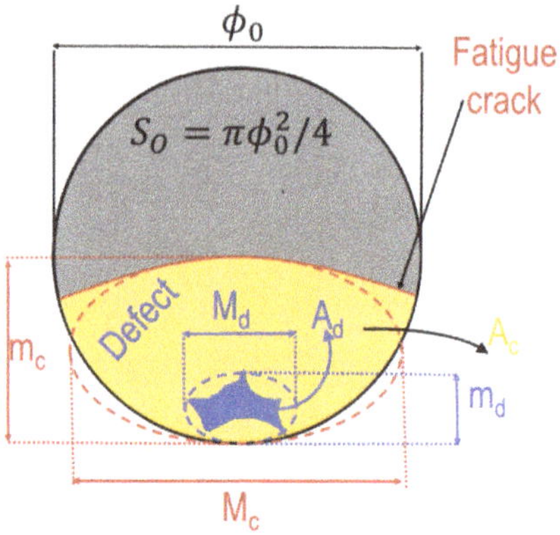


- a - - b -

**Figure 15.** Fracture surface analysis. (**a**) Typical fracture surface with crack initiation at a single defect. (**b**) Different measurements made on each fracture surface to determine the values of the corresponding damage variables.

The cross-sectional area $S_0$ was determined on each sample. The initial defect has an area $A_d$ and the crack an area $A_c$. These areas were measured as the total area of the corresponding pixels. Equivalent ellipses were defined on the initial defect (minor axis $m_d$) and the fatigue crack (minor axis $m_c$). The most commonly used damage parameters are defined as follows:

$$D_1 = \frac{A_c - A_d}{S_0} \quad D_2 = \frac{A_c^{eq} - A_d^{eq}}{S_0} \quad D_3 = \frac{m_c - \sqrt{A_d}}{\varphi_0} \quad D_4 = \frac{m_c - m_d}{\varphi_0} \tag{6}$$

$A_c^{eq}$ and $A_d^{eq}$ designate, respectively the areas of both equivalent ellipses.

$$A_c^{eq} = \frac{\pi M_c m_c}{4} \qquad A_d^{eq} = \frac{\pi M_d m_d}{4} \tag{7}$$

$D_1$ compares the areas of the fatigue crack and the initial defect. $D_2$ corresponds to a similar measure based on the equivalent ellipses. $D_3$ favors the propagation direction of the crack and considers an isotropic influence of the initial defect. Finally, in measure $D_4$ the main weight corresponds to the crack's and initial defect's dimensions in the propagation direction.

For all samples exhibiting a crack initiated at a single defect, the values of the four damage variables have been computed. Table 10 gives the mean values for the three considered strain levels. All damage parameters lead to higher values in the ingot core than the ingot skin. For both locations, the values of all damage variables increase with increasing plastic strain. The two damage parameters based on area measurements ($D_1$ and $D_2$) exhibit exactly the same values. These two damage parameters lead to smaller values than the two damage parameters ($D_3$ and $D_4$) based on length measurements in the propagation direction. Note the microscopic measurements of damage lead to larger values than the shut-off parameter based on the load in tension.

**Table 10.** Damage measurements on fatigue specimens fracture surfaces.

| Damage Variable | Δεp/2 | | | | | |
|---|---|---|---|---|---|---|
| | 0.02% | | 0.1% | | 0.4% | |
| | Core | Skin | Core | Skin | Core | Skin |
| $D_1$ | 28% | 24% | 30% | 29% | 30% | 35% |
| $D_2$ | 28% | 24% | 28% | 29% | 30% | 35% |
| $D_3$ | 34% | 33% | 36% | 38% | 39% | 44% |
| $D_4$ | 35% | 34% | 36% | 38% | 38% | 43% |

4.4.2. Murakami Type Analysis

Based on the previous results, the correlation between the number of cycles at failure with a stress intensity factor based on the initial defect has been tested. *The number of cycles $N_f$ may not be related directly to any stress intensity factor*. However, the number of cycles corresponding to the final stage (stage II), i.e., $(N_f\text{-}N_c)/N_f$, may be related to the stress intensity factor based on $\sqrt{A_d}$ and the stabalized stress $\sigma_{stab}$. Xua et al. [28] successfully modelled fatigue crack propagation (stage II) in G20Mn5QT cast steel by a two parameter model based on a Smith–Watson–Topper approach. The authors introduced a material correction to the classical SWT-parameter. An overview of the different SWT-parameters used in the literature is given in Łagoda et al. [29]. A similar approach is not immediate when considering stage I followed by stage II. In the present work, only the initial defect and the final crack shape could be observed by SEM characterization. Thus, the following classical expression for stress intensity factor $K$, based on Murakami's work, was adopted:

$$K = \sigma_{stab} \cdot \sqrt{\pi \sqrt{A_d}} \tag{8}$$

Figure 16 depicts the stress intensity factor $K$ as a function of the normalized number of cycles in stage II. For Figure 16a the classical logarithmic axes were adopted, whereas Figure 16b shows the same result in cartesian axes. A satisfying linear relationship in the log–log (Figure 16a) was obtained for core samples for all the considered plastic strain increments. For the skin samples, no linear relationship is observed. As only samples with a single initial defect were considered, this might be due to the reduced number of samples. However, the dispersion increases drastically with the plastic strain Δεp or the corresponding stabilized stress $\sigma_{stab}$. Under large plastic strain, the increased number of "smaller" defects in the skin sample leads to an increased number of defects almost perpendicular to the remote stress. For (Δεp/2 < 0.4%), these defects are too small to be activated. In the core samples, this dispersion is not seen, because larger defects (reduced in number) have been activated.

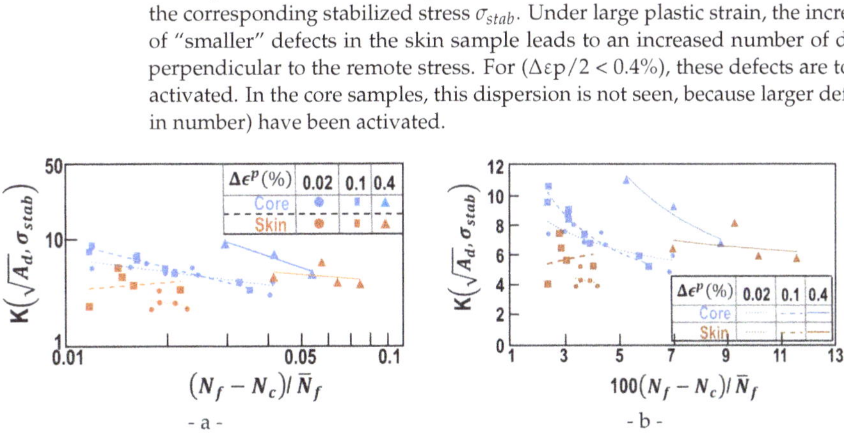


**Figure 16.** Correlation between stress intensity factor based on the initial defect area $A_d$ and the normalized number of cycles in stage II. (**a**) Classical representation with (**b**) the same results in cartesian axes.

## 5. Discussion and Conclusions

The low cycle fatigue behavior of G20Mn5 cast steel was analyzed. Particular castings with large risers were used to eliminate all macro-shrinkages. First, the columnar and equiaxed zones were determined on an as cast ingot. Then, only normalized samples were analyzed. Monotonic tensile and compression curves show that microvoids have no influence on the stress strain curve up to 5%.

The distinction between Core and Skin samples was made based on macrographic observations of the as-cast microstructure. Then, the normalized steel was characterized thoroughly. First, the micro-shrinkage-porosity was characterized on all LCF samples by Micro-CT-scans. Low cycle fatigue tests on normalized G20Mn5 cast steel were conducted at three levels of plastic strain (0.02%, 0.1% and 0.4%).

The main results may be summarized as follows. The microstructure and monotonic mechanical behavior of the ingot skin and core may be considered identical. The sole difference between the core and the skin is controlled by the micro-shrinkage distribution. The core exhibits a widespread volume distribution and the skin one much narrower. However, both core and ingot contain essentially flat (quasi two-dimensional) micro-shrinkages. At this stage, all differences in fatigue life between samples from the ingot skin or ingot core may be attributed to differences of the micro-shrinkage volume distributions.

For all fixed plastic strain increments and both locations (Skin or Core) similar curves are obtained. Depending on the plastic strain, strain hardening or softening may be observed. The stress stabilizes ($\sigma_{stab}$) at about $10^2$ cycles. A Manson–Coffin description with parameters from the literature allows prediction of the numbers of cycles at failure $N_f$. The number of cycles at failure is always smaller in the core than ingot skin.

Introduction of the simple damage variable ($D = \sigma/\sigma_{stab}$), based on values recorded during cycling, shows that the initiation stage ($N < N_c$) spans over more than 90% of the lifetime. The total number of cycles at failure $N_f$ and the number of cycles corresponding to the initiation stage ($N_c$) are smaller in the ingot core than in its skin. This naturally led to a damage characterization based on the observation of the fracture surfaces.

All fracture surfaces were characterized by FEG-SEM observations. All fatigue cracks have been initiated at porosities corresponding to micro-shrinkages. On samples with a single micro-shrinkage responsible for crack initiation, several damage parameters were defined. The most discriminant are based on the characteristic lengths of the initial defect and fatigue crack in the crack propagation direction. Based on these results, a Murakami damage type analysis was done. The stress intensity factor calculated with the initial defect size is related to the number of cycles in the final stage $(N_f\text{-}N_c)/N_f$.

**Author Contributions:** Conceptualization, J.-F.C. and H.K.; Methodology, A.B., J.S., M.L., J.-F.C. and H.K.; validation, H.K.; investigation, A.B., C.R.; data curation, A.B., C.R., writing—original draft preparation, A.B.; writing—review and editing, H.K.; visualization, H.K.; supervision, J.-F.C., H.K.; project administration, J.-F.C., H.K. All authors have read and agreed to the published version of the manuscript.

**Funding:** This research received no external funding.

**Informed Consent Statement:** Not applicable.

**Conflicts of Interest:** The authors declare no conflict of interest.

## References

1. Stefanescu, D.M. *ASM Handbook;* ASM International the Materials Information Company: Materials Park, OH, USA, 2008; Volume 15, Chapter 1, ISBN 978-0-87170-711-6.
2. Suresh, S. *Fatigue of Materials;* Cambridge University Press: Cambridge, UK, 2004; ISBN 978-0-521-57847-9.
3. Pineau, A.; Bathias, C. *Fatigue of Materials and Structures, Fundamentals;* John Wiley & Sons, Inc.: Hoboken, NJ, USA, 2010.
4. Francois, D.; Pineau, A.; Zaoui, A. *Mechanical Behaviour of Materials, Volume II: Fracture Mechanics and Damage, Solid Mechanics and its Applications;* Gladwell, G.M.I., Ed.; Springer: Berlin/Heidelberg, Germany, 2013; Volume 191, pp. 305–362.
5. Mughrabi, H. Dislocations clustering and long range internal stresses in monotonically and cyclically deformed metal crystals. *Rev. Phys Appl.* **1988**, *23*, 367–379. [CrossRef]

6. Murakami, Y. *Metal Fatigue: Effects of Small Defects and Nonmetallic Inclusions*; Elsevier: Amsterdam, The Netherlands, 2002; ISBN 0-08-044064-9.
7. Zerbst, U.; Klinger, C. Material defects as cause for the fatigue failure of metallic components. *Int. J. Fatigue* **2019**, *127*, 312–323. [CrossRef]
8. Dantzig, J.; Rappaz, M. *Solidification*, 2nd ed.; EPFL Press: Lausanne, Switzerland, 2016.
9. Masson, J.-M. Fonderie et Moulage de l'acier-Évolution des Procédés D'élaboration, Techniques de L'ingénieur. 2005. Available online: https://www.techniques-ingenieur.fr/ (accessed on 4 October 2022).
10. Bhadeshia, H.; Honeycombe, R. *Honeycombe, Steels: Microstructure and Properties*, 4th ed.; Butterworth-Heinemann: Amsterdam, The Netherlands, 2017.
11. Niyama, E. A Method of Shrinkage Prediction and Its Application to Steel Casting Practice. *AFS Int. Cast Met. J.* **1982**, *3*, 507–517.
12. Rotella, A. Fatigue d'un Alliage D'aluminium Moulé A357-T6: Rôle de la Morphologie, de la Position des Défauts et Application à Une Structure Pour le Calcul de la Durée de vie en Fatigue. Ph.D. Thesis, ISAE-ENSMA, Chasseneuil-du-Poitou, France, 2017.
13. *Standard E446*; Reference Radiographs for Steel Castings up to 2 inches (51 mm) in Thickness. 2002 Annual Book of ASTM Standards; American Society of Testing and Materials: West Conshohocken, PA, USA, 2002; Volume 3, pp. 189–191.
14. *Standard E186*; Standard Reference Radiographs for Heavy-Walled 4 $\frac{1}{2}$ to 12-in. (114 to 305-mm) Steel Castings. 2002 Annual Book of ASTM Standards; American Society of Testing and Materials: West Conshohocken, PA, USA, 2002; Volume 3, pp. 81–83.
15. *Standard E280*; Standard Reference Radiographs for Heavy-Walled 2 to 4 $\frac{1}{2}$-in. (51 to 114-mm) Steel Castings. 2002 Annual Book of ASTM Standards; American Society of Testing and Materials: West Conshohocken, PA, USA, 2002; Volume 3, pp. 125–127.
16. Beckermann, C.; Hardin, R.A. Effect of Porosity on Mechanical Properties of 8630 Cast Steel. In Proceedings of the 58th SFSA Technical and Operating Conference, no Paper No 4.4, Chicago, IL, USA, 4–6 November 2004.
17. Serrano-Munoz, I.; Jean-Yves Buffiere, J.-Y.; Mokso, R.; Verdu, C.; Nadot, Y. Location, location & size: Defects close to surfaces dominate fatigue crack initiation, Nature. *Sci. Rep.* **2017**, *7*, 45239. [CrossRef]
18. Carlson, K.D.; Beckermann, C. Prediction of Shrinkage Pore Volume Fraction Using a Dimensionless Niyama Criterion. *Metall. Mater. Trans. A* **2009**, *40*, 163–175. [CrossRef]
19. Shouzhu, O.; Carlson, K.D.; Hardin, R.A.; Beckermann, C. Development of New Feeding-Distance Rules Using Casting Simulation: Part II. The New Rules. *Metall. Mater. Trans. B* **2002**, *33*, 741–755.
20. Han, Q.; Guo, Q.; Yin, Y.; Xing, Y. Fatigue Behaviour of G20Mn5QT Cast Steel and Butt Welds with Q345B Steel. *Int. J. Steel Struct.* **2016**, *16*, 139–149. [CrossRef]
21. Wu, S.C.; Qin, Q.B.; Hu, Y.N.; Branco, R.; Li, C.H.; Williams, C.J.; Zhang, W.H. The microstructure, mechanical, and fatigue behaviors of MAG welded G20Mn5 cast steel. *Fatigue Fract. Eng. Mater. Struct.* **2020**, *43*, 1051–1063. [CrossRef]
22. Nagel, S.; Rauber, C.; Veile, I.; Knödel, P.; Ummenhofer, T. Influence of internal imperfections on the fatigue resistance of cast steel–testing methodology. *MATEC Web Conf.* **2019**, *300*, 09001. [CrossRef]
23. CTIF. Méthode Micrographique de Détermination des Inclusions non Métalliques des Aciers Moulés (BNIF 500). Available online: https://ctif-editions.com/ (accessed on 4 October 2022).
24. Manson, S.S. *Behavior of Materials under Conditions of Thermal Stress*; REPORT 1170; National Advisory Committee for Aeronautics: Moffett Field, CA, USA, 1954.
25. Manson, S.S.; Halford, G.R. *Fatigue and Durability of Structural Materials*; ASM International: Materials Park, OH, USA, 2006; ISBN 0-87170-825-6.
26. Lemaitre, J.; Chaboche, J.L. *Mechanics of Solid Materials*; Cambridge University Press: New York, NY, USA, 1990; Chapter 7.
27. Murakami, Y.; Endo, M. Quantitative evaluation of fatigue strength of metals containing various small defects or cracks. *Eng. Fract. Mech.* **1983**, *17*, 1–15.
28. Xua, Y.; Lia, Z.; Xua, J.; Han, Q. Fatigue crack growth of G20Mn5QT cast steel based on a two parameter driving force model. *Eng. Fract. Mech.* **2019**, *208*, 13–26. [CrossRef]
29. Łagoda, T.; Vantadori, S.; Głowacka, K.; Kurek, M.; Kluger, K. Using the Smith-Watson-Topper Parameter and Its Modifications to Calculate the Fatigue Life of Metals: The State-of-the-Art. *Materials* **2022**, *15*, 3481. [CrossRef]

materials

MDPI

Article

# Empirical Solution of Stress Intensity Factors for the Inclined Inner Surface Crack of Pipe under External Pressure and Axial Compression

**Xi-Ming Yao [1,2], Yu-Chen Zhang [1,2], Qi Pei [1,2], Li-Zhu Jin [1,2], Tian-Hao Ma [1,2], Xiao-Hua He [1,2] and Chang-Yu Zhou [1,2,*]**

1 School of Mechanical and Power Engineering, Nanjing Tech University, Nanjing 211816, China
2 Jiangsu Key Lab of Design and Manufacture of Extreme Pressure Equipment, Nanjing 211816, China
* Correspondence: changyu_zhou@163.com; Tel./Fax: +86-25-58139951

**Abstract:** Based on fracture mechanics theory, a finite element method was used to determine the stress intensity factors of the inclined crack on the inner surface of the pipe under axial compression load and external pressure. The effects of different influencing factors on the stress intensity factor along the crack front considering crack closure were systematically explored, which were different to those under internal pressure. The effects of high aspect ratio on $K_{II}$, the crack inclination asymmetry caused by curvature and the effects of the friction coefficient on the stress intensity factors of the pipe with an inclined inner surface crack under axial compression load and external pressure were explored in this paper. To be fit for defect assessment, the solutions for stress intensity factors $K_{II}$ and $K_{III}$ were derived, and new correction factors $f_\theta$ and $f_\mu$ were proposed in the empirical solutions to accommodate the crack inclination asymmetry and the friction coefficient, respectively.

**Keywords:** inclined surface crack; stress intensity factor; pipe; crack closure; external and axial pressure; finite element analysis

**Citation:** Yao, X.-M.; Zhang, Y.-C.; Pei, Q.; Jin, L.-Z.; Ma, T.-H.; He, X.-H.; Zhou, C.-Y. Empirical Solution of Stress Intensity Factors for the Inclined Inner Surface Crack of Pipe under External Pressure and Axial Compression. *Materials* **2023**, *16*, 364. https://doi.org/10.3390/ma16010364

Academic Editor: Roberto Citarella

Received: 25 November 2022
Revised: 19 December 2022
Accepted: 27 December 2022
Published: 30 December 2022



## 1. Introduction

Pipes are widely used in various fields of petroleum, chemical industry, and natural gas. In engineering practice, there are a large number of shells subjected to external pressure loads such as submarines, aerospace simulators, deep-sea pipelines, and buried pipelines. Due to the limitations of the manufacturing process, the structures may have tiny defects such as cracks. Lin and Smith [1] found that arbitrarily shaped cracks become semi-elliptical after a certain period of expansion. The stress intensity factor is an important fracture parameter for judging crack propagation and failure in defect assessment. Therefore, since the last century, scholars have conducted many studies on this subject.

In the early years, Raju and Newman [2] established a stress intensity factor empirical solution for mode I internal and external surface cracks of cylindrical vessels through FEA. The formula gave the influence coefficients based on different shell and crack sizes. Navid and Fenner [3] calculated the stress intensity factors of a thick-walled cylinder with internal pressure for different crack sizes based on the boundary integral equation method. Afterward, various numerical techniques were developed to calculate the stress intensity factor of pipe surface cracks such as the line spring model [4] and the weight function method [5–7]. Kamaya [8] combined the finite element alternating method and the finite element analysis to calculate the stress intensity factors of shell surface cracks. Wallbrink [9] proposed a semi-analytical method of conformal transformation to predict the stress intensity factors of circumferential surface cracks. However, most of these studies were limited to simple loading and the crack direction was axial or circumferential. In fact, in most cases, the loading conditions of the pipelines are complex and the crack direction is inclined.

The research on mixed mode cracks first started from a flat plate. Murakami [10] analyzed $K_I$, $K_{II}$, $K_{III}$ by the body force method, and provided a formula for the maximum value $K_{\theta max}$ at the crack front according to the projected area of the crack in the direction of the maximum principal stress. Zeng and Dai [11] proposed a simplified analytical model of an inclined surface crack under biaxial stress, and proposed a closed solution for $K_I$ and $K_{III}$ at the deepest point of the crack front. Some scholars have also used FEA [12,13], the numerical influence function method [14], and the experimental method [15] to study the mixed mode cracks of flat plates. The study of mixed mode cracks in shell can be divided into two types. First, the crack direction is determined, usually axial or circumferential, and the mixed mode crack is formed by applying a complex far-field load on the structure. Shahani and Habibi [16] considered the stress intensity factor of cylinders with a circumferential crack under the action of axial force, bending moment, and torque. Predan et al. [17] studied the stress intensity factor for circumferential semi-elliptical surface cracks in a hollow cylinder subjected to pure torsion. Ramezani et al. [18] demonstrated the empirical solution of the stress intensity factor of cylinder surface crack under pure torsion. Second, the mixed mode cracks were formed by changing the inclination angle of the surface cracks. For example, Li et al. [19,20] studied the stress intensity factor of the inclined crack in the pipeline under the far field tension and tension-bending, and provided an empirical formula to calculate the complete value of the stress intensity factor at the crack front through the influence coefficient under different aspect ratios. Li and Mao [21] calculated the stress intensity factor of the inclined surface crack of the outer wall of the heat exchanger by numerical simulation.

In the research on cracks, it is inevitable to encounter the problem of complete or partial closure of the cracks. For simplicity, more papers have chosen to ignore the surface contact of the cracks. However, the contact closure of the crack face significantly changes the stress–strain distribution at the crack front, thereby affecting the fracture behaviors of the structure. In fact, since the contact area is time-varying, the contact problem is a typical nonlinear problem with boundary conditions. In static fracture mechanics, the contact friction of the crack face has an obvious influence on the stress intensity factor. At present, there have been few studies on mixed mode cracks on the friction surfaces under compressive loading, which have mainly focused on flat plates [22–24]. Bowie and Freese [25] proposed a crack closure technique to correct the solution of overlapping crack surfaces, but their method does not consider the sliding between crack surface, and is only suitable for the case of a large friction coefficient. Liu and Tan [26] used the boundary element method to study the effect of the interaction between the friction and crack surface on the stress intensity factor. Hammouda et al. [27,28] used finite element analysis to study the effect of crack surface friction and crack inclination on the stress intensity factor of the central cracked plate under unidirectional compressive load. In addition, Dorogoy and Banks-Sills [29] investigated the effects of loading angle and friction coefficient on the stress intensity factor and the crack length of Brazilian disc cracks under concentrated loading using a finite difference solution.

To the best of the authors' knowledge, researchers have not focused on the surface crack stress intensity factor of shell structures subjected to compressive loads, however, in engineering practice, fracture damage occurs in shell structures subjected to simultaneous external and axial pressures. In this paper, the stress intensity factors of the shell subjected to simultaneous external and axial pressures were evaluated by using the three-dimensional finite element analysis method, and the effects of the relative depth of the surface cracks, aspect ratio, crack inclination, and friction coefficient on the stress intensity factor of the crack fronts were investigated, which were different from those under internal pressure. The empirical solutions of the mode II and mode III stress intensity factors along the crack front are given using numerical methods, respectively.

## 2. Finite Element Model of Pipe Crack

### *2.1. Meshing of the Pipe with Inclined Crack on the Inner Surface*

For the study of mixed mode cracks, the most widely used and accurate method is FEA, which involves the extended finite element method (XFEM) and contour integral. Based on the static loading model in this paper, the fluctuation of results caused by inaccurate meshing can be better avoided by using contour integrals. The commercial code ABAQUS [30] was used for FEA. The material used in this paper was TA2 with a modulus of elasticity of 101,901 MPa and a Poisson's ratio of 0.348. The pipe size was fixed at $R_i = 100$ mm, $R_0 = 110$ mm, $t = 10$ mm. In order to avoid the influence of the boundary effect on the stress intensity factor, the pipe length should be six times that of $R_i$ and 20 times that of the crack half-length c, $l = 600$ mm [31,32]. The crack size is described by dimensionless parameters: the relative crack depth ($a/t$) and the ratio of long and minor axes of semi-elliptical cracks ($a/c$). The crack inclination angle is $\theta$. The pipe model is shown in Figure 1a. The external pressure of the pipe $P_0 = 100$ MPa and the axial pressure $p = 525$ MPa. The boundary conditions were that one end of the pipe was completely fixed and the other end restricted its circumferential constraint, as shown in Figure 1b.

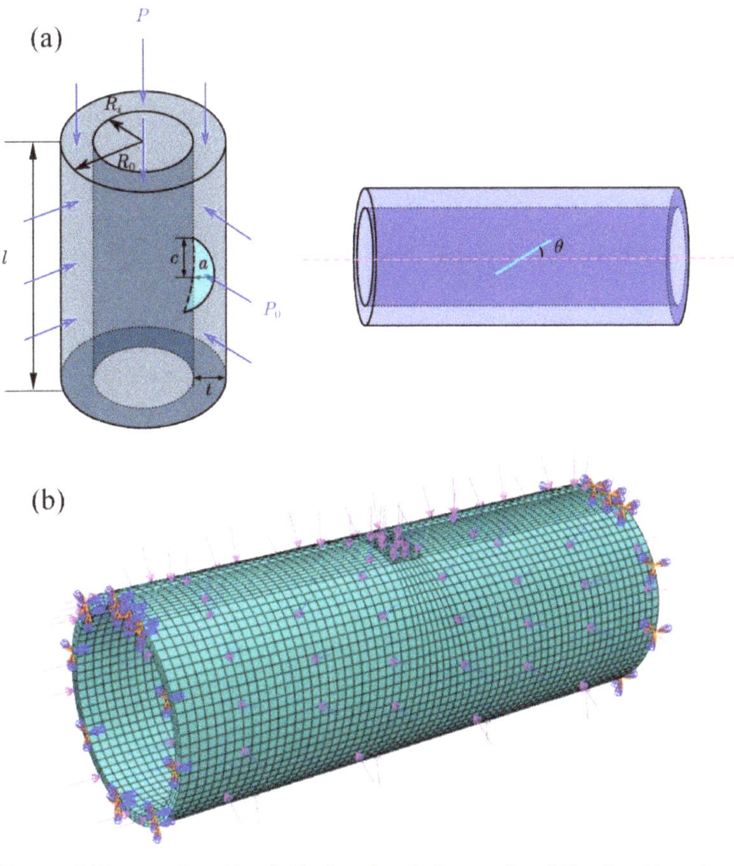


**Figure 1.** (**a**) Pressure pipe with an inclined crack on the inner surface. (**b**) Loads and boundary conditions.

The current general semi-elliptical crack meshing method was used by the authors [31,33]. The six-node triangular prism element C3D6 was used to sweep along the path around the crack tip to simulate the stress–strain singularity at the crack front. C3D6 belongs to a fully integrated element with linear interpolation in all directions, the number of nodes is 6, and the number of integration points is 6. The eight-node hexahedral element C3D8R was used in the nearby area. C3D8R belongs to a linear reduced integral element with eight nodes and only one integral point exists in the center of the cell, which is equivalent to a constant stress cell. The reduced integration element can effectively avoid stress discontinuity problems. The meshes at the crack were subdivided, the mesh size was controlled within 1 mm, and the mesh size of the rest zone was controlled within 10 mm to reduce the calculation time. The mesh division is shown in Figure 2.

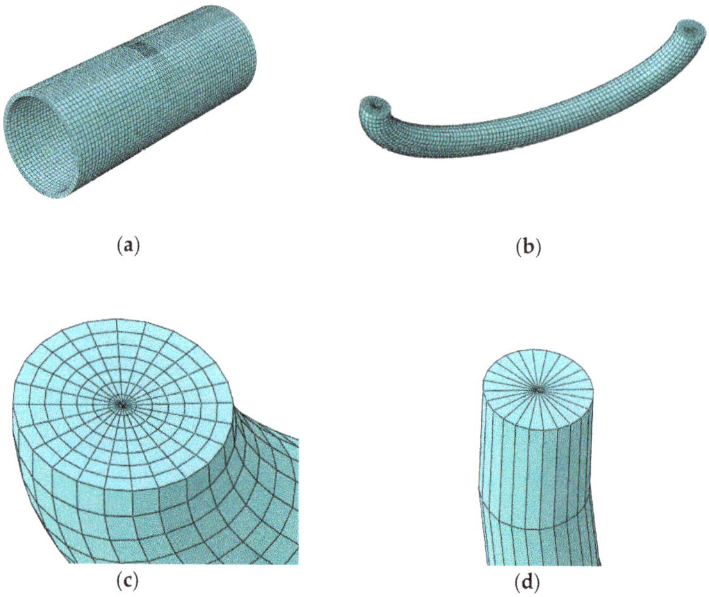

**Figure 2.** Meshing of the pipe with the inclined internal surface crack:(**a**) global view of the cracked pipe; (**b**) equal divisions at the crack front; (**c**) hexahedron elements around the crack front; (**d**) wedge-shaped elements at the crack front.

### *2.2. Penalty Function Method for Contact Problem*

Since the pipe is not only subjected to external pressure but also to axial compressive load, contact should be set between the crack surfaces. Generally speaking, the crack surfaces in the contact state have three characteristics:

(1) The contact surfaces do not penetrate or overlap each other;
(2) The contact surfaces are able to transmit normal pressure and tangential friction;
(3) The contact surfaces generally do not transmit normal tension and can be separated freely.

In ABAQUS, the penalty function method, Lagrange multiplier method, and static friction–kinetic friction index decay method can be used to solve the contact problem. In this paper, a penalty function was used to characterize the tangential friction force between the crack surfaces. The penalty function method requires the normal and tangential friction coefficients, which is similar to setting a "spring" between the contact surfaces, and the "spring" works only when the contact surfaces are closed. In this way, the normal contact pressure can be expressed as:

$$P_n = \begin{cases} 0 & u_n > 0 \\ k_n u_n & u_n \leq 0 \end{cases} \tag{1}$$

According to Coulomb's law of friction, the frictional stress at the contact surface can be written as:

$$\begin{aligned} \tau_s &= k_s u_s - \mu P_n \\ \tau_t &= k_t u_t - \mu P_n \end{aligned} \tag{2}$$

where $t$, $n$, $s$ represent the tangential, normal and sub-normal directions, respectively, as shown in Figure 3. The advantage of the penalty function is that it does not increase the degree of freedom of the problem. The disadvantage is that when the friction coefficient is too large, the convergence will become more difficult. The values of the friction coefficient $\mu$ in this paper were 0, 0.2, and 0.4, respectively.

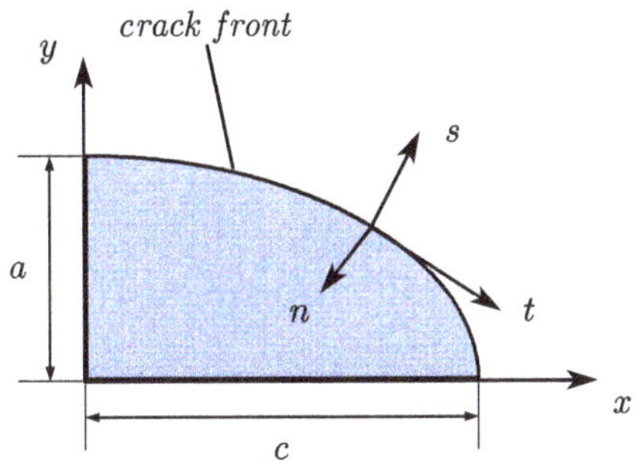


**Figure 3.** Local coordinate system of the quarter-elliptic crack front.

### 2.3. Mesh Independence Verification

Raju and Newman [2] first proposed a formula to calculate the stress intensity factor of a mode I crack in a pipe, and Chun-Qing Li [19] extended this formula to calculate the mixed-mode stress intensity factor of an inclined crack, as follows:

$$\boldsymbol{K} = \sigma_0 \sqrt{\pi \frac{a}{Q}} \boldsymbol{F}\left(\frac{a}{t}, \frac{a}{c}, \frac{t}{R_i}, \xi, \theta\right) \tag{3}$$

where $\sigma_0 = PR/t$ is the average hoop stress of the pipe; $\boldsymbol{K}$ and $\boldsymbol{F}$ are the stress intensity factor and influence coefficient function under different cracking modes, respectively; $Q$ is the shape factor obtained from the second type of elliptic integral; and the empirical formula is given by Shiratori et al. [34].

$$\boldsymbol{K} = \begin{Bmatrix} K_I \\ K_{II} \\ K_{III} \end{Bmatrix} \tag{4}$$

$$\boldsymbol{F}\left(\frac{a}{t}, \frac{a}{c}, \frac{t}{R_i}, \xi, \theta\right) = \begin{Bmatrix} F_I \\ F_{II} \\ F_{III} \end{Bmatrix} \tag{5}$$

$$\begin{aligned} Q &= 1 + 1.464\left(\tfrac{a}{c}\right)^{1.65} & \tfrac{a}{c} \leqslant 1 \\ Q &= \left[1 + 1.464\left(\tfrac{c}{a}\right)^{1.65}\right]\left(\tfrac{a}{c}\right)^2 & \tfrac{a}{c} > 1 \end{aligned} \tag{6}$$

In order to unify the evaluation indicators, it is necessary to normalize the stress intensity factor and the position along the crack front. In this paper, the normalization factor $K_0$ was defined as [18]:

$$K_0 = \sigma\sqrt{\pi \cdot a_0} \quad a_0 = 1 \tag{7}$$

where $\sigma = PR/2t$ is the axial compressive stress. The position along the crack front can be normalized according to the number of equal meshes. The advantage of this normalization method is that $K_0$ becomes a constant, which can truly reflect the variation trend of $K$.

Figure 4 shows the normalized stress intensity factors at the crack front of three kinds of mesh sizes at different angles. The mesh size is shown in Table 1. It can be seen that the results of the first two mesh sizes were basically the same. A comparison with the normalized $K_{II}$ and $K_{III}$ results with Chun-Qing Li [19] after changing the load using mesh2 is given in Figure 5. The average differences were 8.98% and 2.55%, respectively, which demonstrates the accuracy of the results.

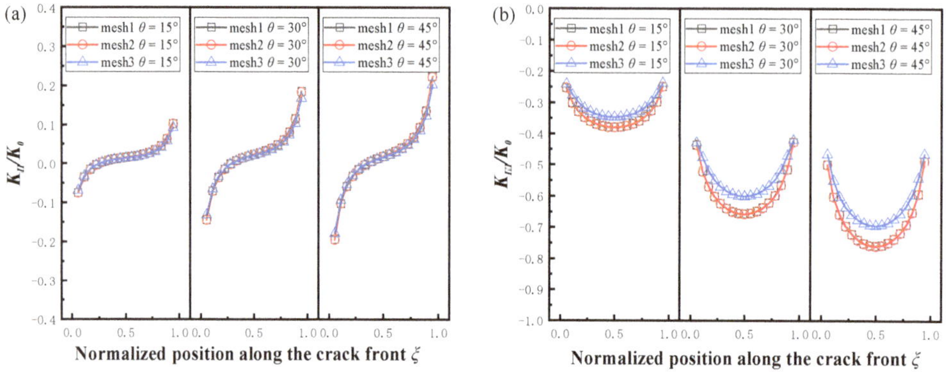


**Figure 4.** Effect of the mesh density on the normalized SIF at different angles, $a/t$ = 0.2, $a/c$ = 0.2. (**a**) Normalized $K_{II}$, (**b**) normalized $K_{III}$.

**Table 1.** Crack front meshing.

| Division Strategy | Mesh1 | Mesh2 | Mesh3 |
|---|---|---|---|
| Wedge element size (mm) | 0.05 | 0.05 | 0.1 |
| Hexahedral element layers | 8 | 6 | 3 |
| Equal divisions at the crack front | 180 | 90 | 45 |

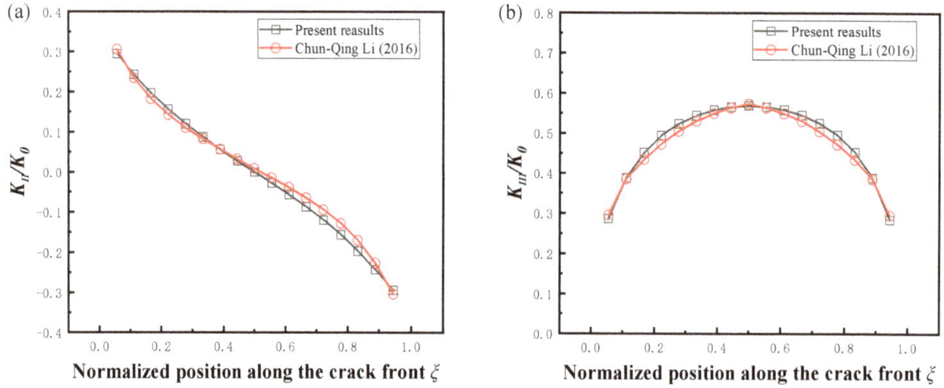


**Figure 5.** Comparison of the stress intensity factors with mixed modes, $a/t$ = 0.2, $a/c$ = 0.4, $\theta$ = 45°. (**a**) Normalized $K_{II}$, (**b**) normalized $K_{III}$ [19].

## 2.4. Comparison of SIFs between Crack Opening and Crack Closing

In order to compare the change in the stress intensity factor for crack opening and crack closing, two sets of cracks with different angles were selected in this paper. The contact surface friction coefficient $\mu$ = 0. The pipe load for crack opening was changed to the internal pressure and axial tension, and the results of the comparison are shown in Figure 6. To facilitate comparison of the differences, the normalized stress intensity factors under external pressure and axial compression were expressed symmetrically about the transverse coordinate. It can be seen that the absolute values of the normalized stress intensity factors at crack closure were obviously higher for both the mode II crack and mode III crack. For this phenomenon, the authors used the stress field calculation method for the plastic zone of the crack tip mentioned by Shlyannikov and Tumanov [35] to convert the three stress states obtained from the finite element calculations to give the sub-normal stresses and tangential stresses at the crack front in these two cases, corresponding to mode II and mode III cracks, respectively, as shown in Figure 7a,b, and it can be seen that $\tau_s$ and $\tau_t$ at the crack front of the pipe under axial compression and external pressure were higher than those under axial tension and internal pressure, which is the fundamental reason for the difference in SIFs. This difference also proves that it is necessary to study the stress intensity factor of cracks under axial compression load and external pressure.

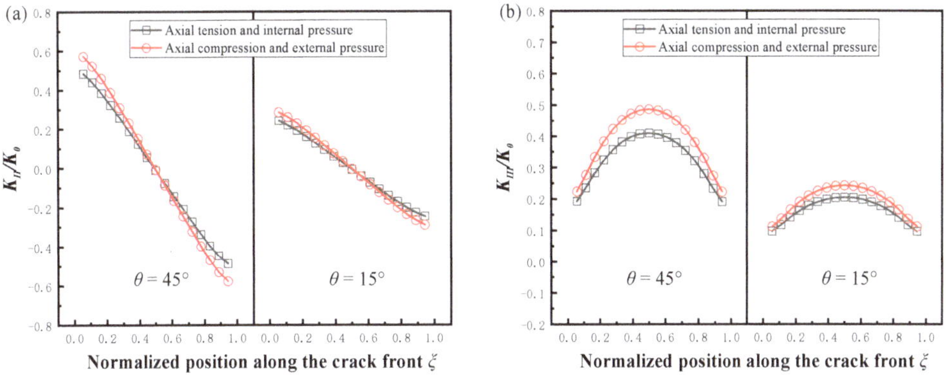


**Figure 6.** Comparison of SIFs under tension and compression loads with different angles, $\mu = 0$, $a/t = 0.2$, $a/c = 0.8$. (**a**) Normalized $K_{II}$, (**b**) normalized $K_{III}$.

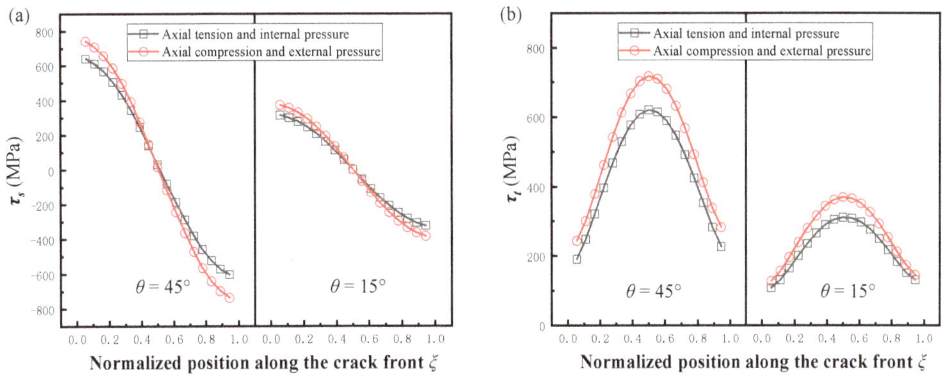


**Figure 7.** Comparison of shear stress along the crack fronts under tension and compression loads. (**a**) Shear stress along the crack front of mode II. (**b**) Shear stress along the crack front of mode III.

## 3. Results and Discussion

The stress intensity factor for mode I crack is 0 under the compression load [36,37]. However, due to the crack inclination and external pressure, the $K_{II}$ and $K_{III}$ generated by the interaction of tangential and frictional forces at the crack front cannot be neglected. In this paper, the SIFs of pipe surface crack fronts were obtained by FEA under different influencing factors, which included the relative depth ($a/t$), the aspect ratio ($a/c$), the crack inclination angle ($\theta$), and the friction coefficient ($\mu$). The specific calculation scheme is shown in Table 2. A total of 336 finite element models were calculated in this paper.

**Table 2.** Finite element calculation scheme.

| Influencing Factors | | | | | | | |
|---|---|---|---|---|---|---|---|
| $a/t$ | 0.2 | 0.4 | 0.6 | 0.8 | | | |
| $a/c$ | 0.2 | 0.4 | 0.8 | 1.6 | | | |
| $\theta$(°) | 0 | 15 | 30 | 45 | 60 | 75 | 90 |
| $\mu$ | 0 | 0.2 | 0.4 | | | | |

### 3.1. Effects of Crack Geometry and Inclination Angle

Figure 8 shows the variation in the mode II and mode III SIFs with relative depth. The long axes of the semi-elliptical cracks were fixed as $c$ = 5 mm and $c$ = 10 mm, respectively, the friction coefficient $\mu$ was 0. For the convenience of presentation, only the comparison results of the maximum value along the crack front are given, and the rules were consistent for the remaining positions. It can be seen that the absolute value of the mode II and mode III stress intensity factors increased with the deepening of the crack, and the increase in the absolute values was more obvious when the crack size was larger. The surface crack was most dangerous at $\theta$ = 45° and became safer with the offset of crack inclination to the axial and circumferential directions.

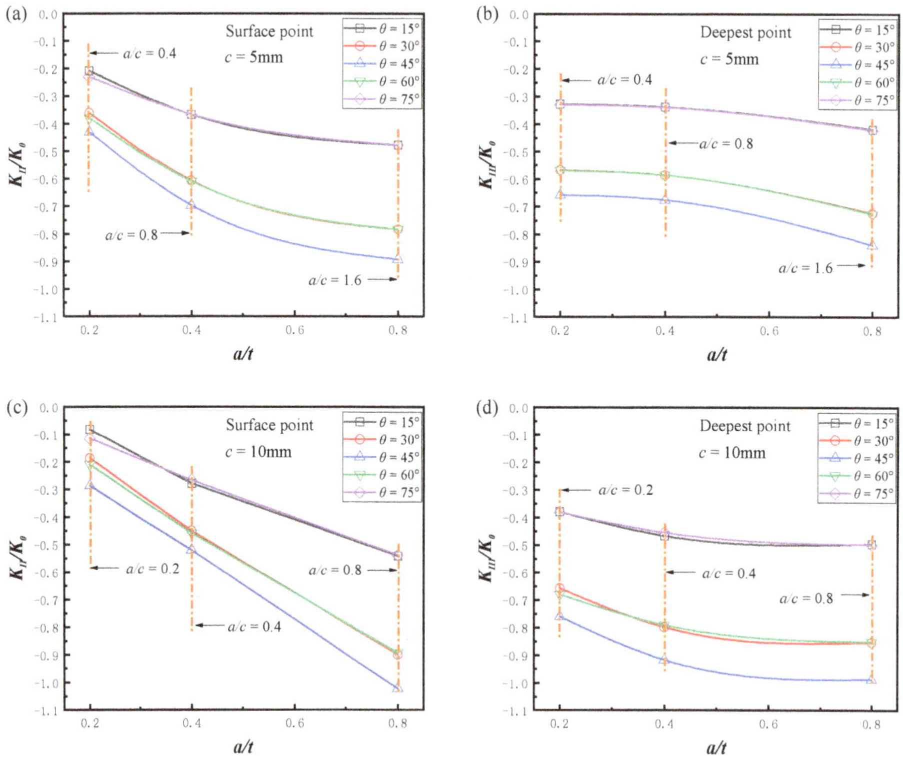


**Figure 8.** Effect of the relative depth on the normalized SIFs, $\mu$ = 0. (**a**) Normalized $K_{II}$ ($c$ = 5 mm, $a/c$ = 0.4, 0.8, 1.6); (**b**) normalized $K_{III}$ ($c$ = 5 mm, $a/c$ = 0.4, 0.8, 1.6); (**c**) normalized $K_{II}$ ($c$ = 10 mm, $a/c$ = 0.2, 0.4, 0.8); (**d**) normalized $K_{III}$ ($c$ = 10 mm, $a/c$ = 0.2, 0.4, 0.8).

The variation in the mode II and mode III SIFs with the aspect ratio is given in Figure 9, and the minor axis of the semi-elliptical crack was fixed as $a$ = 4 mm and $a$ = 6 mm, respectively, the friction coefficient $\mu$ was 0. For the mode II cracks, when $a/c < 1$, the absolute value of the normalized $K_{II}$ increased with the aspect ratio, and when $a/c > 1$, the absolute value of the normalized $K_{II}$ decreased with the increase in the aspect ratio. The $K_{II}$ values of the crack fronts for different aspect ratios at two depths are given in Figure 9e,f. For semi-elliptical cracks with lower aspect ratios, the maximum value of $K_{II}$ always appeared at the surface point of the crack front. However, for semi-elliptical cracks with higher aspect ratios, the maximum value of $K_{II}$ appeared near the surface point, which led to a smaller value at the surface point. These variations are consistent with the results of Yang et al. [38] for high aspect ratio cracks. For mode III cracks, the absolute value of $K_{III}$ decreased monotonically with the increase in aspect ratio, with the most pronounced decrease at $\theta$ = 45°.

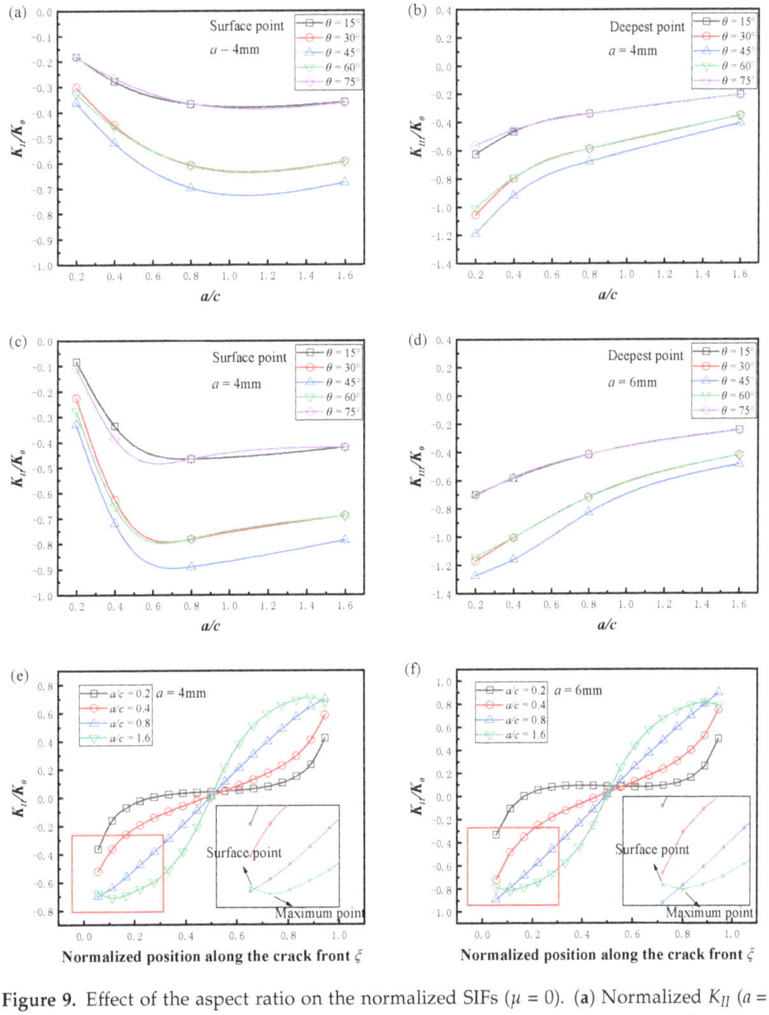


**Figure 9.** Effect of the aspect ratio on the normalized SIFs ($\mu$ = 0). (**a**) Normalized $K_{II}$ ($a$ = 4 mm, $a/t$ = 0.4); (**b**) normalized $K_{III}$ ($a$ = 4 mm, $a/t$ = 0.4); (**c**) normalized $K_{II}$ ($a$ = 6 mm, $a/t$ = 0.6); (**d**) normalized $K_{III}$ ($a$ = 6 mm, $a/t$ = 0.6); (**e**) normalized $K_{II}$ near the crack surface ($a$ = 4 mm, $a/t$ = 0.4); (**f**) normalized $K_{II}$ near the crack surface ($a$ = 6 mm, $a/t$ = 0.6).

The variation in the mode II and mode III SIFs with the crack inclination angle is given in Figure 10, the friction coefficient $\mu$ was 0. It can be seen that in all cases, the absolute values of $K_{II}$ and $K_{III}$ were symmetrical about the deepest point. The maximum value of the stress intensity factor always appeared at the surface point or the deepest point at $\theta = 45°$, and the SIF value was smaller when the crack inclination was closer to the axial and circumferential directions, which was due to the more obvious closure effect under compression load, and the tangential component parallel to the crack surface became smaller. It should be noted that for small-size cracks, $K_{II}$ and $K_{III}$ were almost identical when $\theta$ was symmetric about 45° ($\theta = 30°$, 60° and $\theta = 15°$, 75°), and the pipe surface cracks at this time could be approximated as flat plate cracks, which has been verified in the study of mixed mode cracks in flat plates and pipes [13,20]. However, for large size pipe surface cracks, the difference caused by the angle was more obvious. The crack surface had a different curvature on the inner wall surface of the pipe at different inclination angles, which made the stress intensity factors no longer consistent for $\theta = 30°$, 60° and $\theta = 15°$, 75°. As shown in Figure 10e,f, this difference due to curvature was more pronounced in spherical shells [39].

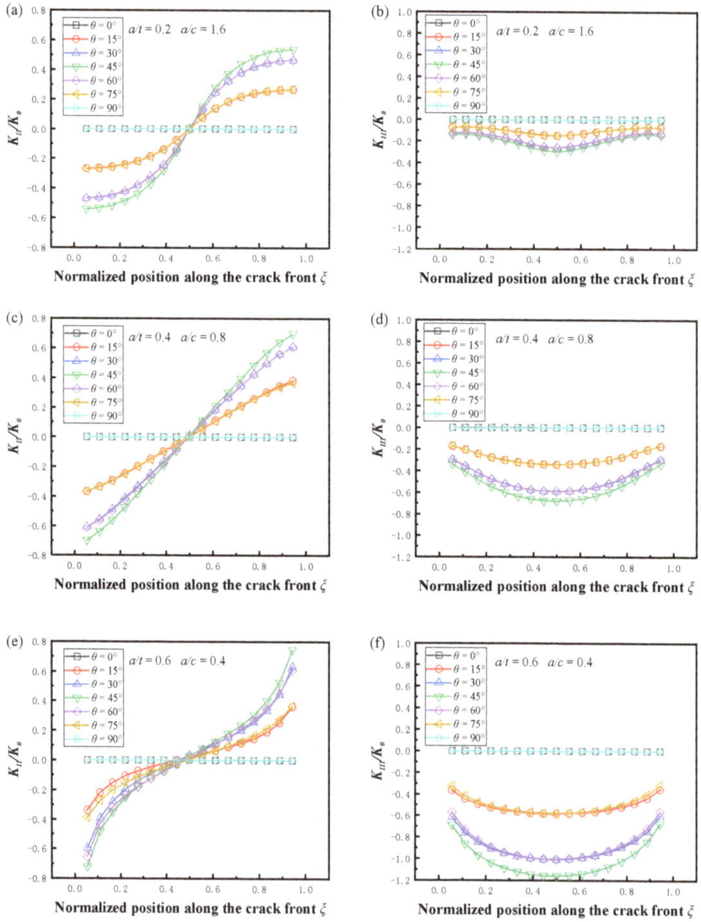


**Figure 10.** Effect of the inclination angle on the normalized SIFs, $\mu = 0$. (**a**) Normalized $K_{II}$ ($a/t = 0.2$, $a/c = 1.6$); (**b**) normalized $K_{III}$ ($a/t = 0.2$, $a/c = 1.6$); (**c**) normalized $K_{II}$ ($a/t = 0.4$, $a/c = 0.8$); (**d**) normalized $K_{III}$ ($a/t = 0.4$, $a/c = 0.8$); (**e**) normalized $K_{II}$ ($a/t = 0.6$, $a/c = 0.4$); (**f**) normalized $K_{III}$ ($a/t = 0.6$, $a/c = 0.4$).

### 3.2. Effect of Friction Coefficient on Contact Surface

The variation in the mode II and mode III SIFs with the friction coefficient is given in Figure 11. Due to the symmetry of the curves, only the results for the half-length along the crack front are shown in the figure. The absolute values of $K_{II}$ and $K_{III}$ decreased with an increasing friction coefficient for different crack sizes and crack inclination angles. At $\theta = 45°$, the friction coefficient had the greatest effect on the SIFs, and this effect also increased as the crack size increased. For the mode II cracks, the friction coefficient had the greatest effect on the SIF at the surface point, and this effect decreased as the location of the crack front moved forward, reaching a minimum at the deepest point. In contrast to mode II cracks, the friction coefficient significantly changed the SIF at the deepest point of mode III, but had a dropping effect on the surface point.

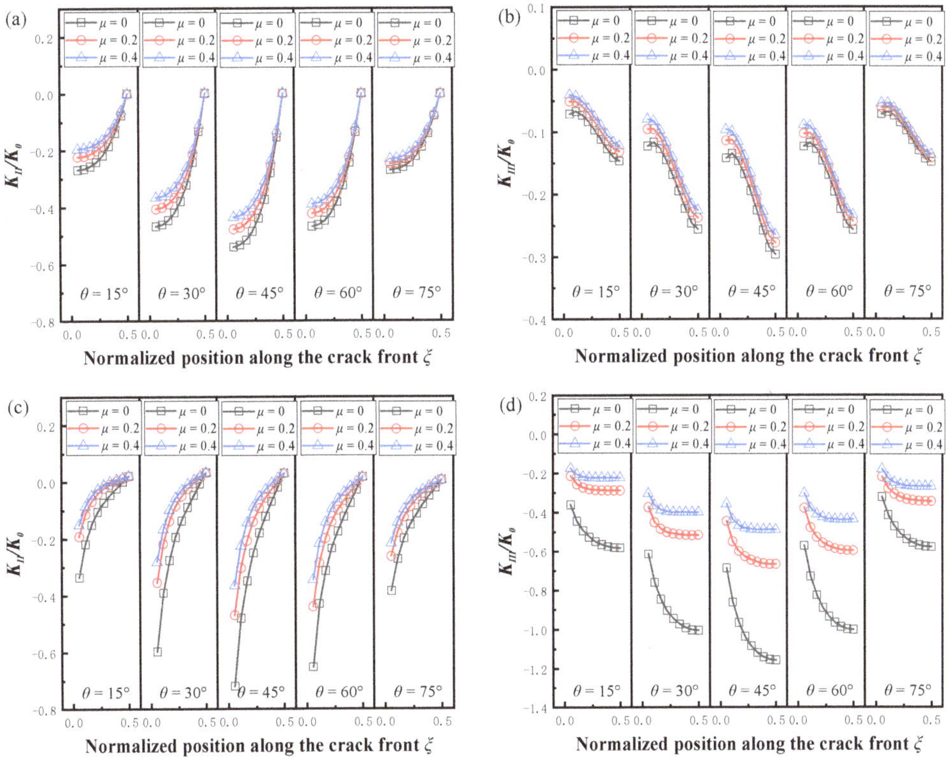


**Figure 11.** Effect of the coefficient of friction on normalized SIFs. (**a**) Normalized $K_{II}$ ($a/t$ = 0.2, $a/c$ = 1.6); (**b**) normalized $K_{III}$ ($a/t$ = 0.2, $a/c$ = 1.6); (**c**) normalized $K_{II}$ ($a/t$ = 0.6, $a/c$ = 0.4); (**d**) normalized $K_{III}$ ($a/t$ = 0.6, $a/c$ = 0.4).

## 4. Empirical Solution of SIFs for the Pipe with Inclined Inner Surface Cracks under External Pressure and Axial Compression

Although FEA is one of the most effective methods for calculating SIFs, it is difficult to apply in engineering practice due to the complexity of modeling and time-consuming calculations. In this paper, a new form of empirical solution was proposed by least-squares fitting based on the FEA results to give the influence coefficients at different $a/t$ with the normalized crack front position $\xi$ as the basis function. A new correction factor for the inclination angle, $f_\theta$, was proposed based on the effect caused by the curvature of the pipe above-mentioned in the effect of the crack inclination angle. The friction coefficient influence coefficient, $f_\mu$, was proposed because the pipe is subjected to external pressure and axial pressure. The influence coefficients are shown in Tables 3 and 4, and the fitted equations are shown in Equations (8)–(16).

$$\boldsymbol{K} = \sigma\sqrt{\pi\frac{a}{Q}}\boldsymbol{F}\left(\frac{a}{t},\frac{a}{c},\xi,f_\theta,f_\mu\right) \tag{8}$$

where $\sigma$ is the far-field compressive stress, $\boldsymbol{K} = \left\{ \begin{array}{c} K_{II} \\ K_{III} \end{array} \right\}$, $\boldsymbol{F}\left(\frac{a}{t},\frac{a}{c},\xi,f_\theta,f_\mu\right) = \left\{ \begin{array}{c} F_{II} \\ F_{III} \end{array} \right\}$.

**Table 3.** The sub-curve fitting constants in Equation (10) $h_i$.

| | a/t = 0.2 | | | | a/t = 0.4 | | | |
|---|---|---|---|---|---|---|---|---|
| Constants | $H_1$ | $H_2$ | $H_3$ | $H_4$ | $H_1$ | $H_2$ | $H_3$ | $H_4$ |
| $h_1$ | 0.161 | −0.323 | 0.159 | −0.117 | −0.121 | 1.404 | −3.037 | 1.865 |
| $h_2$ | −2.105 | 10.376 | −18.905 | 12.616 | −0.753 | 3.320 | −7.455 | 5.953 |
| $h_3$ | 1.795 | −14.836 | 34.186 | −22.826 | 0.285 | −6.828 | 21.492 | −15.793 |
| $h_4$ | −0.492 | 4.623 | −11.073 | 7.399 | 0.020 | 1.850 | −6.686 | 5.024 |
| | a/t = 0.6 | | | | a/t = 0.8 | | | |
| Constants | $H_1$ | $H_2$ | $H_3$ | $H_4$ | $H_1$ | $H_2$ | $H_3$ | $H_4$ |
| $h_1$ | 0.338 | 0.317 | −2.580 | 1.678 | 0.385 | 1.134 | −5.297 | 3.484 |
| $h_2$ | −2.987 | 9.044 | −10.429 | 7.042 | −3.309 | 7.181 | −2.875 | 2.012 |
| $h_3$ | 3.129 | −14.121 | 24.981 | −16.777 | 3.648 | −13.768 | 20.741 | −13.951 |
| $h_4$ | −0.999 | 4.641 | −8.408 | 5.654 | −1.206 | 4.955 | −8.080 | 5.433 |

**Table 4.** The sub-curve fitting constants in Equation (14) $h_i$.

| | a/t = 0.2 | | | a/t = 0.4 | | |
|---|---|---|---|---|---|---|
| Constants | $H_1$ | $H_2$ | $H_3$ | $H_1$ | $H_2$ | $H_3$ |
| $h_1$ | −0.448 | −0.582 | 0.587 | −0.497 | −1.179 | 1.171 |
| $h_2$ | 0.599 | −1.998 | 2.023 | 0.727 | 1.049 | −0.907 |
| $h_3$ | −0.322 | 1.544 | −1.596 | −0.491 | −2.190 | 1.959 |
| $h_4$ | 0.053 | −0.313 | 0.335 | 0.116 | 1.001 | −0.910 |
| | a/t = 0.6 | | | a/t = 0.8 | | |
| Constants | $H_1$ | $H_2$ | $H_3$ | $H_1$ | $H_2$ | $H_3$ |
| $h_1$ | −0.451 | −0.187 | 0.210 | −0.375 | −1.135 | 1.159 |
| $h_2$ | 0.494 | −3.521 | 3.478 | 0.154 | −0.353 | 0.306 |
| $h_3$ | −0.214 | 3.819 | −3.781 | 0.200 | 0.413 | −0.373 |
| $h_4$ | 0.021 | −1.192 | 1.181 | −0.126 | −0.058 | 0.045 |

Influence coefficient functions for $K_{II}$:

$$F_{II} = \left(H_1 + H_2\xi + H_3\xi^2 + H_4\xi^3\right) f_\theta f_\mu \tag{9}$$

$$H_{i(i=1,2,3,4)} = h_1 + h_2\left(\frac{a}{c}\right) + h_3\left(\frac{a}{c}\right)^2 + h_4\left(\frac{a}{c}\right)^3 \tag{10}$$

$$f_\theta = \left\{1 + \frac{ac}{600}[\sin(\theta - 45)]\right\}\sin 2\theta \tag{11}$$

$$f_\mu = (1 + 0.575\mu)(ac)^{-0.505\mu} \tag{12}$$

Influence coefficient functions for $K_{III}$:

$$F_{III} = \left(H_1 + H_2\xi + H_3\xi^2\right) f_\theta f_\mu \tag{13}$$

$$H_{i(i=1,2,3)} = h_1 + h_2\left(\frac{a}{c}\right) + h_3\left(\frac{a}{c}\right)^2 + h_4\left(\frac{a}{c}\right)^3 \tag{14}$$

$$f_\theta = \left\{1+\frac{ac}{1500}[\sin(45-\theta)]\right\}\sin 2\theta \tag{15}$$

$$f_\mu = (1+0.281\mu)(ac)^{-0.511\mu} \tag{16}$$

The fitting results for $\theta = 45°$ and $\mu = 0$ at the three sizes are given in Figure 12a,b. The results of the fit at different angles for the same crack size are given in Figure 12c,d, where $\mu = 0$. The fitting results for the same crack size with different angles and different friction coefficients are given in Figure 12e,f; due to the symmetry of the curves, only the results for the half-length of the crack front are shown in the figure. It can be seen that the fitting results achieved good agreement with the FEA results, which can be applied for the defect assessment.

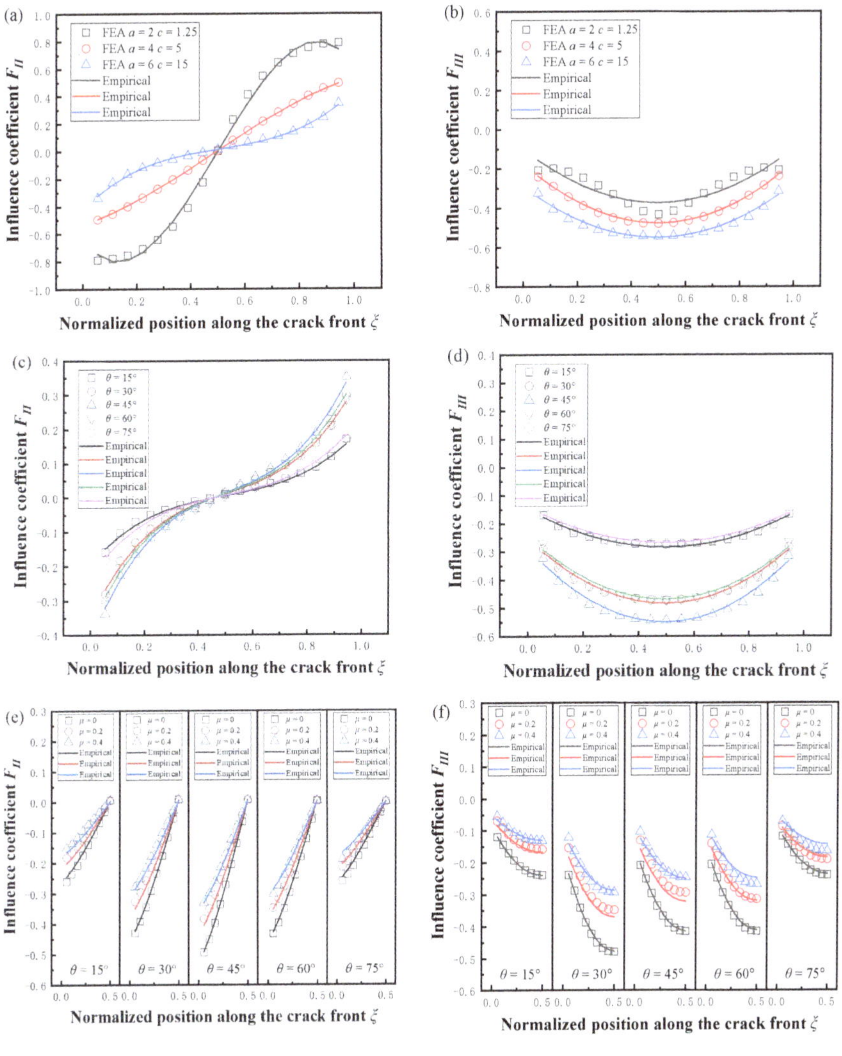


**Figure 12.** Comparison of SIFs fitting results with (**a**,**b**) different crack sizes ($a/t$ = 0.2, 0.4, 0.6, $a/c$ = 1.6, 0.8, 0.4, $\theta = 45°$, $\mu = 0$); (**c**,**d**) different inclination angles ($a/t$ = 0.6, $a/c$ = 0.4, μ = 0); (**e**,**f**) different friction coefficients ($a/t$ = 0.4, $a/c$ = 0.8, $\mu$ = 0, 0.2, 0.4).

## 5. Conclusions

In this paper, the SIFs of the inclined crack on the inner surface of the pipe under combined axial pressure load and external pressure were investigated by the finite element method, and the main conclusions are as follows.

(1) By comparing the SIF values along the crack fronts, it can be found that $K_{II}$ and $K_{III}$ for the closed surface cracks under external and axial pressure were higher than $K_{II}$ and $K_{III}$ for the open surface cracks under internal pressure and axial tension.

(2) The effects of the relative depth and aspect ratio on the SIFs of the inclined inner crack were analyzed. The results showed that the larger the $a$ and $c$, the easier it is for the crack to expand, and when $a/c$ >1, $K_{IImax}$ does not appear at the surface point of the crack, but near the surface point.

(3) The effects of the crack inclination angle on the SIFs of the inclined inner crack were analyzed. The results showed that the SIFs at the crack front were the largest at $\theta = 45°$, and their values decreased with an inclination angle toward the axial and circumferential directions. The larger the crack size, the more obvious the asymmetry of the SIFs about $\theta = 45°$.

(4) The effects of the friction coefficient on the SIFs of surface crack were analyzed. The larger the friction factor, the smaller the SIFs along the crack front. The friction coefficient had the greatest effect on the surface point of mode II cracks and the deepest point of mode III cracks, and the greatest effect on cracks with $\theta = 45°$.

(5) Based on the above results, a new solution for stress intensity factors $K_{II}$ and $K_{III}$ were proposed, and the corresponding coefficients for different crack sizes, an angle correction factor $f_\theta$, and a friction correction factor $f_\mu$ are given.

**Author Contributions:** X.-M.Y.: Conception, methodology, software, writing—manuscript, and designed the research. Y.-C.Z., Q.P. and L.-Z.J.: Software. T.-H.M.: Investigation. X.-H.H.: Funding acquisition. C.-Y.Z.: Formal analysis, conception and funding acquisition. All authors have read and agreed to the published version of the manuscript.

**Funding:** This paper was funded by the National Natural Science Foundation of China (51975271).

**Institutional Review Board Statement:** Not applicable.

**Informed Consent Statement:** Not applicable.

**Data Availability Statement:** The data that support the findings of this study are available from the corresponding author upon reasonable request.

**Acknowledgments:** The authors gratefully acknowledge the financial support of the National Natural Science Foundation of China (51975271).

**Conflicts of Interest:** The authors declare no conflict of interest.

## Abbreviations

| | |
|---|---|
| $a$ | Depth of a semi-elliptical surface crack |
| $c$ | Half-length of a semi-elliptical surface crack |
| $F_I$ | Influence coefficient functions for $K_I$ |
| $F_{II}$ | Influence coefficient functions for $K_{II}$ |
| $F_{III}$ | Influence coefficient functions for $K_{III}$ |
| $f_\theta$ | Angle effect correction factor |
| $f_\mu$ | Frictional influence correction factor |
| $H_1, H_2, H_3, H_4$ | Sub-curve-fitting functions for $F_I$, $F_{II}$, $F_{III}$ |
| $h_1, h_2, h_3, h_4$ | Sub-curve fitting constants for $H_1$, $H_2$, $H_3$, $H_4$ |
| $K_I$ | Stress intensity factors for Modes I crack |
| $K_{II}$ | Stress intensity factors for Modes II crack |
| $K_{III}$ | Stress intensity factors for Modes III crack |

| | |
|---|---|
| $K_0$ | Normalizing factor for the stress intensity factors |
| $k_n$ | Main normal stiffness of crack front |
| $k_s$ | Sub-normal stiffness of crack front |
| $k_t$ | Tangential stiffness of crack front |
| $l$ | Length of a pipe |
| $P$ | Axial pressure of a pipe |
| $P_0$ | External pressure of a pipe |
| $P_n$ | Main normal pressure of crack front |
| $Q$ | Shape factor for elliptical crack |
| $R_i$,$R_0$ | Internal and external radius of a pipe |
| $t$ | Wall thickness of a pipe |
| $u_n$ | Main normal displacement of crack front |
| $u_s$ | Sub-normal displacement of crack front |
| $u_t$ | Tangential displacement of crack front |
| $\theta$ | Inclination angle of a surface crack |
| $\xi$ | Normalized position along the crack front |
| $\sigma$ | Far-field compressive stress |
| $\tau_s$ | Crack front sub-normal shear stress |
| $\tau_t$ | Crack front tangential shear stress |
| $\mu$ | Friction coefficient of the cracked surface |

## References


1. Lin, X.B.; Smith, R.A. Shape Growth Simulation of Surface Cracks in Tension Fatigued Round Bars. *Int. J. Fatigue* **1997**, *19*, 461–469. [CrossRef]
2. Raju, I.S.; Newman, J.C., Jr. Stress-Intensity Factors for Internal and External Surface Cracks in Cylindrical Vessels. *J. Press. Vessel Technol.* **1982**, *104*, 293–298. [CrossRef]
3. Navid, H.; Fenner, R.T.; Nadiri, F.; Webster, G.A. Stress Intensity Factors for Internal and External Cracks in Pressurised Thick-Walled Cylinders. *Int. J. Press. Vessel Pip.* **1985**, *18*, 241–254. [CrossRef]
4. Kumar, V.; German, M.D.; Schumacher, B.I. Analysis of Elastic Surface Cracks in Cylinders Using the Line-Spring Model and Shell Finite Element Method. *J. Press. Vessel Technol.* **1985**, *107*, 403–411. [CrossRef]
5. Zheng, X.J.; Kiciak, A.; Glinka, G. Weight Functions and Stress Intensity Factors for Internal Surface Semi-Elliptical Crack in Thick-Walled Cylinder. *Eng. Fract. Mech.* **1997**, *58*, 207–221. [CrossRef]
6. Jones, I.S.; Rothwell, G. Reference Stress Intensity Factors with Application to Weight Functions for Internal Circumferential Cracks in Cylinders. *Eng. Fract. Mech.* **2001**, *68*, 435–454. [CrossRef]
7. Eshraghi, I.; Soltani, N. Stress Intensity Factor Calculation for Internal Circumferential Cracks in Functionally Graded Cylinders Using the Weight Function Approach. *Eng. Fract. Mech.* **2015**, *134*, 1–19. [CrossRef]
8. Kamaya, M.; Nishioka, T. Analysis of Surface Crack in Cylinder by Finite Element Alternating Method. *J. Press. Vessel Technol.* **2005**, *127*, 165–172. [CrossRef]
9. Wallbrink, C.D.; Peng, D.; Jones, R. Assessment of Partly Circumferential Cracks in Pipes. *Int. J. Fract.* **2005**, *133*, 167–181. [CrossRef]
10. Murakami, Y. Analysis of Stress Intensity Factors of Modes I, II and III for Inclined Surface Cracks of Arbitrary Shape. *Eng. Fract. Mech.* **1985**, *22*, 101–114. [CrossRef]
11. Zhao-Jing, Z.; Shu-Ho, D. Stress Intensity Factors for an Inclined Surface Crack under Biaxial Stress State. *Eng. Fract. Mech.* **1994**, *47*, 281–289. [CrossRef]
12. Ayhan, A.O. Mixed Mode Stress Intensity Factors for Deflected and Inclined Corner Cracks in Finite-Thickness Plates. *Int. J. Fatigue* **2007**, *29*, 305–317. [CrossRef]
13. Şahin, H.; Ayhan, A.O. Three-Dimensional Mixed Mode Stress Intensity Factors for Inclined Elliptical Surface Cracks in Plates under Uniform Tensile Load. *Procedia Struct. Integr.* **2019**, *21*, 38–45. [CrossRef]
14. Phyo Myat, K.; Osawa, N.; Gadallah, R.; Tanaka, S. Accurate and Efficient Method for Analyzing Mixed-Mode SIFs for Inclined Surface Cracks in Semi-Infinite Bodies by Using Numerical Influence Function Method. *Theor. Appl. Fract. Mech.* **2020**, *106*, 102471. [CrossRef]
15. Bhat, M.A.; Shaikh, A.A. Effect of Specimen Parameters on Mixed-Mode I/II Stress Intensity Factors for Additive Manufactured Slant Edge Crack Plate. *Mater. Today Proc.* **2021**, *44*, 4305–4308. [CrossRef]
16. Shahani, A.R.; Habibi, S.E. Stress Intensity Factors in a Hollow Cylinder Containing a Circumferential Semi-Elliptical Crack Subjected to Combined Loading. *Int. J. Fatigue* **2007**, *29*, 128–140. [CrossRef]
17. Predan, J.; Močilnik, V.; Gubeljak, N. Stress Intensity Factors for Circumferential Semi-Elliptical Surface Cracks in a Hollow Cylinder Subjected to Pure Torsion. *Eng. Fract. Mech.* **2013**, *105*, 152–168. [CrossRef]
18. Ramezani, M.K.; Purbolaksono, J.; Andriyana, A.; Ramesh, S.; Putra, I.S. Empirical Solutions for Stress Intensity Factors of a Surface Crack in a Solid Cylinder under Pure Torsion. *Eng. Fract. Mech.* **2018**, *193*, 122–136. [CrossRef]

19. Li, C.-Q.; Fu, G.; Yang, W. Stress Intensity Factors for Inclined External Surface Cracks in Pressurised Pipes. *Eng. Fract. Mech.* **2016**, *165*, 72–86. [CrossRef]
20. Fu, G.; Yang, W.; Li, C.-Q. Stress Intensity Factors for Mixed Mode Fracture Induced by Inclined Cracks in Pipes under Axial Tension and Bending. *Theor. Appl. Fract. Mech.* **2017**, *89*, 100–109. [CrossRef]
21. Li, D.; Mao, Z. Experimental and Numerical Simulations on Compound Stress Intensity Factor of Semi-Elliptical Cracks on the Exchanger Outer Walls with Inclined Angles. *Alex. Eng. J.* **2022**, *61*, 5065–5072. [CrossRef]
22. Woo, C.W.; Cheung, Y.K.; Chen, Y.Z.; Wang, Y.H. A Simple Model for the Contact Problem of a Finite Cracked Plate in Bending. *Eng. Fract. Mech.* **1988**, *29*, 227–231. [CrossRef]
23. Beghini, M.; Bertini, L. Effective Stress Intensity Factor and Contact Stress for a Partially Closed Griffith Crack in Bending. *Eng. Fract. Mech.* **1996**, *54*, 667–678. [CrossRef]
24. Thiagarajan, S.; Alwar, R.S. Influence of Crack Closure in the Case of an Angled Crack. *Eng. Fract. Mech.* **1986**, *24*, 533–537. [CrossRef]
25. Bowie, O.L.; Freese, C.E. On the "Overlapping" Problem in Crack Analysis. *Eng. Fract. Mech.* **1976**, *8*, 373–379. [CrossRef]
26. Liu, S.B.; Tan, C.L. Two-Dimensional Boundary Element Contact Mechanics Analysis of Angled Crack Problems. *Eng. Fract. Mech.* **1992**, *42*, 273–288. [CrossRef]
27. Hammouda, M.M.I.; Fayed, A.S.; Sallam, H.E.M. Mode II Stress Intensity Factors for Central Slant Cracks with Frictional Surfaces in Uniaxially Compressed Plates. *Int. J. Fatigue* **2002**, *24*, 1213–1222. [CrossRef]
28. Hammouda, M.M.I.; Fayed, A.S.; Sallam, H.E.M. Stress Intensity Factors of a Shortly Kinked Slant Central Crack with Frictional Surfaces in Uniaxially Loaded Plates. *Int. J. Fatigue* **2003**, *25*, 283–298. [CrossRef]
29. Dorogoy, A.; Banks-Sills, L. Effect of Crack Face Contact and Friction on Brazilian Disk Specimens—A Finite Difference Solution. *Eng. Fract. Mech.* **2005**, *72*, 2758–2773. [CrossRef]
30. DASSAULT SYSTEMES. 2021. Available online: https://www.3ds.com/support/hardware-and-software/simulia-system-information/abaqus-2021/ (accessed on 19 December 2022).
31. Zhang, T.; Lu, K.; Katsuyama, J.; Li, Y. Stress Intensity Factor Solutions for Surface Cracks with Large Aspect Ratios in Cylinders and Plates. *Int. J. Press. Vessel Pip.* **2021**, *189*, 104262. [CrossRef]
32. Randeniya, C.; Robert, D.J.; Fu, G.; Li, C.Q. The Effect of Corrosion Patch Geometry on Stress Intensity Factors for External Surface Cracks in Cast Iron Water Mains. In Proceedings of the Fourth International Conference on Sustainable Construction Materials and Technologies, Las Vegas, NV, USA, 7–11 August 2016.
33. Teh, S.; Andriyana, A.; Ramesh, S.; Putra, I.S.; Kadarno, P.; Purbolaksono, J. Tetrahedral Meshing for a Slanted Semi-Elliptical Surface Crack at a Solid Cylinder. *Eng. Fract. Mech.* **2021**, *241*, 107400. [CrossRef]
34. Shiratori, M.; Miyoshi, T.; Sakai, Y.; Zhang, G.R. Analysis of Stress Intensity Factors for Surface Cracks Subjected to Arbitrarily Distributed Surface Stresses. *Trans. Jpn. Soc. Mech. Eng.* **1987**, *53*, 1651–1657. [CrossRef]
35. Shlyannikov, V.N.; Tumanov, A.V. An Inclined Surface Crack Subject to Biaxial Loading. *Int. J. Solids Struct.* **2011**, *48*, 1778–1790. [CrossRef]
36. Cordes, R.D.; Joseph, P.F. Crack Surface Contact of Surface and Internal Cracks in a Plate with Residual Stresses. *Int. J. Fract.* **1994**, *66*, 1–17. [CrossRef]
37. Zhu, F.; Liu, H.; Yao, L.; Mei, G. Stress Field Analysis of an Infinite Plate with a Central Closed Inclined Crack under Uniaxial Compression. *Theor. Appl. Fract. Mech.* **2021**, *116*, 103111. [CrossRef]
38. Yang, S.T.; Ni, Y.L.; Li, C.Q. Weight Function Method to Determine Stress Intensity Factor for Semi-Elliptical Crack with High Aspect Ratio in Cylindrical Vessels. *Eng. Fract. Mech.* **2013**, *109*, 138–149. [CrossRef]
39. Ayhan, A.O.; Kurt, E. Three-Dimensional Mixed-Mode Stress Intensity Factors for Deflected External Surface Cracks in Thin and Midsize-Thick-Walled Spherical Pressure Vessels. *Int. J. Press. Vessel Pip.* **2022**, *195*, 104596. [CrossRef]

*materials* MDPI

*Article*

# Analysis of the Vehicle Chassis Axle Fractures

**Živilė Decker [1], Vitalijus Rudzinskas [1], Kazimierz Drozd [2], Jacek Caban [2,*], Jurijus Tretjakovas [3], Aleksander Nieoczym [2] and Jonas Matijošius [4]**

1 Department of Mechanical and Material Engineering, Faculty of Mechanics, Vilnius Gediminas Technical University, J. Basanavičiaus g. 28, LT-03224 Vilnius, Lithuania
2 Faculty of Mechanical Engineering, Lublin University of Technology, 20-618 Lublin, Poland
3 Department of Applied Mechanics, Faculty of Civil Engineering, Vilnius Gediminas Technical University, Saulėtekio al. 11, LT-10223 Vilnius, Lithuania
4 Mechanical Science Institute, Vilnius Gediminas Technical University, J. Basanavičiaus g. 28, LT-10223 Vilnius, Lithuania
* Correspondence: j.caban@pollub.pl

**Abstract:** With the rapid development of the road transport industry, trucks with semi-trailers have become the main means of transporting goods by road. High quality, durability and reliability of the construction are the main requirements for the production of trailers. Trailer and semi-trailer axles are one of the main and most important components of a truck. Due to the fact that semi-trailer axles are subjected to additional static and dynamic loads during operation, their proper construction is extremely important, therefore they should be carefully designed and tested. The durability of the suspension components refers to the duration of the onset of fatigue. This article presents an analysis of damage to the rear axle of the semi-trailer using macroscopic observations of the damage site and dynamic FEA of stress distribution in the axle material. In order to identify the probable cause of the damage, eight cases of loading the semi-trailer axle were considered. Analytical solutions have shown that in various cases the yield point is exceeded and the strength limit of the modeled semi-trailer axle is reached. The risk of damage to the vehicle's suspension system components increases on poor roads (bumps and winding road sections).

**Keywords:** failure analysis; FEA; fracture mechanics; macroscopic research; semi-trailers

**Citation:** Decker, Ž.; Rudzinskas, V.; Drozd, K.; Caban, J.; Tretjakovas, J.; Nieoczym, A.; Matijošius, J. Analysis of the Vehicle Chassis Axle Fractures. *Materials* **2023**, *16*, 806. https://doi.org/10.3390/ma16020806

Academic Editors: Grzegorz Lesiuk and Dariusz Rozumek

Received: 9 December 2022
Revised: 5 January 2023
Accepted: 6 January 2023
Published: 13 January 2023



## 1. Introduction

With the rapid development of the road transport industry, trucks with semi-trailers have become the main means of transporting goods by road. The economic goal of all transport companies is to transport as many goods as possible and with as few journeys as possible. High quality, durability, and structural reliability are the main requirements for the production of trailers. The axles of trailers and semi-trailers are one of the main and most important elements, which must be carefully designed and tested experimentally under static and dynamic loads, as the axle is subjected to additional loads in the event of road roughness or off-road [1]. The durability of the suspension parts refers to the duration of the onset of fatigue, defined as the number of cycles up to a certain component cracking length under cyclic loads [2,3].

The reliability of individual safety systems of a given means of transport translates directly into road traffic safety [4–8]. Much attention is also paid to the diagnostics of individual vehicle systems that affect safety. It is worth mentioning here the research presented by Savchenko et al. [9] and Gnap et al. [10], who demonstrated the possibility of using MEMS sensors in vehicles. Hudec et al. [11], Gajek [12] and Tucki et al. [13] paid attention to the monitoring of the technical condition of vehicles. On the other hand, research on the impact of heavy goods vehicle load during braking is presented in detail in [14–17].

The topic of scientific information is the determination of the causes of vehicle chassis axle defects and fractures, which can be found in a wide variety of ways. The literature begins to address this problem along with the first defects in the chassis axles. A review of recent research shows that researchers are looking to find the most optimal design solutions based on durability, lightness of construction, strength, and low cost. It also examines the causes of defects by evaluating the chemical composition of the metal, microstructure, mechanical properties, acting on loads, the location of the object under investigation in the whole system, and so on.

One of the most popular test methods used to detect cracks, fractures, and other defects is fractographic testing. Such investigations determine the complex causes of decay: the influence of the shape of the part, the load conditions, the interaction of the elements, the peculiarities of the structure of the material, and other factors are determined from the fractures.

In addition to traditional qualitative fractography, quantitative fractography aims to measure the topographic features of the fracture surface, revealing significant fracture surface characteristics. Modern fracture image analysis systems play an important role in advancing and successfully achieving these goals, not only to speed up measurement procedures but also to perform operations that would not be possible in other ways [18].

Fractography is a method in failure analysis for studying the fracture surface of materials. The nature of decay and fracture is characterized by specific characteristics of the mechanisms of decomposition (brittleness, plasticity, etc.). Each substance has the property of decomposing. As the material is exposed to various loads and environmental influences, the nature of decomposition is usually multifaceted. The nature of fracture of a part and the loading conditions has an interface that is characterized by: the nature of the fracture, the mechanism of crack formation, and the relief of the fracture surface [19].

Due to the high load on the rear axle, especially on the tractor, its service life is shortened. Fractographic studies show that the main cause of axle failure is fatigue. Fatigue cracks have been observed near welds. The results show that the axles break due to bending fatigue caused by improper welding [20,21]. Improper welding in the heat-affected zone (HAZ) reduces the plasticity of the material, resulting in structural stress concentration points and inclusions that subsequently affect the cracks. Pre-treatment of pre-weld and post-weld heat treatment of medium carbon steel is necessary to control the hardness level of HAZ and to reduce residual stress [20].

In the literature, it is often recommended to start the engineering calculations of axes from an analytical theoretical model, which can be used to estimate displacements, specific deformations, stresses, and the self-frequency spectrum [1,22]. Theoretical calculations of axle strength are one of the most important tasks in vehicle design. The axles are exposed to different external loads with a certain frequency, which depends on the speed, the actual load, the road conditions, and many other factors. At the same time, resonant phenomena are possible, which can lead to higher than nominal stresses and many other adverse phenomena. Variable external loads cause periodic changes in stresses, which contribute to the growth of fatigue cracks leading to fatigue fractures [1,23].

During the operation of the axle, the load acting on the axle housing in the vertical direction has a significant effect on the fatigue life of the components [24]. Cracks are caused by a constant stress concentration in the axle housing, resulting in fatigue at the concentration point. If the load exceeds a certain threshold, microscopic cracks will begin to form at the stress concentration point. Later, the cracks grow due to the cyclic load of fatigue. Eventually, the cracks reach a critical limit and then the structure suddenly breaks [21,25–28]. Axle housing failures are also affected by factors such as uneven load effect, housing slope, and eccentricity.

The paper presents an analysis of damage to the rear axle of a truck semi-trailer. For this purpose, FEA preceded by a theoretical introduction was used. Several load cases were considered in order to determine the most likely point of damage initiation to the semi-trailer axle. The structure of the work is as follows: the part covering materials and methods is presented in Section 2. Then, in Section 3, the results of analytical calculations and FEM numerical simulations for axle loads with various excitation (forces or displacement) are presented. Macroscopic analysis of the damaged element and material properties tests are also presented in Section 3. As a result of the tests and analyses carried out, conclusions were developed, which are included in Section 4.

## 2. Materials and Methods

### 2.1. Analytical Calculations

The first step in the analysis of the damaged semi-trailer axle was analytical calculations, which were then used as input data for dynamic stress analysis, which was then performed using the finite element method. The calculated values of individual parameters were obtained using appropriate mathematical relationships, taking into account the assumptions of engineering knowledge and the experience contained in the available professional literature was used. The results of the analysis are presented in Section 3.1.

### 2.2. FEM Simulations

Simulations of the operation of the semi-trailer's non-driven axle were carried out using the Abaqus Explicit module. The model (Figure 1), apart from the axle itself, included two hinges, two bushings, two spring-damping elements and two points reflecting the contact points of the wheels with the road. The distribution of forces adopted for the simulation assumed an even position of the nominal load over the entire space of the platform.

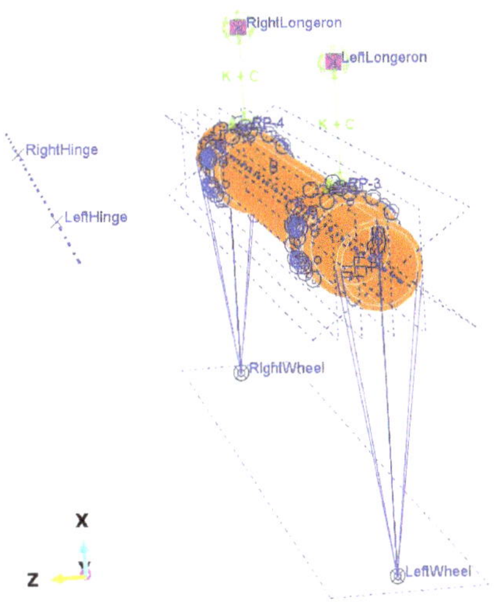


**Figure 1.** Complete model of the axle for simulations (front in Z direction).

The prepared model contains simplifications compared to the actual structure [29,30], i.e., the possibility of deformation of the wishbones, which were modeled with Rigid Body elements, was omitted. All degrees of freedom were fixed but with rotation around the Y axis. Each of the wishbones is coupled with a bushing that is mounted on the axle. This coupling also does not take into account the possibility of deformation of these two parts (rocker arm and bush) relative to each other. The bushings are connected to the side members by means of integrated spring/dashpot elements, while in reality the shock absorber works between the swingarm and the front side member, and the air bellows between the swing arm and the side member behind the axle. The model does not take into account the stiffness and damping of the road wheels. The contact points of the wheels with the road have been connected with Coupling elements to the ends of the axles.

In the presented model, the connection of the bushing with the axle tube was modeled using welds with a shape and location similar to the real structure. In addition, contact was modeled between the inner surfaces of the bushing and the outer cylindrical surface of the axle with a friction coefficient of 1.0. It was assumed that the spring elements of the suspension have a stiffness of 4.5 kN/mm and the damping coefficient of each shock absorber is 10 N s/mm. These parameters were selected in such a way that the wheel does not lose its grip with the ground and that the deflection of the suspension after its load corresponding to the permissible axle load is in the range of 15–20 mm.

The mesh for the axles and bushing was created using linear elements of the C3D8R type, which are bricks with an integration point reduced to its center. In the area of contact between these two parts, the mesh of elements was densified (Figure 2). The bushing has 11,496 elements and 17,943 nodes, while the axle has 10,120 elements and 20,424 nodes.

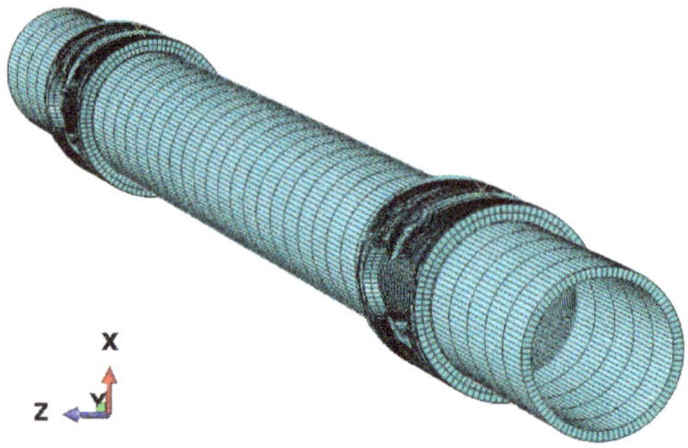


**Figure 2.** Axle meshed for simulations (front in Z direction).

Simulations were carried out for eight axle load cases. For the load with the force of the wheels (wheels) in each case, the value of 44.1 kN was used, resulting from the nominal load capacity of the axle, which increases in 8 ms. A vehicle traveling at a speed of 50 km/h covers about 110 mm of road in this time.

One of the ways of loading concerned only the left wheel, i.e., as if the right wheel fell into a hole in the road. In the second case, both wheels were loaded simultaneously. Four consecutive cases consisted of vertical loading of both wheels, followed by: braking of the left wheel, braking of both wheels simultaneously, side loading of the left wheel, side loading of both wheels. The adopted values of forces (44.1 kN for each wheel) reflect the situation as if the coefficient of friction between the tires and the road surface was 1.0 both when braking and when cornering. One-wheel braking and lateral force on only one wheel can occur when the other wheel loses grip with the road surface.

Simulations were also carried out of the case when the left wheel overruns a small triangular mogul with a height of 30 mm, and then both wheels simultaneously overcome such an obstacle. Each of these cases was simulated as a kinematic excitation occurring in 8 ms.

The simulation results were analyzed in terms of the distribution of stresses and deformations of the axle material and compared with analytical calculations. Next, the results of the simulations (HMH max stress value, point/ node of max stress generation, number of stress cycles in the simulation time for max stressed nodes) were validated with the initiation point of the fracture observed on the real object.

### 2.3. *Fracture Analysis and Fractography*

The damaged semi-trailer axle was inspected. For the fractographic analysis, a fragment of the material from the damaged place was taken, then this element was cleaned of dirt and rust using the ultrasonic method. Fractographic analysis of the crack surface was performed using the MBS-2 stereoscopic microscope. Photographic documentation of the breakthrough was prepared using the Optikam Microscopy (Ponteranica, Italy) digital USB camera. The test sample was cut using a liquid-cooled Optimum (Long Island City, NY, USA) band saw. The sample was then subjected to further analyses.

### 2.4. Strength Tests

The obtained material samples from the damaged semi-trailer axle were subjected to a static tensile test. The tests of the mechanical properties of the steel were carried out in the Laboratory of Strength Mechanics of the Vilnius University of Technology in accordance with the procedure contained in the ISO 6892-1 standard. Young modulus was calculated as a regression coefficient. Tests for obtaining the curve were conducted according to ISO 6892-1:2019 annex G. The Instron 8801 servohydraulic fatigue testing system (with dynamic extensometer) was used for the mechanical characterization of the sample material.

## 3. Results and Discussion

### 3.1. Static Analyses of the Strength of the Axle Material

Bending around the horizontal axis was described in [31] and the results concerning bending around horizontal axis stresses are used here.

Additional bending due to potholes around the horizontal axis is investigated.

If the truck goes to the left (or to the right), the wheel could be in way of the pothole. The most dangerous case in Figure 3 was found to be when the trajectory of the truck has less curvature. Therefore, the speed of the track is minimal. In this case there could be an additional bending moment acting around the horizontal axis, which could cause the bending moment to increase or decrease.

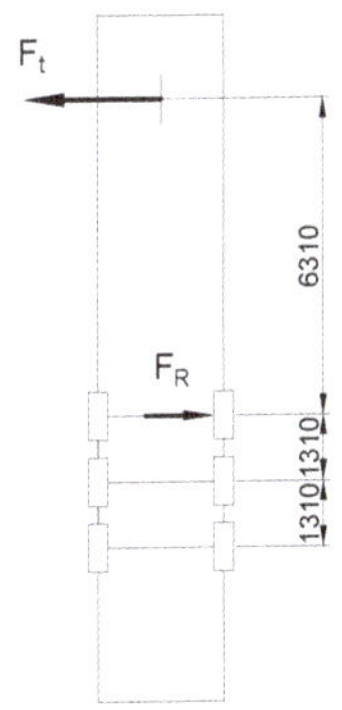


**Figure 3.** The scheme of the trailer turning.

Pull force is the force that a truck or prime mover can exert onto a transporter, or any type of trailer for that matter.

To go from pull force, many terms and conversion factors are thrown on the table, including the number of driven axles, gearbox ratio, rear end ratio, tire size, truck weight, and fifth wheel capacity.

The track pool force according to [32] is 11.5 Mg for truck axles and for truck trailer axles it is no more than 9.0 Mg.

$$F_R = 9.0 \cdot 1000 \cdot 9.81 \cdot \frac{1.0}{2} = 44.1 \text{ kN}, \quad (1)$$

The reaction force of the wheel (Figure 3) is derivable from the equation of equilibrium. The additional bending moment acting around the horizontal axis is shown in (Figure 4).

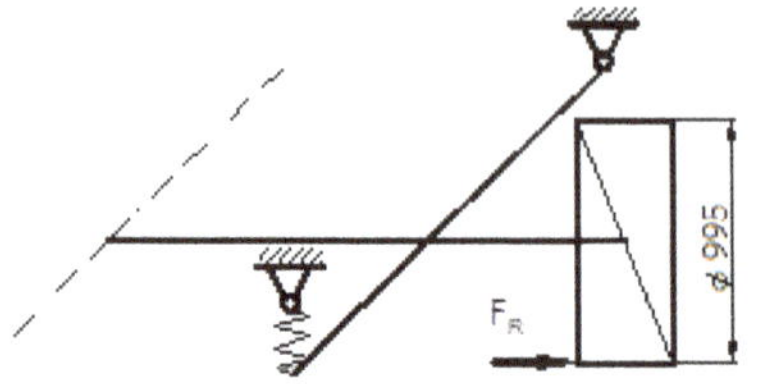


**Figure 4.** Reaction force.

$$M_H = 44.1 \cdot \frac{0.995}{2} = 22.0 \text{ kNm}, \quad (2)$$

Load bending for one side of the axle:

$$M_L = 9.0 \cdot 1000 \cdot 9.81 \cdot 262.5 = 23.2 \text{ kNm}, \quad (3)$$

$$\sigma_L = \frac{M_L}{W} = \frac{23.2 \cdot 10^3}{125 \cdot 10^{-6}} = 185.0 \text{ MPa}, \quad (4)$$

The full stresses due to bending moment acting around the horizontal axis:

$$\sigma_A = \frac{M_H + M_L}{W} = \frac{(22.0 + 23.1) \cdot 10^3}{125 \cdot 10^{-6}} = 360.7 \cdot 10^6 \text{ Pa} = 360.7 \text{ MPa}, \quad (5)$$

**Bending around the vertical axis**

The bending moment acting around the vertical axle is the result of the sudden braking of the truck. The important fact is that the most significant movement occurred just after the first intensive braking [31].

The inertia force of the truck for longitudinally forward, when braking, according to standard EN 12195-1 [32], is $F_i = m{\cdot}g{\cdot}0.8$.

For a semi-trailer, it is approximately 55% load [31] and one axle inertia load is equal:

$$F_{1\text{side}} = \frac{m \cdot g}{2} = \frac{9000 \cdot 9.81}{2} = 44.1 \cdot 10^3 \text{ N} = 44.1 \text{ kN}, \quad (6)$$

Especially, this internal force is decreasing in dynamic shock–image a wheel in a pothole during the braking of the truck. The authors accepted the dynamics coefficient according to [33] as being approximately equal to 1.2.

Bending moment acting around a vertical axis during braking:

$$M_V = 44.1 \cdot 0.515 \cdot 1.2 = 27.3 \text{ kNm}, \quad (7)$$

where: 0.515 m from the center of the wheel to the center of the rocker arm, (49.2 kNm taking into account the braking force).

Stresses due to bending moment around the vertical axis during braking:

$$\sigma_V = \frac{M_V}{W} = \frac{27.3 \cdot 10^3}{125 \cdot 10^{-6}} = 218.2 \cdot 10^6 \text{ Pa} = 218.2 \text{ MPa}, \quad (8)$$

This value will be used for the calculation of compound stresses of the axle for finding the critical region. Torque from load and braking and stress

$$M_{LV} = \sqrt{M_L^2 + M_V^2} = \sqrt{(23.1 \cdot 10^3)^2 + (27.3 \cdot 10^3)^2} = 35.8 \text{ kNm}$$

$$\sigma_V = \frac{M_{LV}}{W} = \frac{35.8 \cdot 10^3}{125 \cdot 10^{-6}} = 286.1 \cdot 10^6 \text{ Pa} = 286.1 \text{ MPa}. \quad (9)$$

**Bending stresses**

The bending moments for tubular section may be summed up superimposed.

$$\sigma_B = \frac{\sqrt{M_H + M_V}}{W} = \frac{\sqrt{(22.0 \cdot 10^3)^2 + (27.3 \cdot 10^3)^2}}{125 \cdot 10^{-6}} = 280.1 \cdot 10^6 \text{ Pa} = 280.1 \text{ MPa}, \quad (10)$$

The bending stresses are maximal in the front down quarter (Figure 3) of the axle.

**Torsion of the axle**

In a scenario with potholes, the wheels of the truck are on different levels. According to [34], air cushion throw is up to 410 mm, while in [35], the nominal ride height is, on average, 300 mm.

When one wheel is in the background and another is in the pothole, this results in torsion of the axle (Figure 5). The angle of twist is:

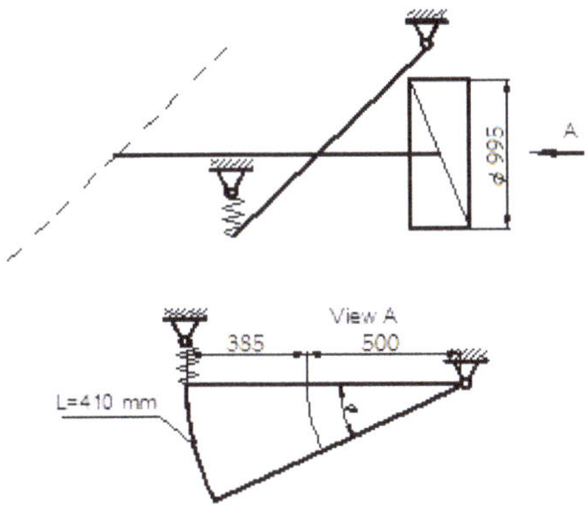


**Figure 5.** Air cushion throw, (L ≤ 300 mm [35]).

$$\phi = \frac{200}{500 + 385} = 0.339 \text{ rad}, \quad (11)$$

The angle of twist, internal torque, polar moment, length, and shear modulus are related by the formula:

$$\Delta\phi = \frac{T \cdot l}{G \cdot I_p}, \quad (12)$$

where: $I_p = 2 \cdot I = 2 \cdot 912 = 1824 \text{ cm}^4$.

Shear modulus is obtained from a relationship that exists between $G$, $E$ and $v$ (Poisson's ratio).

$$G = \frac{E}{2 \cdot (1 + v)} = \frac{204}{2 \cdot (1 + 0.30)} = 78.5 \text{ GPa}, \quad (13)$$

where $E$—modulus of elasticity ($E$ = 204 GPa); $v$—Poisson's ratio ($v$ = 0.30).

Internal torque:

$$T = F_R \cdot 0.500 = 44.1 \cdot 0.5 = 22.1 \cdot 10^3 \text{Nm}. \quad (14)$$

Shearing stresses in slow motion:

$$\tau = \frac{T}{W_p} = \frac{T}{\frac{I_p}{d_e/2}} = \frac{22.1 \cdot 10^3}{250058} = 88.3 \cdot 10^6 \text{ Pa} = 88.3\text{MPa}. \quad (15)$$

Shearing stresses in fast motion:

$$\tau = 1.2 \cdot 88.3 = 105.9 \text{ MPa}, \quad (16)$$

Stress in axle when trailer crosses potholes:

$$\sigma_p = \sqrt{\sigma_L^2 + 3 \cdot \tau_F^2} = \sqrt{185.0^2 + 3 \cdot 105.9^2} = 260.6 \text{ MPa}, \quad (17)$$

**Compound stresses**

The influence of bending and torsion with a truck's turning in slow motion is of an equation:

$$\sigma_d = \sqrt{\sigma_B^2 + 3 \cdot \tau^2} = \sqrt{280.1^2 + 3 \cdot 105.9^2} = 334.9 \text{ MPa}, \quad (18)$$

A compound dynamic stress of 335 MPa is calculated for driving at a speed 90 km/h on roads with potholes and more turns.

### *3.2. Effects of Simulations*

As a result of the simulation, it was found that most often the area of maximum stress (HMH) in the axle material occurs in the place of its cooperation with the sleeve, where it is connected by a weld. This maximum lies closer to the wheel, on the outside of the axle, not between the wishbones. In addition, it usually occurs in the rear part of the axle, which, as it is further away from the axis of rotation of the wishbones, is exposed to higher torsional stresses when the deflection of the axle ends is different.

One of the ways of loading (Figure 6) concerned only the left wheel, i.e., as if the right wheel fell into a hole in the road. Four subsequent cases concerned the vertical load on both wheels (Figure 7), followed by braking the left wheel (Figure 8), braking both wheels (Figure 9), side load on the left wheel (Figure 10), and side load on both wheels (Figure 11). The adopted values of forces ($F_{1side}$ = 44.1 kN for each wheel) reflect the situation as if the coefficient of friction between the tires and the road surface was 1.0, both when braking and when cornering. One-wheel braking and lateral force on only one wheel can occur when the other wheel loses grip with the road surface.

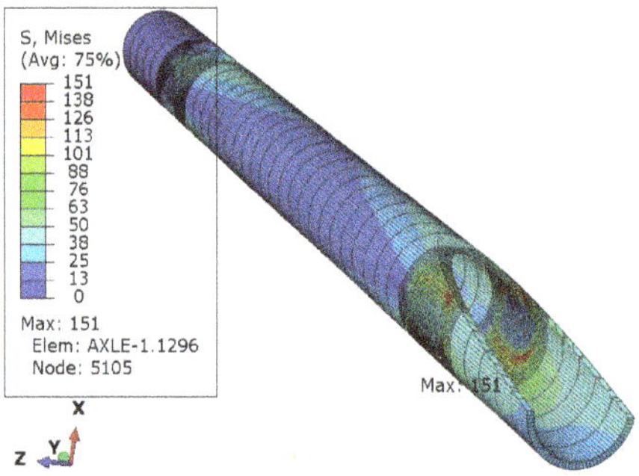


**Figure 6.** HMH stress distribution in the axle while only left wheel loaded with $F_{1side}$ force. Trail front in Z direction.

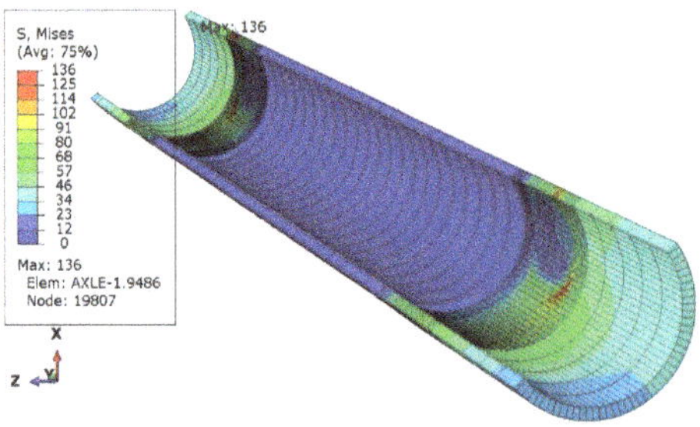


**Figure 7.** Stress distribution in the axle while both wheels loaded with $F_{1side}$ force.

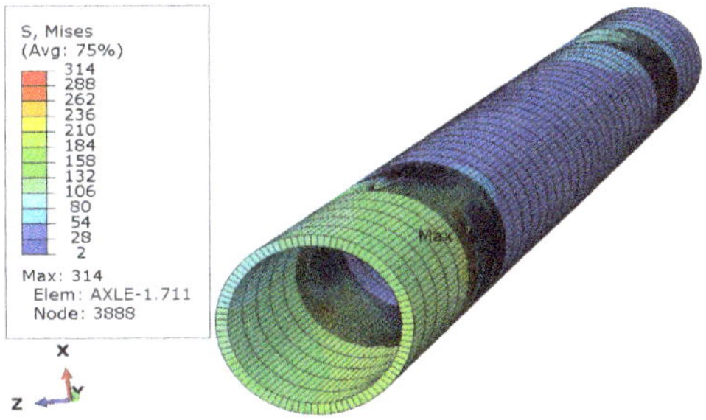


**Figure 8.** Stress distribution in the axle while only left wheel break is active.

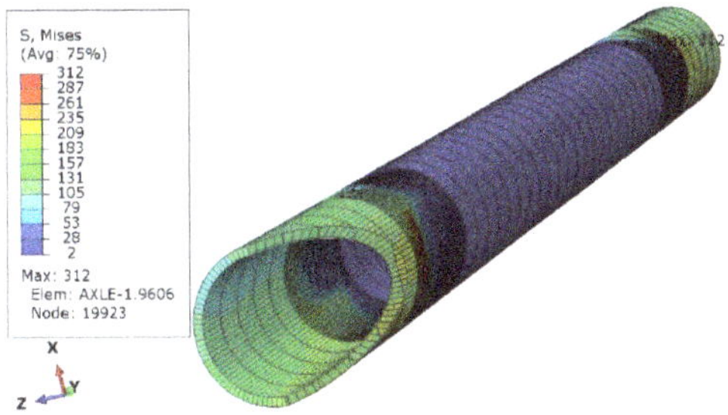


**Figure 9.** Stress distribution in the axle while only both wheels break.

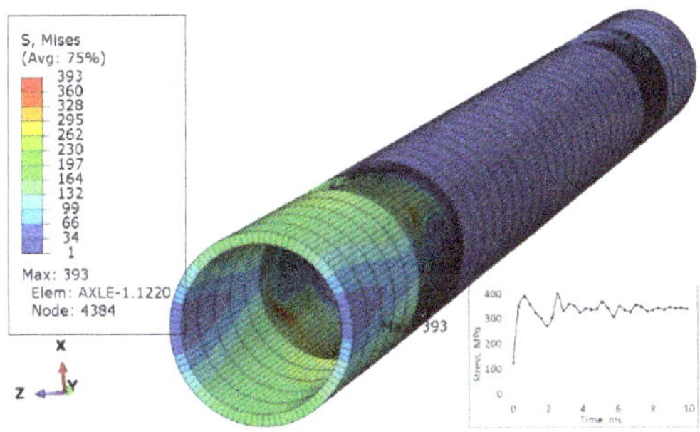


**Figure 10.** Stress distribution in the axle and max strengthen node 4384 while side force during turning acts to the left wheel.

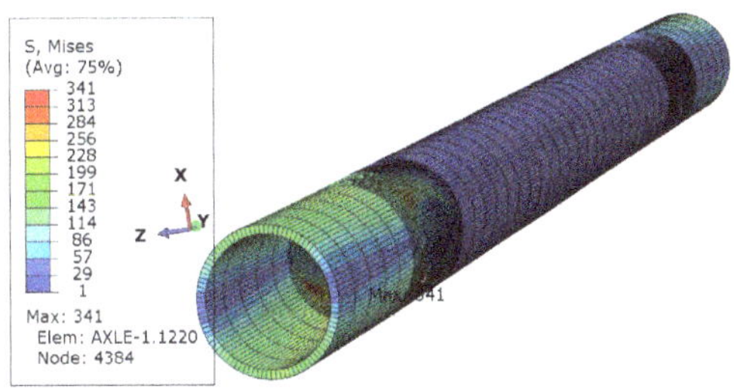


**Figure 11.** Stress distribution in the axle while side force during turning acts to both wheels.

The analysis of the distribution of stress in Figures 6–11 demonstrates that the max values in bushing were always near the welding. For the max strengthened node, the lowest value 136 MPa was observed for the symmetrical loading of two wheels (Figure 7). A slightly larger value (151 MPa) was found for the load on one (left) axle wheel. After that, when side (Figures 8 and 9) or break (Figures 10 and 11) force was added to the simulations, for the most loaded nodes, HMH stress was found to be at least twice as big (312 MPa).

The last simulations concerned the case where the left wheel overruns a small trough (Figure 12) of a triangle shape with a height of 30 mm, and then both wheels simultaneously overcome such an obstacle (Figure 13). Each of these cases was simulated as a kinematic excitation occurring in 8 ms.

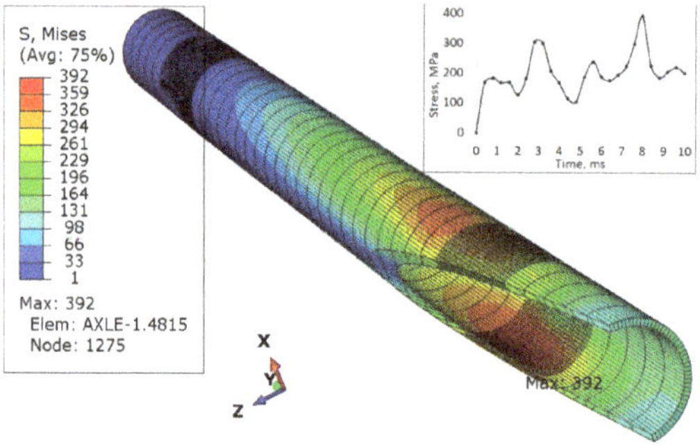


**Figure 12.** Stress distribution in the axle and max strengthened node 1275 while left wheel passes the triangular road deformation.

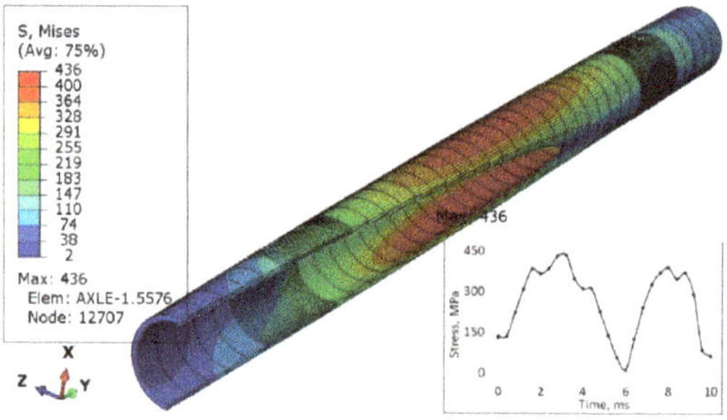


**Figure 13.** Stress distribution in the axle and max strengthened node 12,707 while both wheels pass the triangular road deformation.

When forcing the vertical displacement of the left wheel, the maximum stresses occur at the point of contact between the sleeve and the axle in the lower part of the axle (Figure 12). It can be seen from the graph that apart from the average stresses in the range of 100–200 MPa, there are sudden jumps in their values (peaks) by about 200 units to over 300 MPa and many load cycles.

The value of stresses in the axle (436 MPa in Figure 13) for the case of symmetrical excitation of both wheels is higher than for the displacement of only one wheel (392 MPa in Figure 12). However, their course over time is less abrupt for node 12,707 (Figure 13) than for node 1275 (Figure 12).

For the excitation carried out as displacements of both wheels, the highest stresses of 436 MPa were recorded. They occurred in the central part of the axis, with its deformation corresponding to the first form of natural vibrations.

For analytical calculations, each case can be considered separately: load, braking, side forces.

### 3.3. Visual Fractographic Analysis

The test object is a broken axle of a semi-trailer chassis. The axle broke in February 2016. The total mileage of the axle is 326,516 km. The axle breaks near the weld (near the mounting location). The broken axle is analyzed in the condition in which it was delivered for service under warranty. After inspecting the axle, a fracture of the axle is visible near the coupling joint in which the axle is fastened with the bracket. Non-destructive methods show only part of the fracture (Figure 14). Fractographic analysis of the fracture surface was performed using the MBS-2 stereo microscope and an Optikam Microscopy digital USB camera.

**Figure 14.** Axis fracture location.

For fractographic analysis, a specimen is excised from the fracture site. The ultrasonic bath removes dirt and rust from the fracture surface before inspecting the specimen. The specimen was cut using an Optimum liquid-cooled band saw. Liquid cooling during cutting was used to prevent the structural transformation of the steel.

Mechanical properties of steel tests were performed in the Laboratory of Strength Mechanics of Vilnius Gediminas Technical University according to standard ISO 6892-1. For further analysis using FEM (Finite Element Method), a numerical model of the considered trailer axle was prepared and simulations were carried out using the Abaqus Explicit module.

Fractures of a complex nature predominate in the fracture as the structural material is subjected to environmental influences and various deformations (Figure 15). In the manufacturing process, for example in the welded joints, areas of different mechanical properties are formed. Where there is a decrease in the plasticity of the material or in the area of higher stresses, individual voids may appear and a process of long-term plastic deformation may begin. By inspecting the fracture surface, three areas of fracture can be distinguished: fatigue before fracture, the onset of fracture, and major fracture.

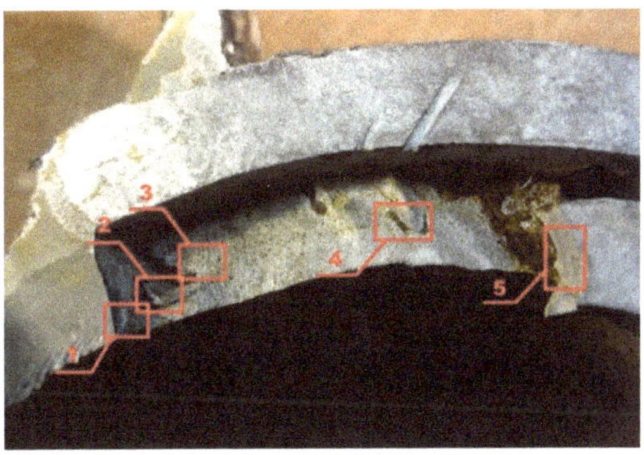


**Figure 15.** Fracture surface: Areas 1 to 5 are used for fractographic analysis.

The first is the area of fatigue before fracture or decay; it has a wavy relief (Figures 16 and 17).

**Figure 16.** Fracture area: the area of the focal point of decay.

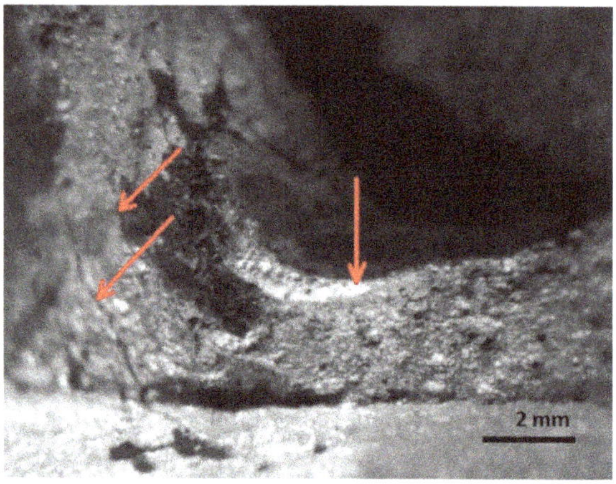


**Figure 17.** Fracture area: wider view of the area of the focal point of decay.

In the area of fatigue before the fracture, in the base and at the weld metal, near the weld where the highest stresses are applied, cracks are observed and a fragile fine-grained structure is visible. The formation and propagation of a crack depending on the type of deformation, the structure of the material, the level of the load, the shape of the part, and many other factors. The fracture is thought to have been initiated by long-term loading, as cracks and ribbed fracture morphology are characteristic of the effects of long-term loading.

Figures 16–20 show the results of fractographic studies of individual fracture areas (Figure 15). Fracture surfaces: Figure 16—an area no. 1; Figure 17—an area no. 2; Figure 18—an area no. 3; Figure 19—an area no. 4; Figure 20—an area no. 5.

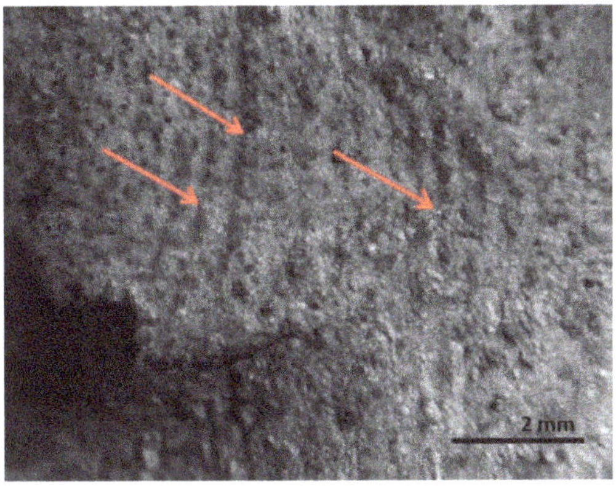


**Figure 18.** Fracture area: the area of gradual crack growth.

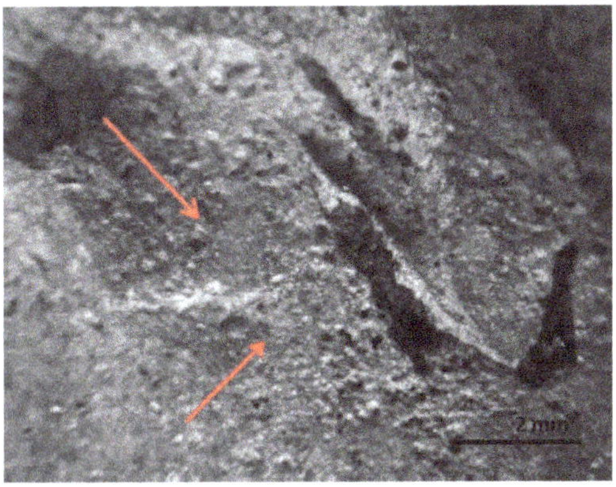


**Figure 19.** Fracture area: The area of the final fracture.

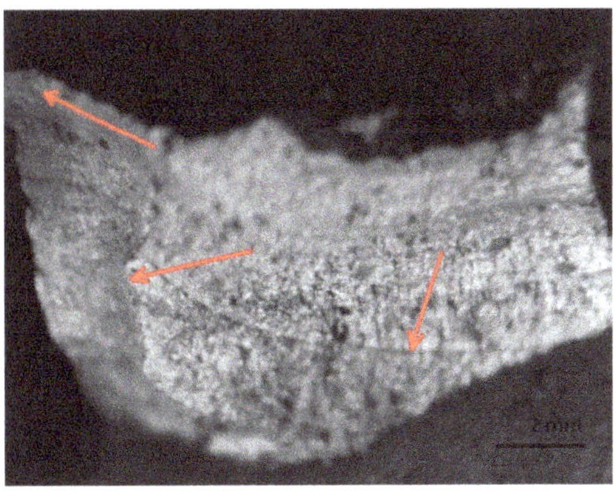

**Figure 20.** Fracture area: area of plastic deformation.

The accumulation of plastic deformation occurs during long-term operation. It is known that during plastic deformation, the density of dislocations in the metal increases. New dislocations are due to new internal sources, the best known of which is the Franco-Reed source. The increase in the density of dislocations affects the constant increase in the hardness of the metal. To achieve this, the process of increasing the hardness of the metal is accompanied by an increase in the brittleness of the metal. As a result, micro-cracks appear on the surface in the area of maximum hardness. During subsequent operations, these cracks turn into macro-cracks.

Fatigue fracture occurs when the surface of the fracture is perpendicular to the direction of the maximum stresses and has characteristic areas: the first—the decay foci, the second—the gradual increase in the crack, the third—the final fracture. In the area of the decay foci, a fine crystalline relief structure is visible, in the second area, the relief is crystalline in structure, and the third area is also crystalline in structure but fragile. Cracks are noticeable on the surface of the part.

The second is the area at the beginning of the fracture. Grooves and fatigue thresholds are observed, which are characteristic of low-cycle fatigue fractures (Figure 18). Fatigue thresholds and grooves are among the signs of macroscopic decay. These signs indicate the direction of crack propagation, which is associated with plastic deformations, a decrease or increase in the rate of crack propagation, depending on the effects of the environment.

The relief of the main or final fracture area is crystalline in structure, brittle, and the surface is rough and porous (Figure 19).

If all materials were absolutely plastic or absolutely brittle, plastic, or brittle fracture would occur during the tensile tests. Since there are no absolutely plastic and absolutely brittle materials, structural plastic materials are fragile when they decompose.

At the edges of the fracture, the shear characteristic of plastic deformation is seen, as well as grooves and fatigue thresholds characteristic of fatigue fracture (Figure 20).

Fractographic analysis shows that it is a fatigue fracture characterized by three fracture areas and the fracture surface oriented perpendicular to the direction of maximum stresses. It can be assumed that the fracture may have been initiated by a long-term load.

### 3.4. *Strength Properties of the Axle Material*

The mechanical properties of the steel of the axle were obtained experimentally according to standard ISO 6892-1. The middle part of the axle was used for testing (Figure 21).

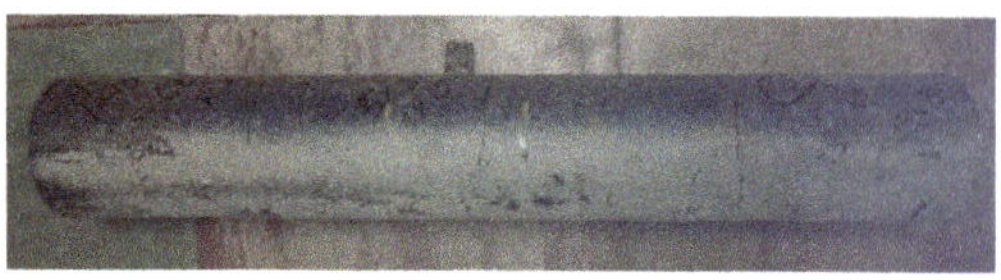

**Figure 21.** Tested middle part of axel.

The blanks for specimens were cut from the middle. It was important to know the numbers of blanks and their position in the axle (Figure 22a,b). The specimen for the tension test was produced by the milling process (Figure 22c).

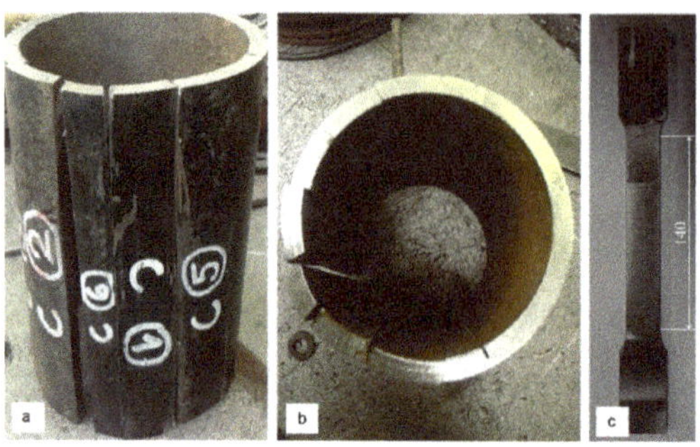


**Figure 22.** The blanks for specimens: (**a**) number of blanks; (**b**) orientations of blanks (**c**) specimen.

To identify the cross sections area of the specimens (Table 1), the cross sections parameters such as length $a$ and thickness $b$ were measured. The Young's modulus $E$ = 204 GPa, yield stress $\sigma_y = 581$ MPa and ultimate stress $\sigma_u = 663$ MPa (Table 1) were obtained as average from the experimental curves.

**Table 1.** Mechanical characteristics of axle.

| Specimen | $E$, GPa | $\sigma_y$, MPa | $\sigma_u$, MPa |
|---|---|---|---|
| C01 | 196 | 567 | 657 |
| C02 | 203 | 592 | 662 |
| C03 | 208 | 572 | 670 |
| C04 | 207 | 592 | 664 |
| Average | 204 | 581 | 663 |

The axis in this place puts the difficult working conditions of the surface in contact with the sleeve. This can cause corrosion of the axle surface (compare Figures 14 and 15) and fretting. In addition, the end of the axle is located near the wheel, from which sand and gravel can be thrown onto the surface of the material and damage the paint coating. Corrosion pits cause additional weakening of the material and surface notches, which are dangerous, especially when the material is subjected to fatigue, i.e., in the conditions of axle operation. The safety factor related to the min yield stress value obtained from strength tests is 567/392 = 1.45. It should be noted that the result of 567 MPa was obtained for a standardly prepared tensile sample (Figure 22c). On the surface of the axis, near the fracture (Figure 14), corrosion pits are clearly visible, reducing the strength of the material, and causing geometrical and structural stress concentrations.

Poisson's ratio $\nu$ = 0.30, which is standard for steel was taken.

After tensions, tests of the specimens were measured and machine diagrams force F- displacement L were transformed to stress σ–strain ε diagrams. Finally, the behavior of the steel used for the axle is described by the non-linear stress σ–strain ε diagram (Figure 23), which is an average of results.

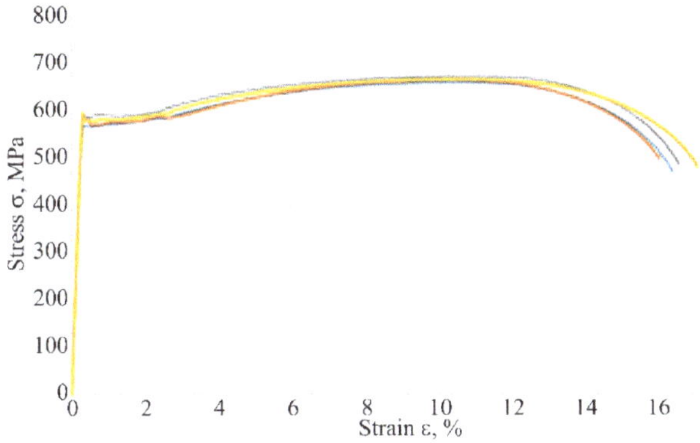


**Figure 23.** Experimental stress–strain diagrams.

In any case, this result has shown that about 60% (335/567) of elastic behavior of steel is used. Note that the analysis Equation (18) did not evaluate possible stress concentration factors due to welding impact on an axle and damage to its surface by corrosion (compare Figure 14).

In addition, the surface of the central part of the pipe is well protected with an anti-corrosion coating (Figure 21) and there is no danger of its abrasion in contact with the sleeve (compare Figure 1) or the impact of gravel thrown by a wheel rolling nearby. For these reasons, a safety factor of 567/436 = 1.30 (Figure 13) may be sufficient to achieve permanent fatigue strength.

Table 2 shows that only in the case of the simplest analytical calculations (bending stress from permissible load for one side of the axle $\sigma_L$) were the calculated stresses greater than those resulting from the simulation. In all other cases, it was the alternative: the simulation resulted in stresses higher than those calculated analytically. The reason for this may be that the simulation takes into account all types of axle loads, while the analytical calculations (except $\sigma_p$) did not take into account possible torsion (different deflection of the wishbones).

**Table 2.** Comparison of maximum stresses calculated analytically (A) and at simulations (S).

| Result of | Stress, MPa | | | |
|---|---|---|---|---|
| | $\sigma_L$ | $\sigma_A$ | $\sigma_{LV}$ | $\sigma_p$ |
| A | 185 | 361 | 286 | 261 |
| S* | 151 | 393 | 314 | 391 |
| S | 136 | 341 | 312 | 436 |
| A-max(S *,S) | 34 | −32 | −28 | −173 |

* load (wheel displacement) of left side only.

Apart from the case of overcoming a hole in the road, the absolute difference between the stress values calculated and those resulting from the simulation was about 30 MPa, i.e., about 5% of the yield stress value.

## 4. Conclusions

On the basis of the non-destructive tests as well as analyses and simulations, the following conclusions were made.

1. Fractographic analysis shows that it is a fatigue fracture characterized by three fracture areas and the fracture surface oriented perpendicular to the direction of maximum stresses.
2. It can be assumed that the fracture may have been initiated by a long-term load, as cracks and ribbed fracture morphology are characteristic of the effects of long-term loading.
3. The mechanical characteristics of the steel of the axle are obtained experimentally: the Young's modulus $E$ = 204 GPa, lowest yield stress $\sigma_y = 567$ MPa and average ultimate tensile stress $\sigma_u = 663$ MPa.
4. In the critical region of the axle during slow turning of the track, bending stresses reach 507 MPa in a dangerous quarter of the cross section. It consists 87% of yielding of steel. Respectively, in fast drive, it consists of 49%.
5. When driving the truck fatly on roads with potholes, due to the resulting torque, the compound stresses reach 335 MPa. However, 40% of the reserve is left up to the yield stress of the steel, but corrosion pits in the contact area of the sleeve and bushing may decrease the value.
6. Analytical solutions have shown that even when the truck is turning on bumpy roads, the yield strength is exceeded and the 93% strength limit is reached. This inevitably raises the fracture in the critical load impact region of the axle.
7. Stress obtained during simulations for the max loaded nodes was usually bigger than calculated and more realistic. It may be caused by the fact that the analytical calculations did not take into account possible torsion as an effect of different deflection of the wishbones while driving.

**Author Contributions:** Conceptualization, Ž.D., K.D. and J.C.; methodology, Ž.D., V.R., K.D. and A.N.; software, K.D.; validation, V.R., A.N. and J.M.; formal analysis, K.D., J.T. and A.N.; investigation, Ž.D., V.R., K.D. and J.T.; resources, Ž.D., K.D. and J.C.; data curation, K.D. and J.T.; writing—original draft preparation, Ž.D., K.D., J.C., J.T. and A.N.; writing—review and editing, Ž.D., K.D., J.C., A.N. and J.M.; visualization, J.C. and J.M.; supervision, K.D., J.C. and J.M.; project administration, J.C. and J.M.; funding acquisition, J.C. All authors have read and agreed to the published version of the manuscript.

**Funding:** This research received no external funding.

**Institutional Review Board Statement:** Not applicable.

**Informed Consent Statement:** Not applicable.

**Data Availability Statement:** Not applicable.

**Acknowledgments:** This work was partly prepared as part of Jacek Caban and Kazimierz Drozd scientific internship at the Institute of Mechanical Science of Vilnius Gediminas Technical University which took place from 26 July to 6 August 2021.

**Conflicts of Interest:** The authors declare no conflict of interest.

## References


1. Mahmoodik, M.; Davoodabadi, I.; Višnjić, V.; Afkar, A. Stress and dynamic analysis of optimized trailer chassis. *Tech. Gaz.* **2014**, *21*, 599–608.
2. Jilek, P.; Berg, J.; Tchuigwa, B.S.S. Influence of the Weld Joint Position on the Mechanical Stress Concentration in the Construction of the Alternative Skid Car System's Skid Chassis. *Appl. Sci.* **2022**, *12*, 397. [CrossRef]
3. Vdovin, D.; Levenkov, Y.; Chichekin, I. Prediction of fatigue life of suspension parts of the semi-trailer in the early stages of design. *IOP Conf. Ser. Mater. Sci. Eng.* **2020**, *820*, 012002. [CrossRef]
4. Hudec, J.; Šarkan, B. Effect of periodic technical inspections of vehicles on traffic accidents in the Slovak Republic. *Commun. Sci. Lett. Univ. Zilina* **2022**, *24*, A142–A159. [CrossRef]
5. Orynycz, O.; Tucki, K.; Wasiak, A.; Sobótka, R.; Gola, A. Evaluation of the Brake's Performance Dependence Upon Technical Condition of Car Tires as a Factor of Road Safety Management. *Energies* **2020**, *13*, 9. [CrossRef]
6. Pečeliūnas, R.; Žuraulis, V.; Droździel, P.; Pukalskas, S. Prediction of Road Accident Risk for Vehicle Fleet Based on Statistically Processed Tire Wear Model. *Promet* **2022**, *34*, 619–630. [CrossRef]
7. Samociuk, W.; Krzysiak, Z.; Bukova, B.; Brumercikova, E.; Bąkowski, H.; Krzywonos, L. Analiza międzynarodowego transportu produktów chemicznych w państwach Grupy Wyszehradzkiej. *Przem. Chem.* **2018**, *97*, 829–833. [CrossRef]
8. Tarkowski, S.; Rybicka, I. Distraction of the Driver and Its Impact on Road Safety. *Transp. Res. Procedia* **2020**, *44*, 196–203. [CrossRef]
9. Savchenko, Y.; Mykhalevych, M.; Drozdziel, P.; Verbitskiy, V.; Wrona, R. Accuracy and durability increasing of the body level control systems in the immobile state of the vehicle. *Diagnostyka* **2022**, *23*, 2022310. [CrossRef]
10. Gnap, J.; Jagelčák, J.; Marienka, P.; Frančák, M.; Kostrzewski, M. Application of MEMS Sensors for Evaluation of the Dynamics for Cargo Securing on Road Vehicles. *Sensors* **2021**, *21*, 2881. [CrossRef]
11. Hudec, J.; Šarkan, B.; Czodörová, R. Examination of the results of the vehicles technical inspections in relation to the average age of vehicles in selected EU states. *Transp. Res. Procedia* **2021**, *55*, 2–9. [CrossRef]
12. Gajek, A. Directions for the development of periodic technical inspection for motor vehicles safety systems. *Arch. Automot. Eng.* **2018**, *80*, 37–51. [CrossRef]
13. Tucki, K.; Wasiak, A.; Orynycz, O.; Mruk, R. Computer Simulation as a Tool for Managing the Technical Development of Methods for Diagnosing the Technical Condition of a Vehicle. *Energies* **2020**, *13*, 2869. [CrossRef]
14. Skrucany, T.; Vrabel, J.; Kazimir, P. The influence of the cargo weight and its position on the braking characteristics of light commercial vehicles. *Open Eng.* **2020**, *10*, 154–165. [CrossRef]
15. Marienka, P.; Frančák, M.; Jagelčák, J.; Synák, F. Comparison of Braking Characteristics of Solo Vehicle and Selected Types of Vehicle Combinations. *Transp. Res. Procedia* **2020**, *44*, 40–46. [CrossRef]
16. Vrabel, J.; Skrucany, T.; Bartuska, L.; Koprna, J. Movement analysis of the semitrailer with the tank-container at hard braking-the case study. *IOP Conf. Ser. Mater. Sci. Eng.* **2019**, *710*, 012025. [CrossRef]
17. Jagelčák, J.; Kiktová, M.; Frančák, M. The Analysis of Manoeuvrability of Semi-trailer Vehicle Combination. *Transp. Res. Procedia* **2020**, *44*, 176–181. [CrossRef]
18. Pantazopoulos, G.A. A Short Review on Fracture Mechanisms of Mechanical Components Operated under Industrial Process Conditions: Fractographic Analysis and Selected Prevention Strategies. *Metals* **2019**, *9*, 148. [CrossRef]
19. Krenevičius, A.; Leonavičius, M.K. *Eksperimentinė Medžiagų Mechanika*; Vilnius Tech: Vilnius, Lithuania, 2007; pp. 98–106.
20. Asi, O. Fatigue failure of a rear axle shaft of an automobile. *Eng. Fail. Anal.* **2006**, *13*, 1293–1302. [CrossRef]

21. Chaudhari, P.C.; Sonara, V.D.; Rathod, P.P. Analysis and Design of Tractor Rear Axle using Finite Element Method-A review. *Int. J. Adv. Eng. Res. Dev.* **2015**, *2*, 3.
22. Petru, C.; Raluca, S.; Vladut, V.; Ungureanu, N. *Technical Aspects Regarding Axles Calculation of Road Transport Means*; Omni Scriptum GmbH & Co., Ed.; LAP Lambert Academic Publishing: Port Louis, Mauritius, 2014; ISBN 978-3-659-39526-0.
23. Togaev, A.; Shermukhamedov, A. Method of calculating the stress strain state of frame structures of the tractor trailers at various external influences. *Glob. J. Sci. Eng.* **2020**, *4*, 5. [CrossRef]
24. Qinghua, M.; Huifeng, Z.; Fengjun, L. Fatigue failure fault prediction of truck rear axle housing excited by random road roughness. *Int. J. Phys. Sci.* **2011**, *6*, 1563–1568. [CrossRef]
25. Aloni, S.; Khedkar, S. Comparative evaluation of tractor trolley axle by using Finite Element Analysis approach. *Int. J. Eng. Sci. Technol.* **2012**, *4*, 1351–1360.
26. Ambiger, H.; Reddy, B. Design Optimization of Automobile Rear Axle Housing for Fatigue Loads using Finite Element Analysis. *IJSRD* **2016**, *4*, 5.
27. Kurniawan, A.; Andoko. Stress and Crack Simulation on Axle Housing Mitsubishi L300 Pickup Car using Finite Element Method. *IOP Conf. Ser. Mater. Sci. Eng.* **2019**, *494*, 012107. [CrossRef]
28. Dižo, J.; Blatnicky, M.; Melnik, R.; Semenov, S.; Mikhailov, E.; Kurtulik, J. Static Analysis of Tipping Superstructure of Single-Axle Tractor Trialer. In Proceedings of the 21st International Scientific Conference Engineering for Rural Development, ERD 2022, Jelgava, Letonia, 25–27 May 2022; pp. 13–21.
29. Dębski, H.; Koszałka, G.; Ferdynus, M. Application of FEM in the analysis of the structure of a trailer supporting frame with variable operation parameters. *Eksploat. Niezawodn.* **2012**, *14*, 107–114.
30. Podkowski, K.; Barszcz, Z.; Senko, J. FEM (Finite Element Method) Numeric Analyses of the Syrenka S201 Car Model. In Proceedings of the 1st Renewable Energy Sources-Research and Business (RESRB-2016), Wrocław, Poland, 22–24 June 2016; Springer: Cham, Switzerland, 2017; pp. 415–421.
31. Vrabel, J.; Jagelcak, J.; Zamecnik, J.; Caban, J. Influence of Emergency Braking on Changes of the Axle Load of Vehicles Transporting Solid Bulk Substrates. *Procedia Eng.* **2017**, *187*, 89–99. [CrossRef]
32. *EN 12195-1*; The Standard for Securing the Cargo According to the Directive 2014/47/EU. European Standard: Pilsen, Czech Republic, 2010.
33. Buhari, R.; Rohani, M.M.; Abdullah, M.E. Dynamic load coefficient of tyre forces from truck axles. *Appl. Mech. Mater.* **2013**, *405–408*, 1900–1911. [CrossRef]
34. Zhao, X.; Kang, J.; Lei, T.; Li, X.; Cao, Z.; Wang, Y. Research on air cushion suspension support system with low speed and heavy load. *J. Phys. Conf. Ser.* **2020**, *1549*, 052020. [CrossRef]
35. SAF-Holland Group. Design Manual, Edition 2020-02. Available online: https://safholland.com/cz/en/download-center/document/resource/environment/project1_p/documents/documentationP/2022/2022_07/LG%2520Hercules%2520KH%2520Edition-2020_en-DE.pdf (accessed on 30 November 2022).

*materials*

MDPI

*Article*

# Prediction of Fatigue Crack Initiation of 7075 Aluminum Alloy by Crystal Plasticity Simulation

**Takayuki Shiraiwa *, Fabien Briffod and Manabu Enoki**

Department of Materials Engineering, School of Engineering, The University of Tokyo, 7-3-1 Hongo, Bunkyo-ku, Tokyo 113-8656, Japan
* Correspondence: shiraiwa@rme.mm.t.u-tokyo.ac.jp

**Abstract:** The 7075 aluminum alloy is a promising material for the aerospace industry due to its combination of light weight and high strength. This study proposed a method for predicting fatigue crack initiation of the 7075 aluminum alloy by crystal plasticity finite element analysis considering microstructures. In order to accurately predict the total fatigue life, it is necessary to calculate the number of cycles for fatigue crack initiation, small crack growth, and long crack growth. The long crack growth life can be estimated by the Paris law, but fatigue crack initiation and small crack growth are sensitive to the microstructures and have been difficult to predict. In this work, the microstructure of 7075 aluminum alloy was reconstructed based on experimental observations in the literature and crystal plasticity simulations were performed to calculate the elasto-plastic deformation behavior in the reconstructed polycrystalline model under cyclic deformation. The calculated local plastic strain was introduced into the crack initiation criterion (Tanaka and Mura, 1981) to predict fatigue crack initiation life. The predicted crack initiation life and crack morphology were in good agreement with the experimental results, indicating that the proposed method is effective in predicting fatigue crack initiation in aluminum alloys. From the obtained results, future issues regarding the prediction of fatigue crack initiation were discussed.

**Keywords:** aluminum alloy; fatigue; crystal plasticity; finite element method; crack initiation

**Citation:** Shiraiwa, T.; Briffod, F.; Enoki, M. Prediction of Fatigue Crack Initiation of 7075 Aluminum Alloy by Crystal Plasticity Simulation. *Materials* **2023**, *16*, 1595. https://doi.org/10.3390/ma16041595

Academic Editors: Grzegorz Lesiuk and Dariusz Rozumek

Received: 23 November 2022
Revised: 12 January 2023
Accepted: 13 February 2023
Published: 14 February 2023



## 1. Introduction

The 7000 (Al-Zn-Mg) and 2000 (Al-Cu-Mg) series aluminum alloys are widely used as structural materials in aircraft and space applications. The stable precipitate phases in Al-Zn-Mg alloys are $MgZn_2$ (η phase) and $Mg_3Zn_3Al_2$ (T phase) [1]. The metastable phases (η′, T′ phases) that appear during the transition from the spherical GP zone to the stable phase are fine and produce substantial lattice strains, resulting in excellent precipitation strengthening. The 7075 alloy (Al-5.6Zn-2.5Mg-1.6Cu-0.23Cr in wt%) is a representative 7000 series alloys, further enhanced with the addition of Cu and Cr, and is commonly utilized in a fully precipitated state by T6 heat treatment. While the 7075 alloys have the highest tensile strength among aluminum alloys, they are inferior in fatigue strength and corrosion resistance, making them difficult to use in parts subjected to tensile fatigue loading. Their application is limited to parts subjected to compressive fatigue loading, such as the upper wing surface. Since aircraft are subjected to cyclic loading due to takeoffs, landings, and changes in atmospheric pressure, further improvement in fatigue performance is required.

The fatigue failure of metallic materials occurs when the material is subjected to repeated loads, resulting in irreversible cyclic plastic deformation, crack initiation, the small crack growth, the long crack growth, and final failure. The behavior of long crack growth can be predicted based on linear fracture mechanics. However, the prediction of fatigue crack initiation and the microstructurally small fatigue crack growth remains challenging. Since the work by Forthyth [2], many observations have been made regarding fatigue crack initiation in aluminum alloys. Under the positive stress ratio (tensile fatigue loading), fatigue crack initiations from inclusions [3] and voids [4] have been reported in 7000 series alloys, but the effect of the type and size of the inclusions on crack initiation is not clear. On the other hand, it has been observed that under conditions of stress ratios below zero, as used in practical applications, fatigue cracks are generated by the formation of transgranular facets at the surface of the material [5,6]. In order to predict such fatigue crack initiation within grains, it is necessary to analyze the cyclic plastic deformation behavior of polycrystalline materials considering the grain morphologies and crystallographic orientations.

The crystal plasticity finite element method (CPFEM) is a suitable numerical method for such purposes and has been used to predict fatigue properties in steels [7], titanium alloys [8], and aluminum alloys [9–11]. In these analyses, fatigue crack initiation life is often predicted by quantifying the driving force for fatigue crack initiation as a fatigue indicator parameter (FIP) [12], which is defined by the Tanaka-Mura model based on dislocation theory [13] and/or the Fatemi-Socie model based on critical plane theory [14]. However, methods for quantifying the FIP from the stress-strain field computed by CPFEM are still not well established. Previous papers on fatigue prediction of aluminum alloys have several drawbacks, such as the inability to predict crack shape considering the plastic anisotropy of each grain and the inability to relate the slip band length to grain morphology in the Tanaka-Mura model. Our research group has proposed a method to solve these problems by introducing potential crack paths parallel to the slip bands, and applied the method to pure iron [15], titanium alloys [16], and magnesium alloys [17].

The purpose of this paper is to investigate the applicability of fatigue crack initiation prediction by the crystal plasticity simulation to aluminum alloys. Among the 7000 series alloys, the 7075 alloy was selected as the target of this study because of the abundance of experimental fatigue data. The fatigue prediction results by CPFEM and potential crack paths are compared with the experimental data in the literature. The physical interpretation of the fatigue crack initiation criterion will be also discussed.

## 2. Materials and Methods

### 2.1. Materials

The material investigated in this study is 7075 aluminum alloy. Numerical models in this paper were constructed and calibrated with reference to experimental data of the 7075 aluminum alloy from the literature [5,18]. In the literature, a rolled 7075 aluminum alloy was subjected to T6 heat treatment, i.e., solution heat treatment at 480 °C for 1 h followed by aging at 120 °C for 24 h. The grains were elongated in the rolling direction, with average dimensions of 600 μm in the rolling direction (RD) and 108 μm in the transverse direction (TD). The tensile strength was 623 MPa, and elongation was 16%. The inverse polar figure map obtained by electron back-scatter diffraction (EBSD) experiments [18] are shown in Figure 1a.

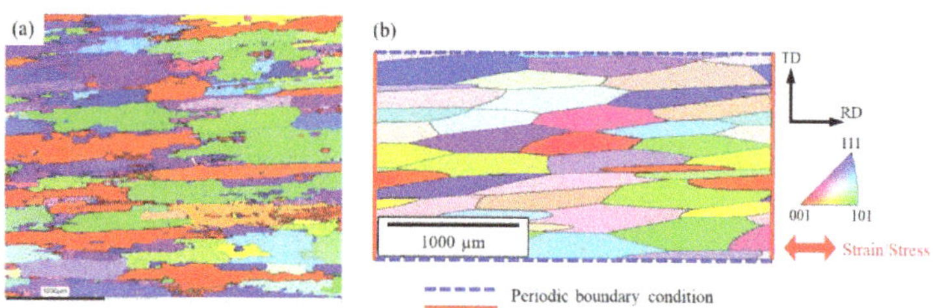


**Figure 1.** (**a**) Inverse pole figure map of 7075 aluminum alloy in the TD-RD plane (reprinted with permission from Ref. [18]), (**b**) finite element model of polycrystalline aggregate and boundary conditions.

*2.2. Generation of Synthetic Microstructures*

As mentioned in the previous section, the material has an anisotropic grain morphology. To reproduce such grain morphology, a polycrystalline aggregate model was constructed by anisotropic weighted Voronoi tessellation [15]. In this tessellation method, ellipses are sampled from assumed statistical distributions of grain shapes and positioned in model space by a relaxed random sequential addition (RSA) algorithm [19]. Then, an anisotropic tessellation is performed using the center positions of the ellipses as seeds, and finally the crystal orientation is assigned. In this study, the statistical distributions of grain shape parameters were assumed based on the experimental data. The major axis length was assumed to be normally distributed with a mean of 600 μm and a standard deviation of 50 μm, the aspect ratio of the major axis to the minor axis was normally distributed with a mean of 0.2 and a standard deviation of 0.05, and the angle between the major axis and RD was normally distributed with a mean of 0° and a standard deviation of 5°. The crystal orientation was assumed uniformly random for simplicity. A synthetic polycrystalline microstructure generated using these statistical distributions is shown in Figure 1b. It is noted that the polycrystalline structure is periodic in the x- and y-directions (RD and TD). Considering the computational cost, the model size and mesh size were set to 1500 μm × 3000 μm and 5 μm, respectively. The number of elements was 180,000. Three-dimensional eight-node hexahedral elements (C3D8) were used for meshing.

*2.3. Constitutive Law and Parameter Calibration*

To analyze the elasto-plastic deformation behavior of the polycrystalline model under cyclic loading, a constitutive law based on crystal plasticity theory was used in the finite element analysis. The total deformation gradient $\mathbf{F}$ can be decomposed into an elastic part $\mathbf{F}_e$ and a plastic part $\mathbf{F}_p$:

$$\mathbf{F} = \mathbf{F}_e\mathbf{F}_p \tag{1}$$

where $\mathbf{F}_e$ represents the elastic deformation and the rigid body rotation, and $\mathbf{F}_p$ represents the plastic deformation of the crystal lattice. The plastic velocity gradient $\mathbf{L}$ is defined as:

$$\mathbf{L}_p = \dot{\mathbf{F}}_p\mathbf{F}_p^{-1} = \sum_{\alpha=1}^{N} \dot{\gamma}^{\alpha}\mathbf{m}^{\alpha} \otimes \mathbf{n}^{\alpha} \tag{2}$$

where $\dot{\gamma}^{\alpha}$, $\mathbf{m}^{\alpha}$ and $\mathbf{n}^{\alpha}$ are the plastic shear rate, slip direction vector and normal to the slip plane of the slip system $\alpha$, respectively. A total of 12 slip systems of $\{111\}\langle 1\bar{1}0\rangle$ was considered. The shear rate of each slip system is modeled by a phenomenological power law:

$$\dot{\gamma}_s^{\alpha} = \dot{\gamma}_0\left|\frac{\tau^{\alpha}}{\tau_c^{\alpha}}\right|^{n} \mathrm{sgn}(\tau^{\alpha} - \chi^{\alpha}) \tag{3}$$

where $\dot{\gamma}_0$ is the reference shear rate ($\dot{\gamma}_0 = 10^{-3}$ s), $\tau^{\alpha}$ is the resolved shear stress of the slip system $\alpha$, $\tau_c^{a}$ is the critical resolved shear stress (CRSS), and $n$ is the strain rate exponent. The work hardening behavior due to slip-slip interaction was formulated as follows:

$$\tau_c^{\alpha} = \tau_0^{\alpha} + \int_t \dot{\tau}_{s\to s}^{\alpha}\mathrm{d}t \tag{4}$$

$$\dot{\tau}_{s\to s}^{\alpha} = \frac{\mathrm{d}\tau_{s\to s}^{\alpha}}{\mathrm{d}\Gamma_s}\sum_{\beta=1}^{N_s} h^{\alpha\beta}\left|\dot{\gamma}_s^{\alpha}\right| \tag{5}$$

$$\tau_{s\to s}^{\alpha} = \tau_1^{\alpha}\left\{1 - \exp\left(-\left|\frac{b_1^{\alpha}}{\tau_1^{\alpha}}\right|\Gamma_s\right)\right\} \tag{6}$$

where $\tau_0^{\alpha}$ is the initial CRSS, $\tau_1^{\alpha}$ is the saturated CRSS, $b_1^{\alpha}$ is the initial hardening rate, $\Gamma_s$ is the total shear strain, and $h^{\alpha\beta}$ is the interaction coefficient of the slip systems $\alpha$ and $\beta$.

To simulate cyclic deformation behavior, the back stress evolution was defined by the Armstrong-Frederick model as follows:

$$\dot{\chi}^{\alpha} = A\dot{\gamma}^{\alpha} - B\left|\dot{\gamma}^{\alpha}\right|\chi^{\alpha} \tag{7}$$

where $A$ and $B$ are constants, each associated with a dynamic hardening and a dynamic recovery.

To calibrate the crystal plasticity parameters, strain-controlled low-cycle fatigue simulations were performed on the polycrystalline model generated in the previous section. The boundary conditions are depicted together with the model in Figure 1b. Periodic boundary conditions were applied in the x- and y-directions (RD and TD), respectively. To provide the periodic boundary conditions, equation constraints in the finite element code Abaqus were used to constrain the displacement of each node on the left end to match the displacement of the corresponding node on the right end. The top and bottom ends were similarly constrained. While maintaining these constraints, the average value of the x-displacement at the left end was fixed, and displacement (or load) was applied on the right end to simulate strain-controlled (or stress-controlled) cyclic fatigue tests. In the strain-controlled tests, the polycrystalline model was subjected to cyclic displacement in the RD with a maximum strain of 1.5%, a strain ratio of $R_{\varepsilon} = -1$, and the number of cycles at 10. The stress-strain hysteresis loop was compared with experimental data [18] and the crystal plasticity parameters were calibrated to minimize the discrepancy. The calibrated crystal plasticity parameters are shown in Table 1.

**Table 1.** Elastic constants and crystal plasticity parameters for aluminum alloy 7075.

| **Property** | $C_{11}$ | $C_{12}$ | $C_{44}$ | $\dot{\gamma}_0$ □ | $n$ | $\tau_0^{\alpha}$ □ | $\tau_1^{\alpha}$ □ | □ $b_1^{\alpha}$ | $A$ | $B$ |
|---|---|---|---|---|---|---|---|---|---|---|
| Unit | GPa | GPa | GPa | $s^{-1}$ | - | MPa | MPa | MPa | MPa | - |
| Value | 106.75 | 60.41 | 28.34 | 0.001 | 7.14 | 150 | 160 | 200 | 8100 | 80 |

### 2.4. *High-Cycle Fatigue Simulation and Crack Initiation Prediction*

Stress-controlled fatigue simulations were performed to predict crack initiation in high-cycle fatigue tests. A periodic boundary condition was applied to the polycrystalline model as in the previous section, and cyclic loading was applied in the RD direction with a maximum stress of 190 to 340 MPa and a stress ratio of R = −1. The number of cycles was set to 10.

To assess the crack initiation life, an analysis based on the Tanaka-Mura model [20] was applied to the results of the crystal plasticity finite element simulations. A schematic of the crack initiation analysis is shown in Figure 2. The Tanaka-Mura model considers that a crack occurs on a slip band when the accumulated strain energy on the slip band exceeds a critical value. To specify the slip bands in the finite element model, the intersecting lines of the 2D finite element model and the slip plane were drawn at regular intervals for each slip system of each grain, as shown in Figure 2a. These lines were defined as potential crack paths. As an example, Figure 2b shows potential crack paths in a certain grain. For each potential crack path, the fatigue indicator parameter (FIP) was defined as a function of the plastic shear strain amplitude at the last cycle averaged over the path, $\Delta\gamma^{\alpha}$, and the length of the path, $d$.

$$FIP = d\left(\frac{\Delta\gamma^{\alpha}}{2}\right)^2 \tag{8}$$

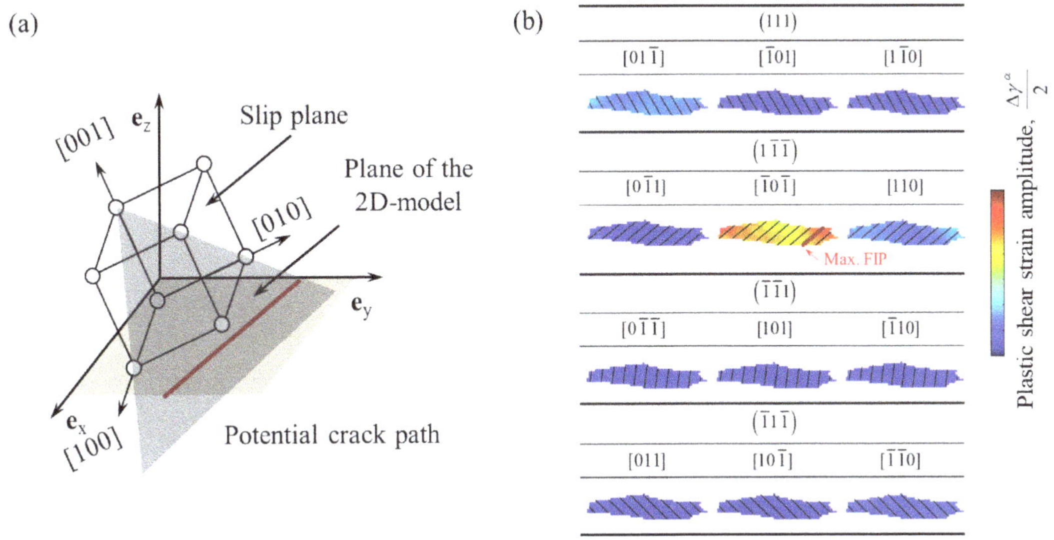


**Figure 2.** (**a**) Schematic of a potential crack path defined as the intersection of the slip plane and the 2D finite element model. (**b**) An example of the potential crack paths for a single crystalline grain. The path with the highest fatigue indicator parameter (FIP) is highlighted with a bold line.

Following the Tanaka-Mura model, the crack initiation life, $N_i$, is given by:

$$N_i = \frac{8GW_c}{\pi(1-\nu)d(\Delta\tau - 2\tau_c)^2} = \frac{A_{\mathrm{TM}}}{FIP} \quad (9)$$

where $G$ is the shear modulus, $W_c$ is the fracture surface energy per unit area, $\nu$ is the Poisson's ratio. Using the FIP defined above, the crack initiation life can be predicted by the inverse FIP and a single material constant, $A_{\mathrm{TM}}$. The material constant was determined by fitting with experimental data on crack initiation life at maximum stress of 270 MPa [5]. For all potential crack paths in the polycrystalline model, $N_i$ was calculated and it was assumed that the initial crack occurs in the path with the smallest value of $N_i$.

## 3. Results and Discussion

### *3.1. Low-Cycle Fatigue Behavior*

Low-cycle fatigue simulations were repeated with varying crystal plasticity parameters until the difference in the stress-strain hysteresis loop from the experimental data was sufficiently small. The first ten hysteresis loops obtained from low-cycle fatigue experiments performed in the rolling direction are shown in Figure 3a. Basically, it is difficult to uniquely determine the solution for multiple crystal plasticity parameters based only on macroscopic stress-strain behavior. To avoid this non-uniqueness problem, among the crystal plasticity parameters, $n$, $\dot{\gamma}_0$, $A$, and $B$ were fixed using literature values [9], while the other parameters $\tau_0{}^{\alpha}$, $\tau_1{}^{\alpha}$, and $b_1{}^{\alpha}$ were varied. The stress-strain hysteresis loops calculated with these crystal plasticity parameters are shown in Figure 3b. The obtained stress-strain behavior agreed well with the experimental data [18], including the yield stress, the maximum stress, the loop shape, the slight cyclic hardening at the beginning, and almost saturation by 10 cycles. Other experimental investigations have also reported that the hysteresis loop reaches a saturated state after about 10 cycles at most strain levels [21–23].

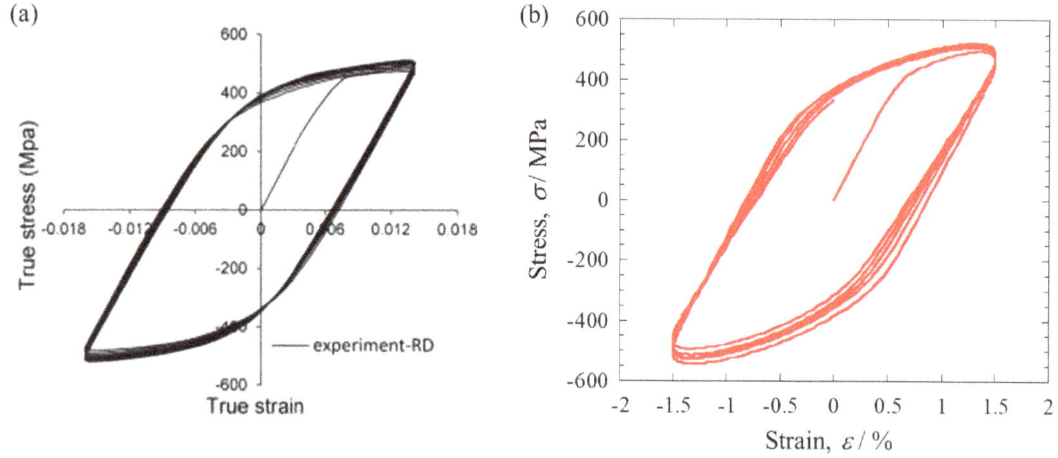


**Figure 3.** (**a**) First ten hysteresis loops obtained from low-cycle fatigue experiments performed in the rolling direction (reprinted with permission from Ref. [18]), (**b**) calibrated stress-strain hysteresis loops in low-cycle fatigue simulations.

### *3.2. High-Cycle Fatigue Behavior*

Stress-controlled fatigue simulations were performed using the crystal plasticity parameters calibrated in the previous section. Based on the results of the low-cycle fatigue simulation, the number of cycles was set to 10, and the stress-strain state in the last cycle was assumed to be the same as the steady state in the high-cycle fatigue test (e.g., half the number of cycles to failure). As an example, the Mises stress distribution at maximum load in the last cycle under the conditions of a maximum stress of 340 MPa and a stress ratio of $R = -1$ is shown in Figure 4a. When the applied stress was 340 MPa, the local stress exceeded 400 MPa due to the anisotropy of the elasto-plastic behavior of the crystal grains. The macroscopic stress-strain curve is shown in Figure 4b. The highest strain appeared in the tensile loading of the first cycle, and the loops after the second cycle overlapped each other. On the compression side, almost the same loops showed from the first cycle. To examine the local deformation behavior in more detail, stress-strain curves at higher and lower stress positions were displayed in Figure 4c. At the higher stress position, the maximum tensile stress in the loading direction was 421 MPa, and at the lower stress position, the stress was 269 MPa, indicating that the stress varied about $\pm 20\%$ from the applied stress. Additionally, the local stress ratio showed slightly different values for each position. For example, the two elements used in Figure 4c had a stress ratio of $-0.95$ at the higher stress position and a stress ratio of $-1.07$ at the lower stress position.

### *3.3. Prediction of Fatigue Crack Initiation*

The plastic shear strain amplitudes for each slip system were extracted from the high-cycle fatigue simulations in the previous section to calculate the FIP values for all potential crack paths. The cumulative plastic shear strain is shown in Figure 5 (the slip system corresponding to each contour plot is listed in Figure 2b). As can be seen from the figure, the plastic strain showed a quite heterogeneous distribution due to the plastic activities on slip systems. The FIP values of all paths were computed and the path with the highest FIP was assumed to be the crack initiation site. The predicted initial crack is indicated by an arrow in the figure. It was oriented close to 45° to the loading direction and occurred in a grain with a significant accumulation of plastic strain. This inclination of the crack to the loading direction is consistent with experimental observations [5]. Fatigue prediction studies using FIP often use volume averaging, but such methods can eliminate the intragranular heterogeneity of plastic strain. Especially for coarse grains with anisotropic geometry, the slip band size is highly dependent on the slip plane and grain shape. The concept of potential crack paths used in this paper can take into account the effect of grain shape anisotropy on crack initiation. The same crack initiation analysis was performed by changing the maximum stress in the range from 190 Mpa to 340 Mpa. The material constant $A_{TM}$ in the Tanaka-Mura model was determined to be 0.66 mm by fitting based on the experimental data of the crack initiation life at an intermediate stress level, the maximum stress of 270 Mpa [5]. This value was used to predict fatigue crack initiation life at other stress levels. The predicted crack initiation life is shown in Figure 6 together with the experimental S-N curve. The predicted crack initiation life at the maximum stress of 340 Mpa was in good agreement with the experimental value of $N_i$. Under loading conditions with no experimental crack initiation data, the predicted results of crack initiations were not contradicted with experimental failure life, $N_f$ (the predicted crack initiation life was smaller than the $N_f$). These results suggest that the crystal plasticity simulation proposed in this study can predict the fatigue crack initiation life in 7075 aluminum alloy. It should be noted that experimental results may be affected not only by the microstructures, but also by various factors such as specimen geometry, surface roughness, and alignment of the fatigue testing machine.

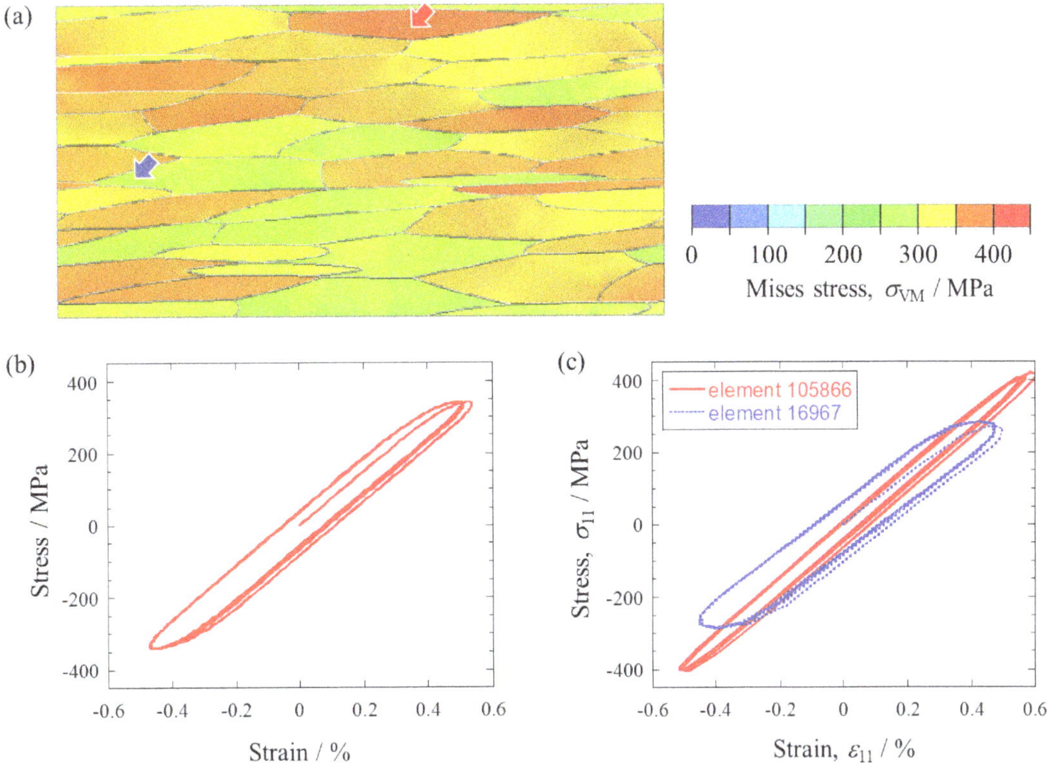


**Figure 4.** (**a**) Mises stress distribution at maximum stress applied ($\sigma_{max}$ = 340 MPa, R = −1). (**b**) Macroscopic stress-strain curve. (**c**) Stress-strain curves at the higher and lower stress locations indicated by arrows in (**a**).

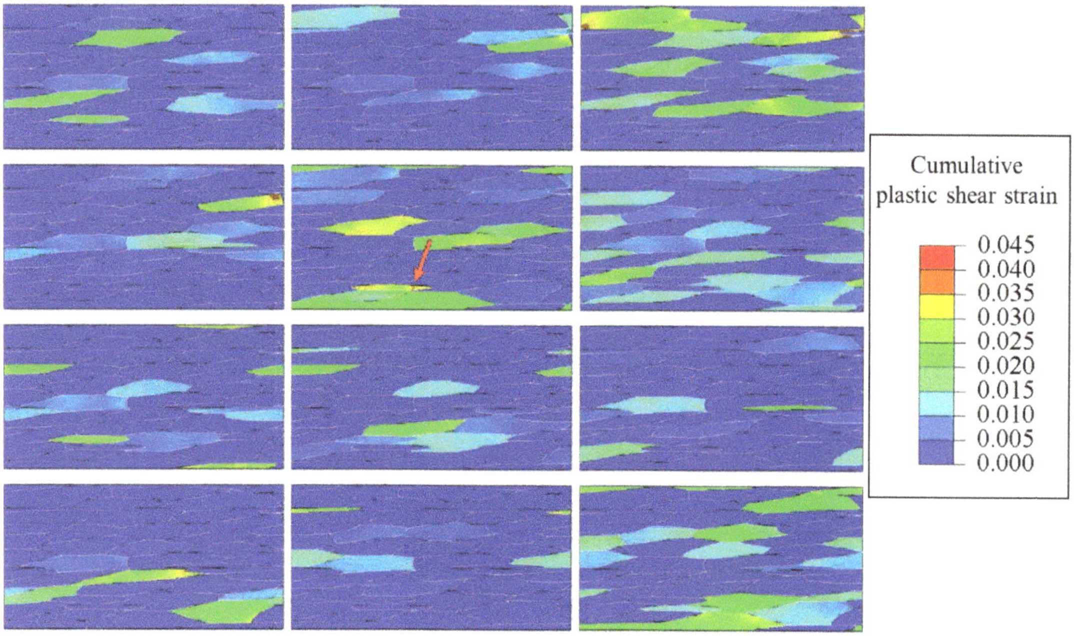


**Figure 5.** Distribution of cumulative plastic shear strain for 12 slip systems in the last calculation step ($\sigma_{max}$ = 340 MPa, R = −1). The predicted crack initiation site is indicated by an arrow and a red solid line.

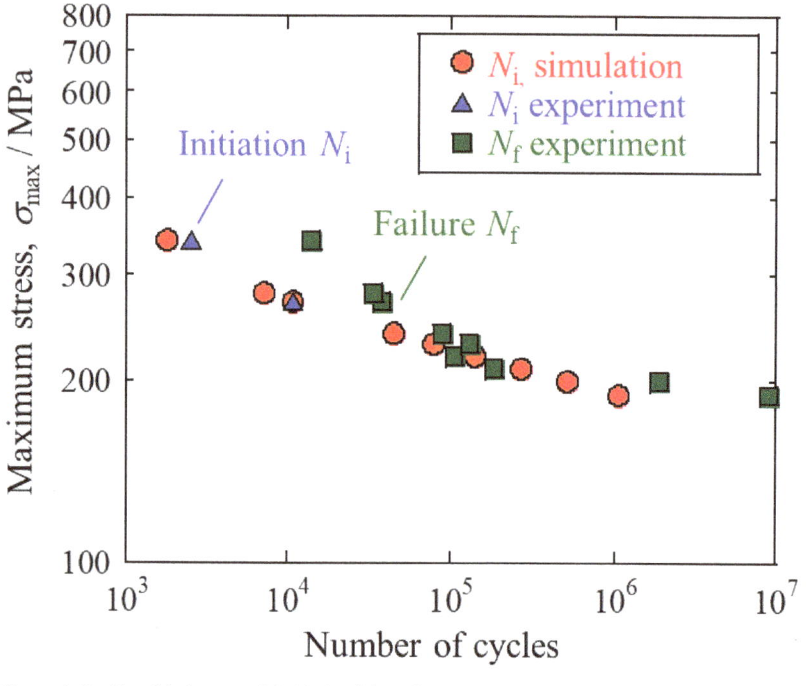


**Figure 6.** Predicted fatigue crack initiation life and experimental fatigue crack initiation and failure life taken from the literature [5].

The proposed method has the advantage of accurately predicting crack initiation with only a single fitting parameter ($A_{TM}$) and is useful for predicting fatigue crack initiation under various loading conditions. On the other hand, several issues remain to be addressed in order to apply the proposed method more universally to various situations. The first issue is the effect of free surfaces. In the experiments, extrusion and intrusion on an unconstrained surface can lead to fatigue crack initiation [2]. Therefore, it is necessary to include surface topology in the numerical simulations. A study that applied a crystal plasticity simulation similar to this paper to a polycrystalline model with a free surface reported that crack initiation was predicted on the surface [24]. The second issue is the crystallographic texture. It has been shown experimentally in aluminum alloys that the crystallographic texture has a significant effect on fatigue propagation and total fatigue life [25,26]. Virtual experiments based on numerical simulations also showed that the effect of the texture on fatigue crack initiation is not negligible [27]. The third issue is to model the various sources of crack initiation. It has been reported that cracks initiate from non-metallic inclusions under conditions where the stress ratio $R = 0$ or higher [3,23]. There are hard inclusions such as $Al_2Cu$ and soft inclusions such as $Mg_2Si$, both of which can be sources of fatigue crack initiation. Another issue is the number of grains included in the polycrystalline model. Since fatigue crack initiation is a phenomenon involving scattering, fatigue crack initiation analysis should be performed on models with sufficient volume and the highest value of the FIP should be adopted. Since computational cost often limits the model size, applying an extreme value analysis [28] can be helpful. To predict total fatigue life, it is also necessary to accurately predict microstructural small fatigue crack propagation [10].

### 3.4. *Physical Meaning of Tanaka-Mura Parameter*

Let us consider the physical meaning of the fitted parameter, $A_{TM}$. This consideration is important to make the proposed method more generally applicable to various materials. In the fatigue analysis proposed in this paper, the plastic shear strain induced by external forces and the microstructure-dependent slip length are quantified in the FIP. Hence, the $A_{TM}$ is considered to involve the nature of the material. Basically, a higher $A_{TM}$ means higher resistance to fatigue crack initiation. Pioneering work by Feltner et al. [29] more than 50 years ago pointed out that the stacking fault energy (SFE) affects dislocation structure changes and cyclic stress-strain responses, which are precursors to fatigue crack initiation. Then, many studies have reported that in materials with high SFE, cross slip is promoted, and subgrain structures of dislocations form under cyclic loading, leading to fatigue crack initiation along subgrain boundaries [30–33]. In their experiments, it has been difficult to separate the effect of SFE from the effect of yield stress on fatigue properties, making quantitative evaluation challenging. In the analysis performed in this paper, the effects of yield stress are separated to the FIP and the effect of SFE is considered to be included in the $A_{TM}$. The $A_{TM}$ values derived for various metallic materials in previous studies are listed in Table 2 together with the SFE in the literature [34,35]. Overall, the $A_{TM}$ tended to increase with decreasing the SFE. This trend is roughly consistent with previous studies, as a lower SFE implies superior fatigue properties. The $A_{TM}$ value of 7075 aluminum alloy obtained in this study appear to deviate from the trend. One of the reasons is that the experimental values for crack initiation life were taken from literature 30 years ago, and the spatial resolution of the observations was lower than for the other three alloys obtained more recently, resulting in an overestimation of $N_i$. Another reason is that the 7075 aluminum alloy is precipitation-hardened, and the finely dispersed particles on the atomic order promote planar slips, as examined in detail by Gerold et al. [36]. As Mughrabi points out [37], explaining fatigue behavior with SFE is not inherently appropriate and should be compared to the ease of cross slip and the nature of dislocation slip (plane or wavy slip) because short-range order and short-range clustering in solid solution have a more significant effect on promoting planar slip than SFE. To this end, it is important to quantitatively evaluate the dislocation behavior of materials subjected to cyclic deformation, which should be discussed with numerical simulations based on discrete dislocation dynamics (DDD) [38]. In DDD, the challenge is how to formulate the dislocation mobility law, which would require MD calculations [39] and inverse analysis of dislocation motion by acoustic emission (AE) [40].

**Table 2.** Tanaka-Mura parameter and stacking fault energy (SFE). The SFE is the literature value of the base metal of each material.

| Material | Tanaka-Mura Parameter/mm | Stacking Fault Energy /mJ/mm$^2$ | Crystal Structure |
|---|---|---|---|
| Ti-6Al-4V alloy | 0.0014 [15] | 310–82 [33] | HCP |
| Fe | 0.14 [14] | 180 [34] | BCC |
| 7075 Al alloy | 0.66 | 166 [34] | FCC |
| Mg-Al-Ca-Mn alloy | 0.245 [16] | 125 [34] | HCP |

It should be noted that the $A_{TM}$ also depends on the distance between the slip bands assumed in the model. Recently, high-resolution digital image correlation (HR-DIC) has made it possible to observe slip bands [41]. This technique would be useful for introducing reasonable slip-band spacing into the numerical model. Additionally, it should be kept in mind that the Tanaka-Mura model (Equation (9)) is derived from strong hypotheses, such as the irreversibility of dislocation movement, uniform shear stress on the slip band, and no work hardening. The assumption regarding slip irreversibility is inconsistent with the recent observations made by atomic force microscopy [42,43]. The observation suggested that the slip irreversibility depends on the plastic strain amplitude. One possible approach to incorporate this phenomenon is to define the fraction of the slip irreversibility as a function of plastic strain amplitude and introduce it into the Tanaka-Mura model, but there is a lack of sufficient experimental knowledge to formulate such a function. Experimental methods to measure slip irreversibility more easily are required. The DIC technique and AE method may be useful for such purposes. Previous studies have suggested that in cyclic testing of pure aluminum, the generated AE signals were associated with dislocations that move during stress reversal [44]. Further quantitative discussion of dislocation behavior during cyclic deformation will be necessary to better understand the physics of the fatigue crack initiation.

## 4. Conclusions

This work proposed a method to predict fatigue crack initiation of 7075 aluminum alloy by calculating the cyclic deformation of a polycrystalline model based on crystal plasticity simulations and substituting the results into a crack initiation criterion of the Tanaka-Mura model. The proposed method requires the crystal plasticity parameters of the material and the critical value of the Tanaka-Mura model. The following conclusions can be drawn:

1. In both strain-controlled and stress-controlled fatigue simulations, the stress-strain hysteresis loop was saturated by 10 cycles. In the high-cycle (stress-controlled) fatigue simulations, local stress amplitudes were heterogeneously distributed in the polycrystalline model due to the anisotropic elasto-plastic deformation of crystalline grains. On the other hand, local stress ratios were almost uniform.
2. To predict crack initiation in high-cycle fatigue experiments, potential crack paths were assumed parallel to the slip planes, and FIPs were derived on the paths and substituted into the fatigue crack initiation criterion of the Tanaka-Mura model. The predicted crack initiation life and initial crack orientation were in good agreement with the experimental results.
3. The physical meaning of the fitting parameter ($A_{TM}$) in the Tanaka-Mura model was discussed, compared with values in other alloys. Although the $A_{TM}$ was weakly correlated with SFE, it was suggested that SFE is not considered the only dominant

parameter on fatigue crack initiation. The importance of analyzing the dynamic dislocation behavior under cyclic loading was pointed out.

**Author Contributions:** Conceptualization, T.S., F.B. and M.E.; methodology, T.S. and F.B.; software, T.S. and F.B.; validation, T.S.; formal analysis, T.S. and F.B.; investigation, T.S.; resources, M.E.; data curation, T.S. and F.B.; writing—original draft preparation, T.S.; writing—review and editing, F.B. and M.E.; visualization, T.S. and F.B.; supervision, M.E.; project administration, T.S., F.B. and M.E.; funding acquisition, T.S. and M.E. All authors have read and agreed to the published version of the manuscript.

**Funding:** This work was partially supported by grant-in-aid from Japan Aluminium Association (JAA), Light Metal Educational Foundation of Japan, JSPS KAKENHI for Scientific Research (B) (Grant Number 21H01648), and JST SICORP (Grant Number JPMJSC21E1).

**Institutional Review Board Statement:** Not applicable.

**Informed Consent Statement:** Not applicable.

**Data Availability Statement:** The raw/processed data are available from the corresponding author upon reasonable request.

**Conflicts of Interest:** The authors declare no conflict of interest.

## References

1. Mondolfo, L.F. 8—Technological Properties. In *Aluminum Alloys*; Mondolfo, L.F., Ed.; Butterworth-Heinemann: Oxford, UK, 1976; pp. 173–210.
2. Forsyth, P.J.E. Fatigue damage and crack growth in aluminium alloys. *Acta Metall.* **1963**, *11*, 703–715. [CrossRef]
3. Payne, J.; Welsh, G.; Christ, R.J.; Nardiello, J.; Papazian, J.M. Observations of fatigue crack initiation in 7075-T651. *Int. J. Fatigue* **2010**, *32*, 247–255. [CrossRef]
4. Przystupa, M.A.; Bucci, R.J.; Magnusen, P.E.; Hinkle, A.J. Microstructure based fatigue life predictions for thick plate 7050-T7451 airframe alloys. *Int. J. Fatigue* **1997**, *19*, 285–288. [CrossRef]
5. Tokaji, K.; Ogawa, T.; Kameyama, Y. The Effect of Stress Ratio on Growth Behaviour of Small Fatigue Cracks in an Aluminum Alloy 7075-T6 with Special Interest in Stage I Crack Growth. *J. Soc. Mater. Sci. Jpn.* **1989**, *38*, 1019–1025. [CrossRef]
6. Li, Y.; Retraint, D.; Xue, H.; Gao, T.; Sun, Z. Fatigue properties and cracking mechanisms of a 7075 aluminum alloy under axial and torsional loadings. *Procedia Struct. Integr.* **2019**, *19*, 637–644. [CrossRef]
7. Wang, W.; Wei, P.; Liu, H.; Zhu, C.; Deng, G.; Liu, H. A micromechanics-based machine learning model for evaluating the microstructure-dependent rolling contact fatigue performance of a martensitic steel. *Int. J. Mech. Sci.* **2022**, *237*, 107784. [CrossRef]
8. Stopka, K.S.; Gu, T.; McDowell, D.L. Effects of algorithmic simulation parameters on the prediction of extreme value fatigue indicator parameters in duplex Ti-6Al-4V. *Int. J. Fatigue* **2020**, *141*, 105865. [CrossRef]
9. Li, L.; Shen, L.; Proust, G. Fatigue crack initiation life prediction for aluminium alloy 7075 using crystal plasticity finite element simulations. *Mech. Mater.* **2015**, *81*, 84–93. [CrossRef]
10. Sun, T.; Qin, L.; Xie, Y.; Zheng, Z.; Xie, C.; Huang, Z. An Approach for Predicting the Low-Cycle-Fatigue Crack Initiation Life of Ultrafine-Grained Aluminum Alloy Considering Inhomogeneous Deformation and Microscale Multiaxial Strain. *Materials* **2022**, *15*, 3403. [CrossRef]
11. Gaur, V.; Briffod, F.; Enoki, M. Micro-mechanical investigation of fatigue behavior of Al alloys containing surface/superficial defects. *Mater. Sci. Eng. A* **2020**, *775*, 138958. [CrossRef]
12. McDowell, D.L. Simulation-based strategies for microstructure-sensitive fatigue modeling. *Mater. Sci. Eng. A* **2007**, *468–470*, 4–14. [CrossRef]
13. Tanaka, K.; Mura, T. A Dislocation Model for Fatigue Crack Initiation. *J. Appl. Mech.* **1981**, *48*, 97–103. [CrossRef]
14. Fatemi, A.; Socie, D.F. A Critical Plane Approach to Multiaxial Fatigue Damage Including Out-of-Phase Loading. *Fatigue Fract. Eng. Mater. Struct.* **1988**, *11*, 149–165. [CrossRef]
15. Briffod, F.; Shiraiwa, T.; Enoki, M. Microstructure modeling and crystal plasticity simulations for the evaluation of fatigue crack initiation in α-iron specimen including an elliptic defect. *Mater. Sci. Eng. A* **2017**, *695*, 165–177. [CrossRef]
16. Briffod, F.; Shiraiwa, T.; Enoki, M. Nucleation and propagation modeling of short fatigue crack in rolled bi-modal Ti–6Al–4V alloy. *Mater. Sci. Eng. A* **2020**, *790*, 139710. [CrossRef]
17. Briffod, F.; Shiraiwa, T.; Enoki, M. Numerical investigation of the influence of twinning/detwinning on fatigue crack initiation in AZ31 magnesium alloy. *Mater. Sci. Eng. A* **2019**, *753*, 79–90. [CrossRef]
18. Li, L.; Shen, L.; Proust, G. A texture-based representative volume element crystal plasticity model for predicting Bauschinger effect during cyclic loading. *Mater. Sci. Eng. A* **2014**, *608*, 174–183. [CrossRef]
19. Tarjus, G.; Schaaf, P.; Talbot, J. Random sequential addition: A distribution function approach. *J. Stat. Phys.* **1991**, *63*, 167–202. [CrossRef]

20. Tanaka, K.; Mura, T. A Theory of Fatigue Crack Initiation at Inclusions. *Metall. Trans. A.* **1982**, *13*, 117–123. [CrossRef]
21. Liu, Z.; Zhang, H.; Yan, Z.; Liaw, P.K.; Dong, P. Cyclic deformation and fatigue behavior of 7075-T651 Al alloy with a gradient structure. *Mater. Sci. Eng. A* **2021**, *822*, 141669. [CrossRef]
22. Branco, R.; Costa, J.D.; Prates, P.A.; Berto, F.; Pereira, C.; Mateus, A. Load sequence effects and cyclic deformation behaviour of 7075-T651 aluminium alloy. *Int. J. Fatigue* **2022**, *155*, 106593. [CrossRef]
23. Ma, J.; Wang, Q.; Yang, Y.; Yang, F.; Dong, B.; Che, X.; Cao, H.; Zhang, T.; Zhang, Z. Anisotropic Low Cycle Behavior of the Extruded 7075 Al Alloy. *Materials* **2021**, *14*, 4506. [CrossRef]
24. Shiraiwa, T.; Briffod, F.; Enoki, M. Development of integrated framework for fatigue life prediction in welded structures. *Eng. Fract. Mech.* **2018**, *198*, 158–170. [CrossRef]
25. Hu, Y.; Liu, Z.; Zhao, Q.; Bai, S.; Liu, F. P-Texture Effect on the Fatigue Crack Propagation Resistance in an Al-Cu-Mg Alloy Bearing a Small Amount of Silver. *Materials* **2018**, *11*, 2481. [CrossRef]
26. Strohmann, T.; Breitbarth, E.; Besel, M.; Zaunschirm, S.; Witulski, T.; Requena, G. Damage Mechanisms and Anisotropy of an AA7010-T7452 Open-Die Forged Alloy: Fatigue Crack Propagation. *Materials* **2022**, *15*, 3771. [CrossRef]
27. Briffod, F.; Shiraiwa, T.; Enoki, M. Numerical investigation of the influence of rolling texture and microstructure on fatigue crack initiation in BCC polycrystals. *Int. J. Fatigue* **2018**, *107*, 72–82. [CrossRef]
28. Przybyla, C.P.; McDowell, D.L. Microstructure-sensitive extreme value probabilities for high cycle fatigue of Ni-base superalloy IN100. *Int. J. Plast.* **2010**, *26*, 372–394. [CrossRef]
29. Feltner, C.E.; Laird, C. Cyclic stress-strain response of F.C.C. metals and alloys—I Phenomenological experiments. *Acta Metall.* **1967**, *15*, 1621–1632. [CrossRef]
30. DH, A.; Backofen, W.; Miller, G. Fatigue-crack growth in some copper-base alloys. *Trans. Metall. Soc. AIME* **1966**, *236*, 1667–1673.
31. Saxena, A.; Antolovich, S.D. Low cycle fatigue, fatigue crack propagation and substructures in a series of polycrystalline Cu-Al alloys. *Metall. Trans. A* **1975**, *6*, 1809. [CrossRef]
32. Awatani, J.; Katagiri, K.; Koyanagi, K. A study on the effect of stacking fault energy on fatigue crack propagation as deduced from dislocation patterns. *Metall. Trans. A* **1979**, *10*, 503–507. [CrossRef]
33. Lukáš, P.; Klesnil, M. Dislocation structures in fatigued single crystals of Cu-Zn system. *Phys. Status Solidi A* **1971**, *5*, 247–258. [CrossRef]
34. Guo, Z.; Miodownik, A.P.; Saunders, N.; Schillé, J.P. Influence of stacking-fault energy on high temperature creep of alpha titanium alloys. *Scr. Mater.* **2006**, *54*, 2175–2178. [CrossRef]
35. Edalati, K.; Hashiguchi, Y.; Iwaoka, H.; Matsunaga, H.; Valiev, R.Z.; Horita, Z. Long-time stability of metals after severe plastic deformation: Softening and hardening by self-annealing versus thermal stability. *Mater. Sci. Eng. A* **2018**, *729*, 340–348. [CrossRef]
36. Gerold, V.; Karnthaler, H.P. On the origin of planar slip in f.c.c. alloys. *Acta Metall.* **1989**, *37*, 2177–2183. [CrossRef]
37. Mughrabi, H. Fatigue, an everlasting materials problem—Still en vogue. *Procedia Eng.* **2010**, *2*, 3–26. [CrossRef]
38. Meng, F.; Ferrié, E.; Déprés, C.; Fivel, M. 3D discrete dislocation dynamic investigations of persistent slip band formation in FCC metals under cyclical deformation. *Int. J. Fatigue* **2021**, *149*, 106234. [CrossRef]
39. Groh, S.; Marin, E.B.; Horstemeyer, M.F.; Zbib, H.M. Multiscale modeling of the plasticity in an aluminum single crystal. *Int. J. Plast.* **2009**, *25*, 1456–1473. [CrossRef]
40. Scruby, C.B.; Wadley, H.N.G.; Hill, J.J. Dynamic elastic displacements at the surface of an elastic half-space due to defect sources. *J. Phys. D Appl. Phys.* **1983**, *16*, 1069. [CrossRef]
41. Stinville, J.C.; Callahan, P.G.; Charpagne, M.A.; Echlin, M.P.; Valle, V.; Pollock, T.M. Direct measurements of slip irreversibility in a nickel-based superalloy using high resolution digital image correlation. *Acta Mater.* **2020**, *186*, 172–189. [CrossRef]
42. Risbet, M.; Feaugas, X.; Guillemer-Neel, C.; Clavel, M. Use of atomic force microscopy to quantify slip irreversibility in a nickel-base superalloy. *Scr. Mater.* **2003**, *49*, 533–538. [CrossRef]
43. Mughrabi, H. Cyclic slip irreversibility and fatigue life: A microstructure-based analysis. *Acta Mater.* **2013**, *61*, 1197–1203. [CrossRef]
44. Kishi, T.; Tanaka, H.; Horiuchi, R.; Obata, Y.; Aoki, K. Acoustic emission peak of bauschinger effect. *Scr. Metall.* **1975**, *9*, 1023–1026. [CrossRef]

materials

MDPI

*Article*

# General Reference and Design S–N Curves Obtained for 1.2709 Tool Steel

**Michał Böhm [1,*], Adam Niesłony [1], Szymon Derda [1], Robert Owsiński [1], Miloslav Kepka, Jr. [2], Ivana Zetkova [2], Miroslav Zetek [2], Šárka Houdková [3] and Mariusz Prażmowski [1]**

1 Department of Mechanics and Machine Design, Faculty of Mechanical Engineering, Opole University of Technology, Mikołajczyka 5, 45-271 Opole, Poland
2 Regional Technological Institute, Faculty of Mechanical Engineering, University of West Bohemia, Univerzitní 8, 301 00 Plzeň, Czech Republic
3 Research and Testing Institute in Plzeň, Tylova 1581/46, 301 00 Plzeň, Czech Republic
* Correspondence: m.bohm@po.edu.pl; Tel.: +48-77-449-8418

**Abstract:** At present, due to advanced fatigue calculation models, it is becoming more crucial to find a reliable source for design S–N curves, especially in the case of new 3D-printed materials. Such obtained steel components are becoming very popular and are often used for important parts of dynamically loaded structures. One of the commonly used printing steels is EN 1.2709 tool steel, which has good strength properties and high abrasion resistance, and can be hardened. The research shows, however, that its fatigue strength may differ depending on the printing method, and may be characterized by a wide scatter of the fatigue life. This paper presents selected S–N curves for EN 1.2709 steel after printing with the selective laser melting method. The characteristics are compared, and conclusions are presented regarding the resistance of this material to fatigue loading, especially in the tension–compression state. A combined general mean reference and design fatigue curve is presented, which incorporates our own experimental results as well as those from the literature for the tension–compression loading state. The design curve may be implemented in the finite element method by engineers and scientists in order to calculate the fatigue life.

**Keywords:** fatigue of materials; S–N curves; 1.2709 steel; 3D-printed materials; SLM 3D printing

**Citation:** Böhm, M.; Niesłony, A.; Derda, S.; Owsiński, R.; Kepka, M., Jr.; Zetkova, I.; Zetek, M.; Houdková, Š.; Prażmowski, M. General Reference and Design S–N Curves Obtained for 1.2709 Tool Steel. *Materials* **2023**, *16*, 1823. https://doi.org/10.3390/ma16051823

Academic Editor: Feng Qiu

Received: 25 January 2023
Revised: 17 February 2023
Accepted: 20 February 2023
Published: 23 February 2023



## 1. Introduction

The proper use of S–N curve data is one of the most important steps in the process of fatigue life assessment. The reliability of the results depends on the engineer's experience, but mostly on the quality of the S–N data. Therefore, it is important to work on reliable data sets when implementing S–N curves in the fatigue life estimation process, usually together with the finite element method (FEM). The process is especially interesting when dealing with 3D-printed materials. EN 1.2709 tool steel can be found in the literature under different descriptions and names such as EOS MS1: US classification 18% Ni Maraging 300, European 1.2709 and German X3NiCoMoTi 18-9-5. It is used in the additive manufacturing process of 3D elements. EOS MS1 is delivered only as a powder. The material is praised for its good strength properties and high abrasion resistance as well as its potential to be hardened. Due to the fact that the material exists under different names, it is sometimes difficult to find the required experimental data in order to predict the fatigue life of designed components printed using it. Due to the 3D printing technology, an increasing number of mechanical elements are being designed and produced with the selective laser melting method. This method is regarded as the most promising in terms of additive manufacturing. The analysis performed within this paper consists of fatigue S–N data that were obtained for specimens produced with the use of the selective laser melting (SLM) method. The influence of different parameters was taken into account during the data analysis. Parameters such as different heat treatments and defect sizes and locations as well as the specimen smoothness and stress concentrators in the specimen shape and geometry were also described for the literature data. All data analyses were performed for the case of cyclic constant amplitude stress loading conditions. This served in the process of the comparison between the S–N curves taken from the literature and the obtained experimental curve for the tension–compression tests.

As we sorted through the literature, we noticed that papers about the fatigue life of EN 1.2709 tool steel are difficult to find, especially due to the fact that the steel can be found under different names. After noticing this, we observed that this topic is being pursued by some researchers, especially for 3D printing with the SLM technology. In this section, a number of related papers are presented, which take into account different parameters that influence the obtained experimental fatigue results; some of them also refer to the powder bed fusion (PBF) process or direct metal laser sintering (DMLS), which are equal to the SLM technology. Interesting papers that discuss the maraging steel fatigue properties for casted specimens are also mentioned. The authors mostly focus on papers that published an S–N curve. We can divide the papers into subcategories that present parameters such as:

- Different heat treatments;
- Defect size and location;
- Powder contamination;
- Specimen shape and build orientation;
- Loading type;
- Non-zero mean stress loading conditions.

As for different heat treatments, we found a large number of papers, such as the paper by Bouzakis et al. [1], where the authors additionally discussed the corrosion fatigue problem. Another group of authors led by Kostic et al. [2] analyzed the effect on heat-treated and untreated specimens under bending loading. Elangeswaran et al. [3] also focused on the heat treatment effects as well as the surface post-treatment processes that might increase the fatigue life. It can be concluded that heat treatment increases the fatigue life of specimens.

As for the defect size and location, we found the paper by Bai et al. [4], where the authors discussed the influence of hole-type defects inside the printed material. No specific S–N curve was presented, which could have been widely discussed or used in the analysis. It may be concluded that laser power has a positive effect on the surface quality, which serves in prolonging the fatigue life of specimens.

As for powder contamination, we found the paper by Gatto et al. [5], where the authors analyzed this influence on fatigue life in comparison with forged specimens. They stated that powder contamination has a great influence on the fatigue life of 3D-printed specimens in terms of its decrease. The presented experimental results were found for the case of non-zero mean stress under tension–compression.

As for the specimen shape, we found the papers by Branco et al., where the authors discussed the effect of notches on the fatigue life [6,7]. This effect of a stress concentrator in the specimen affects the fatigue life performance of SLM-printed materials. Dörfert et al. [8] discussed the fatigue of conventionally obtained and 3D-printed materials. The S–N curves showed that the conventionally obtained material had higher fatigue strength characteristics than the 3D-printed material. Fitzka et al. [9] analyzed the size effect in terms of the fatigue life for two different sizes of sheet specimens that were tested under ultrasonic frequencies obtained from a coil material. Guo et al. [10] presented a review in which they compared the fatigue results of 3D-printed specimens for notched and un-notched cases, as well as for cases using PBF printing technology. They presented an S–N curve for a variety of experimental results under tension–compression taken from the literature. Meneghetti et al. [11] presented a paper focused mostly on the effects of the build orientation of specimens and heat treatments, along with the fatigue results for the tension–compression loading state with the use of DMLS. Solberg et al. [12] discussed the effect of the directional build orientation under tension–compression for plate specimens. The comparison was performed for different build angles, and surfaces were as-built during the printing process or machined. The machined specimens had superior fatigue properties in comparison to the as-built ones. Tshabalala et al. [13] also performed a tension–compression test for cylindrical specimens with different build orientations. They observed that tool steels develop compressive mean stress during axial tests. It can be concluded that the shape, dimensions and distortion in terms of stress concentrators in the form of notches can cause a decrease in the fatigue life.

As for the loading type, we found the paper by Branco et al. [14], where the authors presented fatigue results under constant and variable amplitude loading. The same authors discussed the problem of multiaxial loading conditions concerning the studied material [15]. Crocollo et al. [16] presented experimental results for specimens under bending loading conditions. They focused on the sensitivity of the technology in regard to the build orientation. The paper by Damon et al. [17] presented tension–compression fatigue results under different temperatures and discussed the effect of the vertical and horizontal orientation of the printing procedure. Vilchez et al. [18] presented the fatigue results for maraging steel 300 in the tension–compression state under ultrasonic frequencies. The material was not printed but obtained from solid bars, but the obtained results showed the enormous possibilities in terms of the obtained fatigue life. It can be concluded that variable loads can cause a decrease in the fatigue life of specimens.

As for non-zero mean stress loading conditions, we found the paper by Chang [19] et al., where the authors presented multiple S–N curves for four different cycle asymmetry ratios, R. Their paper also strongly focused on the description of the defects that may influence the fatigue life of the components, as well as the temperature effects. Nevertheless, these tests were performed on a casted material, where the specimens were obtained from casted rods. Schuller et al. [20] presented fatigue S–N curves for two different R values under ultrasonic loading tests. It may be concluded that positive mean stresses influence the fatigue life of printed specimens by decreasing their lifetime.

Summarizing the literature review, it is found that the material has been tested by various scientists under different conditions. The shortcoming is the fact that it is difficult to formulate a "safe" design curve on the basis of individual literature results. It is important to note that the authors analyzed the fatigue results for 3D-printed specimens of this steel under different loading states inter alia for bending and tension–compression. The study presents never-before-published experimental fatigue results for 3D-printed specimens of 1.2709 tool steel under tension–compression on a fatigue test stand under the stress ratio $R = -1$. The most important part of this paper is the presentation of new experimental fatigue results as well as the presentation of a mean reference curve for the obtained experimental data and literature data as well as a reference design curve that was prepared according to the British Standard: BS 7608:1993 for the same data. There are currently no reference curves that represent a data set as large as the one presented within this paper. The design curve may be implemented in FEM by engineers and scientists in order to calculate the fatigue life in a reliable way, especially for simple objects that have been printed, e.g., plates with the dimensions reported by Van Vihn et al. [21].

## 2. Materials and Methods

The material used in the experimental investigation was EN 1.2709 tool steel. The mechanical properties of this steel are presented in Table 1. The powder of this material was used in order to print specimens, whose geometry and dimensions are presented in Figure 1. The specimens were not subject to additional machining, but they were in the state after solution annealing.

**Table 1.** Important mechanical properties of 1.2709 tool steel.

| $R_m$ MPa | $R_{0.2}$ MPa | $E$ GPa | $v$ | |
|---|---|---|---|---|
| 1000–1200 | 900–1100 | 160–200 | 0.3 | As-built |
| 1670–2230 | 1800–2000 | 160–200 | 0.3 | After age hardening |
| 1000–1200 | 800–920 | 160–200 | | After solution annealing |

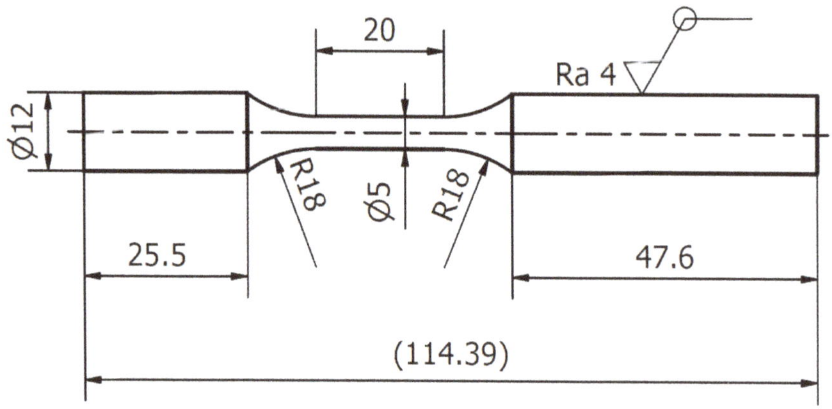


**Figure 1.** Geometry and dimensions of the specimens used in this experimental research for the tension–compression tests.

The machine used during the printing process was the EOS M290 with EOSTATE Exposure OT, MeltPool Monitoring systems and is presented in Figure 2. The default EOS parameter set was used for the printing of the samples in the Z direction (laser power of 258 W, scanning rate of 960 m/s, hatching of 0.11 mm, layer thickness of 40 micrometers, protective atmosphere of nitrogen gas). The samples were tested after solution annealing (temperature of 820 °C/1 h, slow cooling in the furnace).

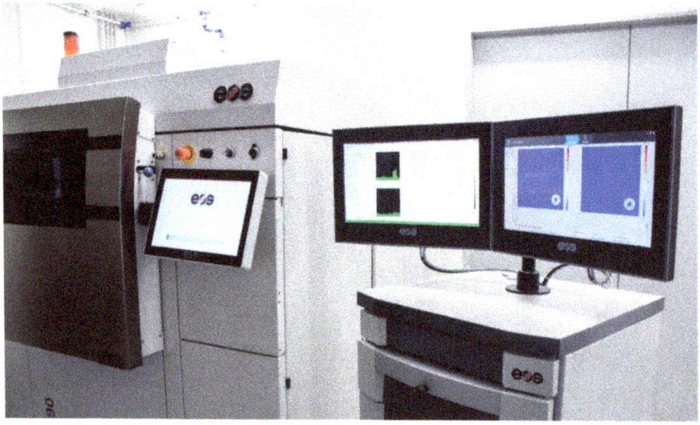

**Figure 2.** Selective laser sintering 3D printing machine EOS M290 used in the process of specimen preparation.

The specimens underwent fatigue tests under tension–compression performed on the Instron 8852 test stand presented in Figure 3. The tests were performed with a zero mean stress value. Therefore, no initial compression or tension was added. In Figure 4, a photograph of the specimens before and after the fatigue tests is shown.

**Figure 3.** Fatigue test stand for tension–compression axial loading, Instron 8852, at Opole University of Technology.

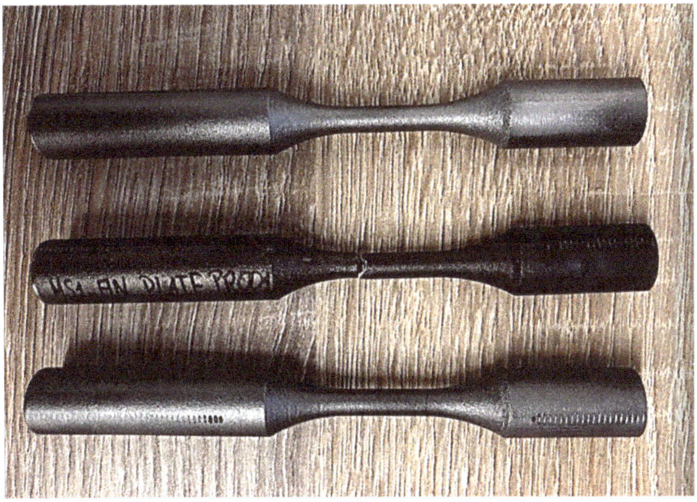

**Figure 4.** Photograph of test specimens before and after axial fatigue testing, OUTech Opole.

The basic formula used for the description of the S–N curve is

$$N = B(\sigma_a)^{-m}, \tag{1}$$

where $N$ is the number of cycles until failure, $B$ is the S–N curve constant and $m$ is the slope of the S–N curve.

An examination of the microstructure was also performed. The metallographic specimens were ground with abrasive papers or diamond pastes with decreasing gradation. Then, they were polished using an aqueous suspension of $Al_2O_3$ and etched using an etching reagent (10 mL HF, 30 mL $HNO_3$, 50 mL $H_2O$).

## 3. Results

In this section, the obtained experimental results and S–N curves are presented, as well as the S–N curves taken from the literature in order to compare other states of loading conditions. The results of the experimental fatigue tests are presented in Table 2. The microstructure of the material is presented in Figure 5. The microstructure resembles a Widmanstätten austenite structure. The Widmanstätten structure is formed in the weld after remelting and cooling. This might be the case for this material as the SLM method works on the basis of this principle. In Figure 6, the fatigue fracture surfaces of three specimens under different loading amplitude values are presented. It can be noticed that the fractures obtained using the low-cycle (MS1_AN_PLATF_5_Pr012 and MS1_AN_PLATF_5_Pr015) and high-cycle regimes (MS1_AN_PLATF_5_Pr004) in the fatigue curve have an expected fracture area of propagation and initiation on the surface of the specimens, which is the expected case for the tension–compression loading state. In Figure 7, the authors present the S–N curves obtained for the tension–compression loading state under R = −1 for smooth cylindrical specimens.

**Table 2.** Fatigue test results under tension–compression loading and R = −1.

| Specimen | $\sigma_a$, MPa | $N_f$, Cycles |
|---|---|---|
| MS1_AN_PLATF_5_Pr001 | 407 | 93,149 |
| MS1_AN_PLATF_5_Pr002 | 407 | 832,264 |
| MS1_AN_PLATF_5_Pr003 | 433 | 2,000,000 |
| MS1_AN_PLATF_5_Pr004 | 459 | 153,479 |
| MS1_AN_PLATF_5_Pr005 | 510 | 111,884 |
| MS1_AN_PLATF_5_Pr006 | 535 | 50,761 |
| MS1_AN_PLATF_5_Pr007 | 535 | 35,282 |
| MS1_AN_PLATF_5_Pr008 | 446 | 192,177 |
| MS1_AN_PLATF_5_Pr009 | 560 | 59,186 |
| MS1_AN_PLATF_5_Pr010 | 700 | 10,921 |
| MS1_AN_PLATF_5_Pr011 | 800 | 2269 |
| MS1_AN_PLATF_5_Pr012 | 800 | 1939 |
| MS1_AN_PLATF_5_Pr013 | 650 | 19,705 |
| MS1_AN_PLATF_5_Pr014 | 750 | 6555 |
| MS1_AN_PLATF_5_Pr015 | 600 | 22,496 |

Beside the experimental results of the fatigue tests, this section also presents an analysis of different S–N curves obtained for additively manufactured specimens reported by other authors. A Wöhler diagram is used to describe the fatigue life. The diagram presents the fatigue life in terms of cycles until failure $N_f$ on the x axis. The constant stress amplitude $\sigma_a$ on the y axis represents the value at which the material is loaded during the constant tests. The literature sometimes presents the S–N diagrams in relation to the maximum stress $\sigma_{max}$. In Figure 8, the experimental results of the tension–compression tests under R = −1 performed by Branco et al. [14] for specimens created with the use of the SLM technology are shown.

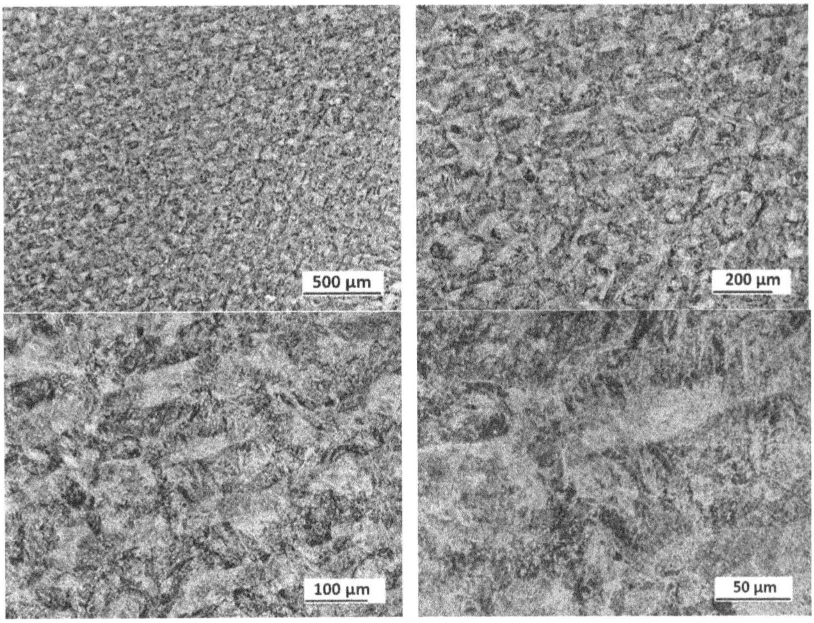


**Figure 5.** Microstructure of the 3D-printed specimens under different magnifications.

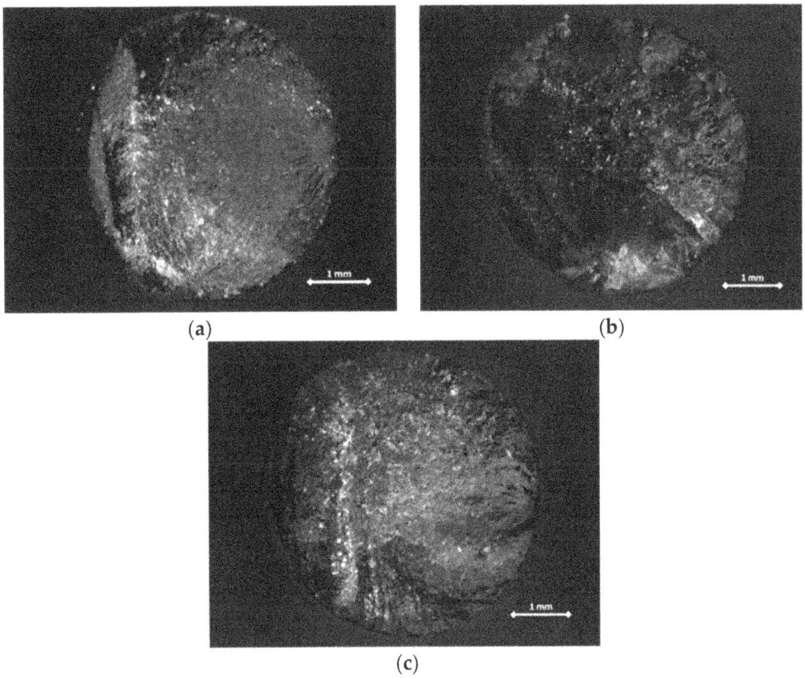


**Figure 6.** Fatigue fracture surfaces of three specimens under different loading amplitudes: (**a**) MS1_AN_PLATF_5_Pr004; (**b**) MS1_AN_PLATF_5_Pr012; (**c**) MS1_AN_PLATF_5_Pr015.

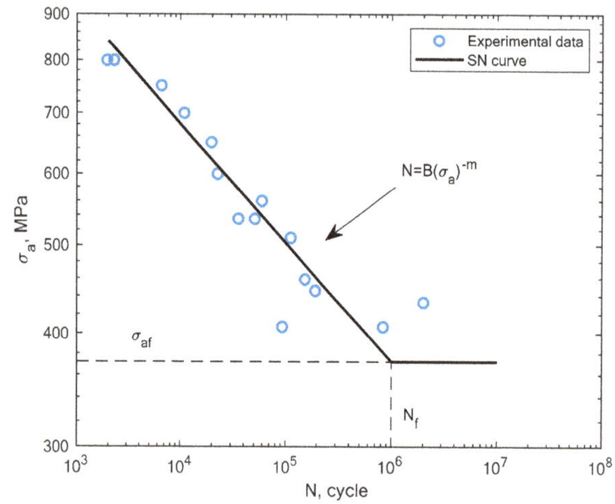


**Figure 7.** S–N curve for the experimental tension–compression fatigue test results of 1.2709 steel obtained by the authors under R = −1.

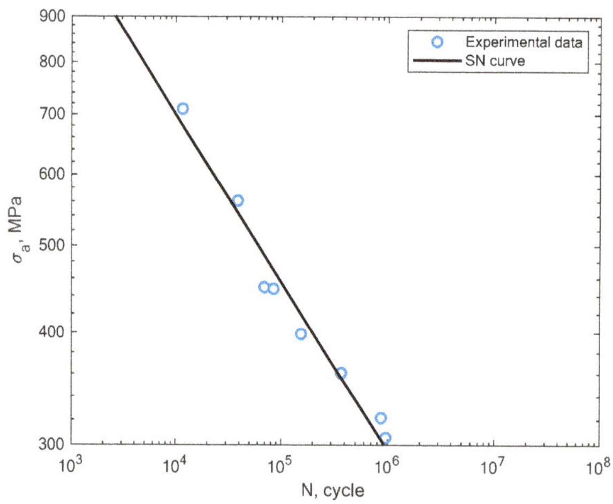


**Figure 8.** S–N curve for the experimental tension–compression fatigue test results (R = −1) reported by Branco et al. [14].

In Figure 9, the experimental results obtained in the work of Croccolo et al. for specimens created with the direct metal laser sintering method and tested under a rotating bending load (R = −1) are shown [16]. The authors tested the sensitivity of the material in terms of the build direction. In Figure 10, the experimental results obtained by Cruces et al. for the tension–compression (R = −1) of smooth specimens with hollow cylinders are shown [7]. The authors analyzed smooth specimens with two differently sized circular stress concentrators. It can be noticed that their results apply to notched specimens with 0.4 and 1 mm notches.

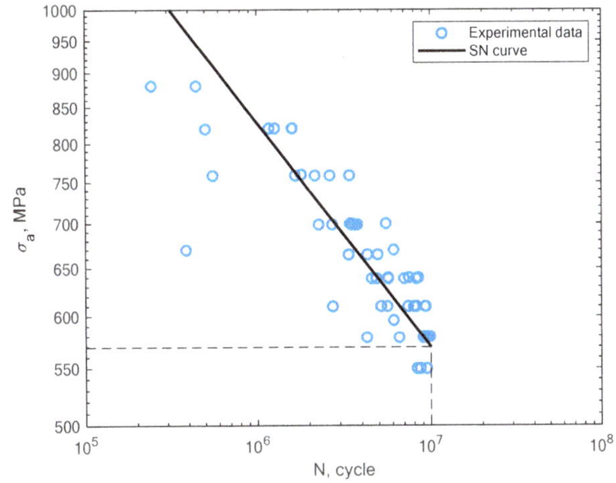


**Figure 9.** S–N curve for the experimental rotating bending fatigue test results reported by Croccolo et al. [16].

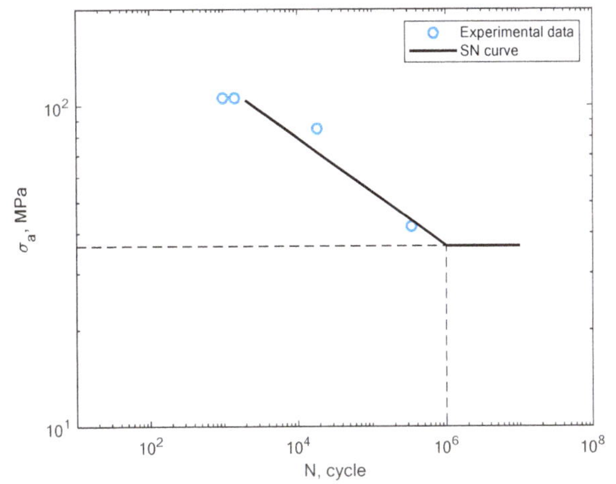


**Figure 10.** S–N curve for the experimental tension–compression fatigue test results reported by Cruces et al. [7].

In Figures 11–14, the results of the bending fatigue tests performed by Kostic et al. are presented [2]. The figures show the results for four different material conditions. The first is the condition of no heat treatment and no machining of the surface; the second is the condition of heat treatment but no machining; the third is the condition of no heat treatment and machining; and the fourth is the condition of heat treatment and machining. The experimental tension–compression S–N curves for the tests performed by Damon et al. [17] under $R = -1$ with two different temperature conditions, namely, room temperature (25 °C) and 400 °C, are shown in Figures 15–18. The specimens were created with the use of the SLM technology in the horizontal and vertical directions. In Figure 19, the case of an S–N curve with non-zero mean stress under a stress ratio of $R = 0$ is presented. In Figure 20, the case of non-zero mean stress with a value of $R = 0.55$ is presented. Both these cases apply to tension–compression loading.

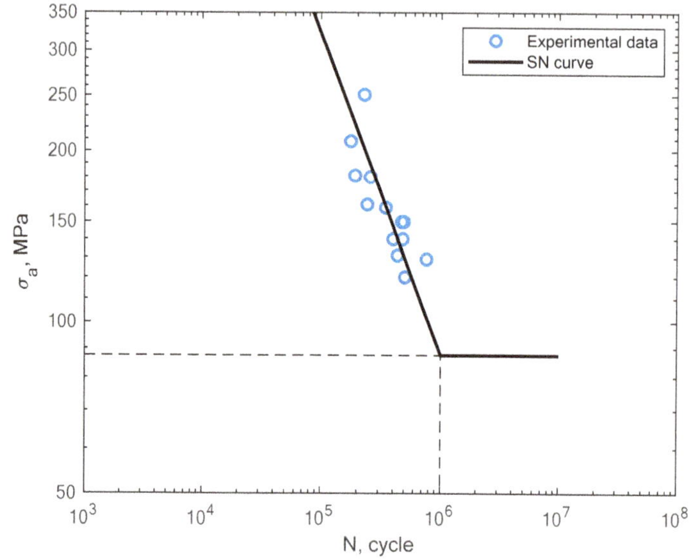


**Figure 11.** S–N curve for the experimental bending fatigue test results reported by Kostic et al. [2] for the case of no heat treatment and no machining.

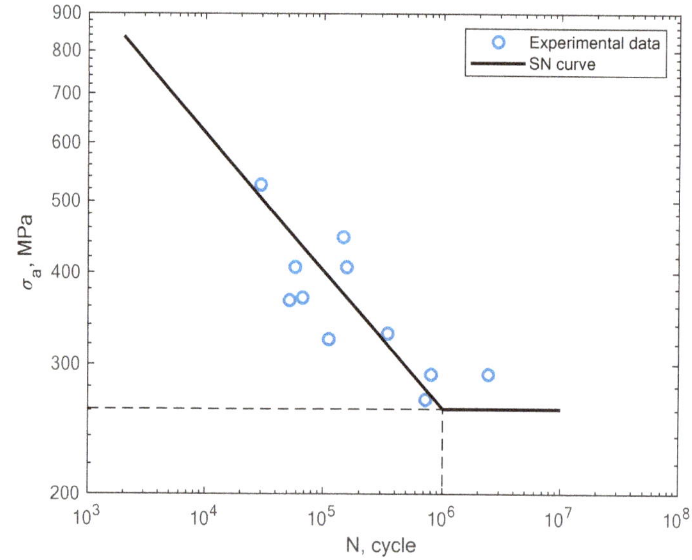


**Figure 12.** S–N curve for the experimental bending fatigue test results reported by Kostic et al. [2] for the case of heat treatment and no machining.

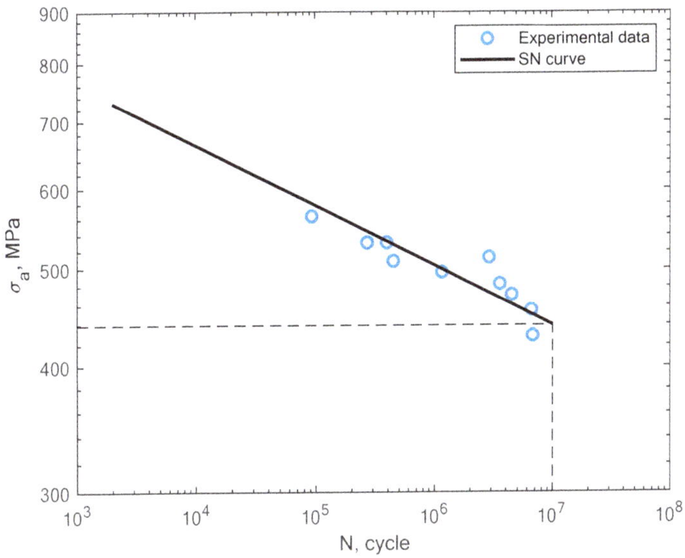


**Figure 13.** S–N curve for the experimental bending fatigue test results reported by Kostic et al. [2] for the case of no heat treatment and machining.

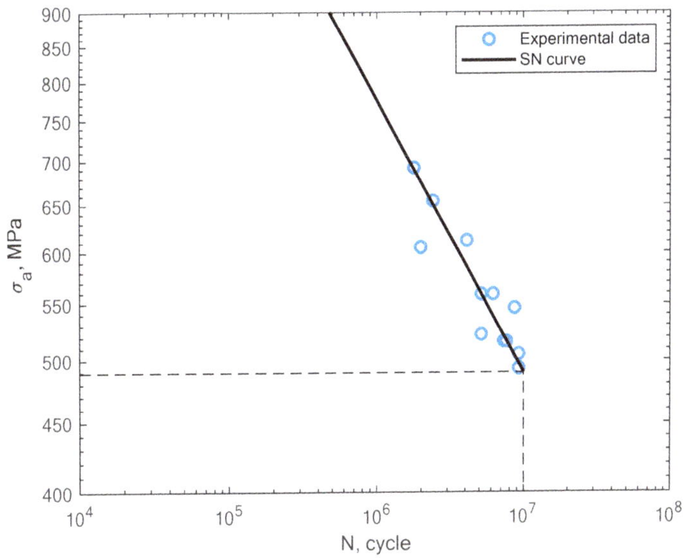


**Figure 14.** S–N curve for the experimental bending fatigue test results reported by Kostic et al. [2] for the case of heat treatment and machining.

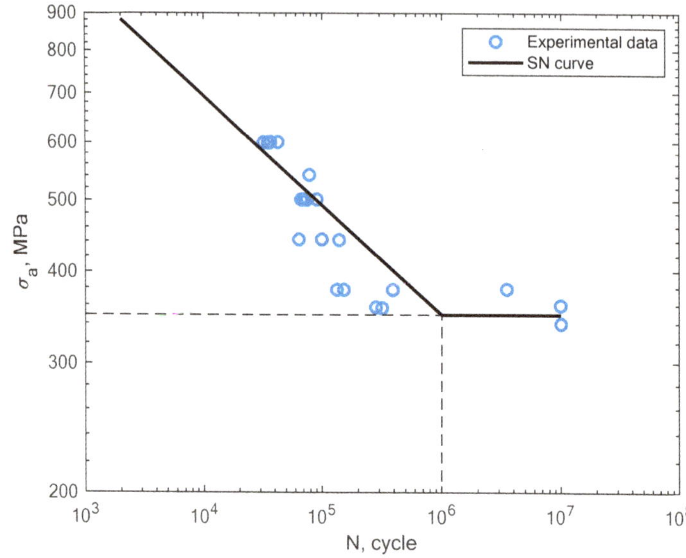


**Figure 15.** S–N curve for the experimental tension–compression (R = −1) fatigue test results reported by Damon et al. [17] under the conditions of room temperature and a horizontal direction of the printing orientation.

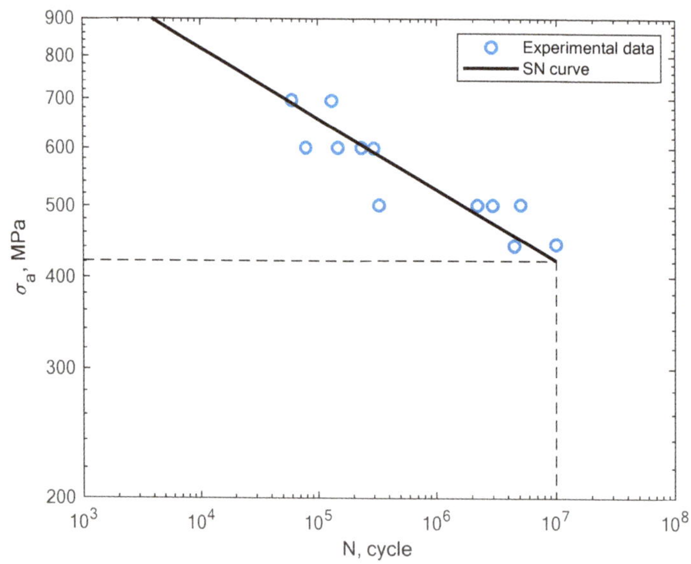


**Figure 16.** S–N curve for the experimental tension–compression (R = −1) fatigue test results reported by Damon et al. [17] under the conditions of room temperature and a vertical direction of the printing orientation.

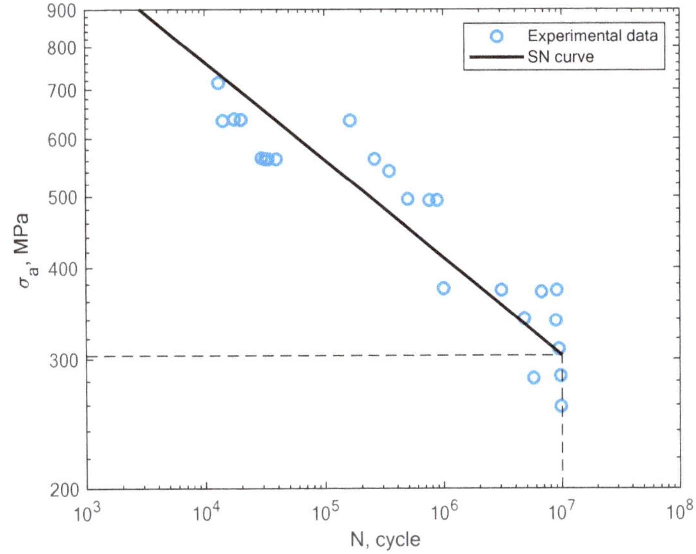


**Figure 17.** S–N curve for the experimental tension–compression (R = −1) fatigue test results reported by Damon et al. [17] under the conditions of a high temperature of 400 °C and a horizontal direction of the printing orientation.

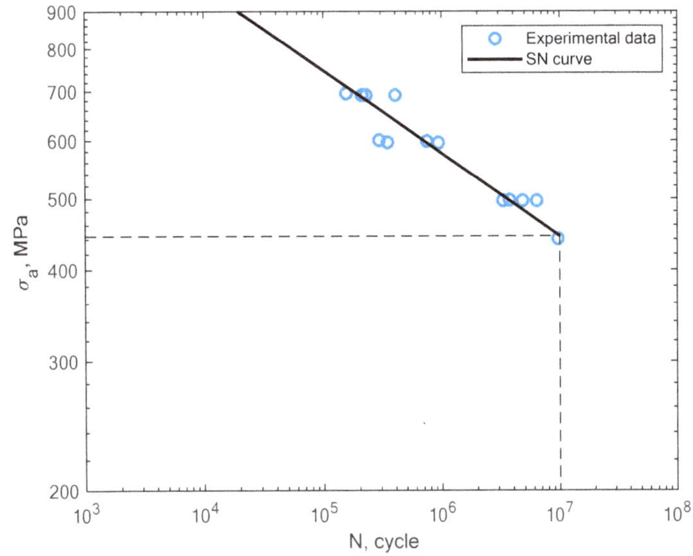


**Figure 18.** S–N curve for the experimental tension–compression (R = −1) fatigue test results reported by Damon et al. [17] under the conditions of a high temperature of 400 °C and a vertical direction of the printing orientation.

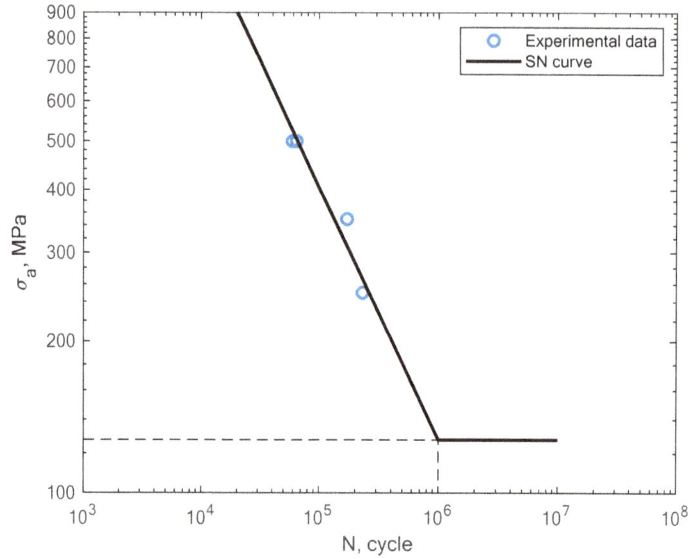


**Figure 19.** S–N curve for the experimental tension–compression (R = 0) fatigue test results reported by Gatto et al. [5] for the case of non-zero mean stress.

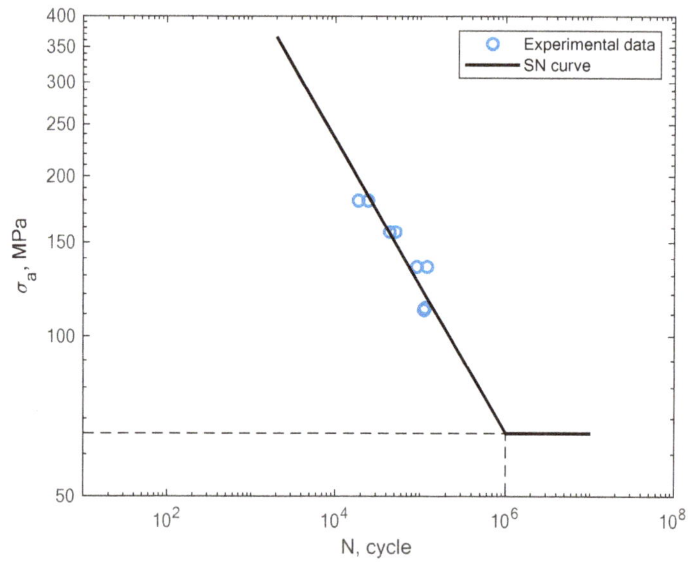


**Figure 20.** S–N curve for the experimental tension–compression (R = 0.55) fatigue test results reported by Tshabalala et al. [13] for the case of non-zero mean stress.

## 4. Discussion

In terms of practical use and the determination of parameters that might be used by engineers and designers, we can notice that there are some specific types of S–N curves that might be applicable during their praxis. Most of them incorporate simple constant amplitude tension–compression or bending states. These types of S–N curves should be divided in terms of heat treatment, loading type, specimen smoothness and stress concentrators. From the variety of S–N curves that can be found in the literature, some specific discussion points may arise. We can notice that simple fatigue tests performed under rotating bending or pure bending allow the reader to gain more information about the behavior of printed maraging steel. However, due to the fact that these results are very rare, it is currently difficult to create a reference fatigue curve for a large amount of data. On the other hand, it is possible to form such a curve for tension–compression data. On the basis of this research, including the authors' own experimental data and the literature data, we can try to create a reference fatigue curve for the stress asymmetry ratio R = −1 with no temperature effect. This type of curve can be used in the FEM calculation procedure in order to simulate simple as well as more sophisticated loading states. Such a reference curve is presented in Figure 21. The curve was calculated on the basis of the 55 experimental points taken from our own results and the literature results. For the bending or other specific loading cases, we can notice that the results may still be insufficient to present a reference fatigue curve, which would be based on the data from only one or two papers. What can be noticed while comparing the stress amplitudes in terms of the loading state is that the bending and other states have lower curves than the tension–compression S–N curves in terms of the stress amplitude values. Table 3 presents the fatigue curve values obtained in the experimental and literature results. It also presents the reference tension–compression fatigue curve data.

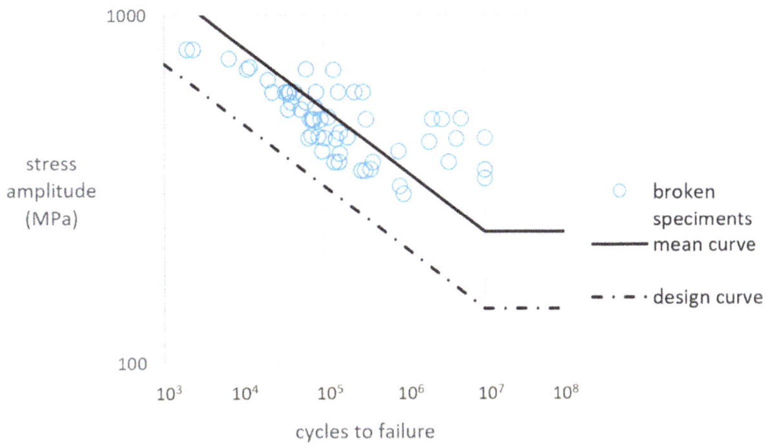


**Figure 21.** A combined reference mean and design fatigue curve on the basis of the tension–compression experimental and literature results presented in the previous section for the case of R = −1 with no high temperature effects.

**Table 3.** Fatigue test results under tension–compression loading and R = −1.

| Reference | B | m |
|---|---|---|
| Experimental results | $4.0531 \times 10^{25}$ | 7.6277 |
| Branco et al. [14] | $1.6762 \times 10^{19}$ | 5.3522 |
| Crocollo et al. [16] | $9.0667 \times 10^{23}$ | 6.1542 |
| Cruces et al. [7] | $1.8687 \times 10^{15}$ | 5.9366 |
| Kostic et al. [2], no heat treatment and no machining | $2.7356 \times 10^{9}$ | 1.7701 |
| Kostic et al. [2], heat treatment and no machining | $7.9276 \times 10^{18}$ | 5.3371 |
| Kostic et al. [2], no heat treatment and machining | $2.0597 \times 10^{51}$ | 16.7676 |
| Kostic et al. [2], heat treatment and machining | $2.8920 \times 10^{20}$ | 5.0046 |

**Table 3.** *Cont.*

| Reference | B | m |
|---|---|---|
| Damon et al. [17], room temperature and horizontal direction of the printing orientation | $1.0819 \times 10^{23}$ | 6.6992 |
| Damon et al. [17], room temperature and vertical direction of the printing orientation | $1.3550 \times 10^{34}$ | 10.3383 |
| Damon et al. [17], high temperature of 400 °C and horizontal direction of the printing orientation | $4.5591 \times 10^{25}$ | 7.5174 |
| Damon et al. [17], high temperature of 400 °C and vertical direction of the printing orientation | $2.9195 \times 10^{30}$ | 8.8621 |
| Gatto et al. [5] | $1.6320 \times 10^{10}$ | 2.0002 |
| Tshabalala [13] | $3.7732 \times 10^{12}$ | 3.6185 |
| Reference mean curve for tension–compression | $3.2076 \times 10^{20}$ | 5.6825 |
| Reference design curve with 97.7% certainty of survival | $1.7861 \times 10^{19}$ | 5.6825 |

As an addition to the mean curve based on the above data, there is a calculated design S–N curve with a certainty of survival equal to 97.7%. A lognormal distribution and therefore a relationship between the standard deviation and probability are assumed. The design fatigue curve is shifted by two standard deviations below the generic mean curve. This is carried out using a similar method to that of the British Standard [22].

## 5. Conclusions and Observations

On the basis of the obtained results, we can formulate the following conclusions and observations:

- The literature review has shown that there is a deficit in terms of design S–N curves for 1.2709 steel in 3D-printed form for both the tension–compression and bending loading states.
- On the basis of 55 experimental data points taken from the literature as well as our own experiments, general mean and design S–N curves for the state of tension–compression are presented.
- The presented reference mean and reference design fatigue curves, which were calculated according to the BS 7608:1993 standard for the tension–compression loading state under $R = -1$, may be used by engineers in their fatigue estimation of 1.2709 steel printed elements.
- The design curve has a certainty of survival equal to 97.7%.
- On the basis of the literature results, it can be noticed that the fatigue curves for the bending loading conditions are lower in comparison to the tension–compression loading state in terms of the stress amplitude values, which may be an effect of the build orientation.
- It can be noticed that the printing orientation together with the surface quality and heat treatment may influence the fatigue life of maraging steels.
- This paper serves as a reference to analyze the current state of fatigue conditions that would affect the fatigue life of 1.2709 steel used in the process of 3D printing with the use of the SLM technology.
- The next stage of this project is to present the experimental results of 3D-printed specimens with and without special surface layers that influence the durability of the material, and the authors will focus on the bending loading state in order to extend the fatigue knowledge of this material.

**Author Contributions:** Conceptualization, A.N. and M.B.; methodology, M.B. and A.N.; software, M.B.; validation, M.B. and M.K.J.; formal analysis, A.N. and M.B.; investigation, S.D.; resources, I.Z. and M.Z.; data curation, M.B. and M.K.J.; writing—original draft, M.B. and A.N.; writing—review and editing, M.B., R.O., M.K.J., S.D., I.Z., M.Z. and M.P.; visualization, M.B. and R.O.; supervision, A.N.; project administration, Š.H.; funding acquisition, Š.H.; microstructure, M.P. All authors have read and agreed to the published version of the manuscript.

**Funding:** This research was co-financed by: (1) National Science Centre, Poland, 2020/02/Y/ST8/00093 within the M-ERA NET 2 program, project DePriSS; (2) Technology Agency of the Czech Republic (TA ČR) under the EPSILON Programme within the M-Era.Net 2 co-fund call.

**Institutional Review Board Statement:** Not applicable.

**Informed Consent Statement:** Not applicable.

**Data Availability Statement:** The data presented in this study are available on request from the corresponding author.

**Conflicts of Interest:** The authors declare no conflict of interest.

## References

1. Bouzakis, E.; Arvanitidis, A.; Kazelis, F.; Maliaris, G.; Michailidis, N. Comparison of Additively Manufactured vs. Conventional Maraging Steel in Corrosion-Fatigue Performance after Various Surface Treatments. *Procedia CIRP* **2020**, *87*, 469–473. [CrossRef]
2. Ćirić-Kostić, S.; Croccolo, D.; De Agostinis, M.; Fini, S.; Olmi, G.; Paiardini, L.; Robusto, F.; Šoškić, Z.; Bogojević, N. Fatigue Response of Additively Manufactured Maraging Stainless Steel CX and Effects of Heat Treatment and Surface Finishing. *Fatigue Fract. Eng. Mater. Struct.* **2022**, *45*, 482–499. [CrossRef]
3. Elangeswaran, C.; Gurung, K.; Koch, R.; Cutolo, A.; Van Hooreweder, B. Post-Treatment Selection for Tailored Fatigue Performance of 18Ni300 Maraging Steel Manufactured by Laser Powder Bed Fusion. *Fatigue Fract. Eng. Mater. Struct.* **2020**, *43*, 2359–2375. [CrossRef]
4. Bai, Y.; Zhao, C.; Wang, D.; Wang, H. Evolution Mechanism of Surface Morphology and Internal Hole Defect of 18Ni300 Maraging Steel Fabricated by Selective Laser Melting. *J. Mater. Process. Technol.* **2022**, *299*, 117328. [CrossRef]
5. Gatto, A.; Bassoli, E.; Denti, L. Repercussions of Powder Contamination on the Fatigue Life of Additive Manufactured Maraging Steel. *Addit. Manuf.* **2018**, *24*, 13–19. [CrossRef]
6. Branco, R.; Costa, J.D.; Jesus, J.; Berto, F.; Ferreira, J.A.M.; Capela, C. Prediction of Multiaxial Fatigue Life of Notched Maraging Steel Components Manufactured by Selective Laser Melting. *Procedia Struct. Integr.* **2022**, *39*, 273–280. [CrossRef]
7. Cruces, A.S.; Exposito, A.; Branco, R.; Borrego, L.P.; Antunes, F.V.; Lopez-Crespo, P. Study of the Notch Fatigue Behaviour under Biaxial Conditions of Maraging Steel Produced by Selective Laser Melting. *Theor. Appl. Fract. Mech.* **2022**, *121*, 103469. [CrossRef]
8. Dörfert, R.; Zhang, J.; Clausen, B.; Freiße, H.; Schumacher, J.; Vollertsen, F. Comparison of the Fatigue Strength between Additively and Conventionally Fabricated Tool Steel 1.2344. *Addit. Manuf.* **2019**, *27*, 217–223. [CrossRef]
9. Fitzka, M.; Pennings, B.; Karr, U.; Schönbauer, B.; Schuller, R.; Tran, M.-D.; Mayer, H. Influence of Cycling Frequency and Testing Volume on the VHCF Properties of 18Ni Maraging Steel. *Eng. Fract. Mech.* **2019**, *216*, 106525. [CrossRef]
10. Guo, L.; Zhang, L.; Andersson, J.; Ojo, O. Additive Manufacturing of 18% Nickel Maraging Steels: Defect, Structure and Mechanical Properties: A Review. *J. Mater. Sci. Technol.* **2022**, *120*, 227–252. [CrossRef]
11. Meneghetti, G.; Rigon, D.; Cozzi, D.; Waldhauser, W.; Dabalà, M. Influence of Build Orientation on Static and Axial Fatigue Properties of Maraging Steel Specimens Produced by Additive Manufacturing. *Procedia Struct. Integr.* **2017**, *7*, 149–157. [CrossRef]
12. Solberg, K.; Hovig, E.W.; Sørby, K.; Berto, F. Directional Fatigue Behaviour of Maraging Steel Grade 300 Produced by Laser Powder Bed Fusion. *Int. J. Fatigue* **2021**, *149*, 106229. [CrossRef]
13. Tshabalala, L.; Sono, O.; Makoana, W.; Masindi, J.; Maluleke, O.; Johnston, C.; Masete, S. Axial Fatigue Behaviour of Additively Manufactured Tool Steels. *Mater. Today: Proc.* **2021**, *38*, 789–792. [CrossRef]
14. Branco, R.; Silva, J.; Ferreira, J.M.; Costa, J.D.; Capela, C.; Berto, F.; Santos, L.; Antunes, F.V. Fatigue Behaviour of Maraging Steel Samples Produced by SLM under Constant and Variable Amplitude Loading. *Procedia Struct. Integr.* **2019**, *22*, 10–16. [CrossRef]
15. Branco, R.; Costa, J.D.; Martins Ferreira, J.A.; Capela, C.; Antunes, F.V.; Macek, W. Multiaxial Fatigue Behaviour of Maraging Steel Produced by Selective Laser Melting. *Mater. Des.* **2021**, *201*, 109469. [CrossRef]
16. Croccolo, D.; De Agostinis, M.; Fini, S.; Olmi, G.; Robusto, F.; Ćirić-Kostić, S.; Morača, S.; Bogojević, N. Sensitivity of Direct Metal Laser Sintering Maraging Steel Fatigue Strength to Build Orientation and Allowance for Machining. *Fatigue Fract. Eng. Mater. Struct.* **2019**, *42*, 374–386. [CrossRef]
17. Damon, J.; Hanemann, T.; Dietrich, S.; Graf, G.; Lang, K.-H.; Schulze, V. Orientation Dependent Fatigue Performance and Mechanisms of Selective Laser Melted Maraging Steel X3NiCoMoTi18-9-5. *Int. J. Fatigue* **2019**, *127*, 395–402. [CrossRef]
18. Ruiz Vilchez, J.A.; Domínguez Almaraz, G.M.; Sánchez Miranda, M.A. Ultrasonic Fatigue Endurance of the Maraging 300 Steel. *Procedia Struct. Integr.* **2021**, *33*, 658–664. [CrossRef]

19. Chang, D.-K.; Wang, Q.-Q.; Chen, H.; Zhang, X.-C.; Wen, J.-F.; Tu, S.-T. Effects of Defect Size and Location on High Cycle Fatigue Life of a Maraging Stainless Steel at Ambient and Cryogenic Temperatures. *Int. J. Fatigue* **2022**, *161*, 106906. [CrossRef]
20. Schuller, R.; Fitzka, M.; Irrasch, D.; Tran, D.; Pennings, B.; Mayer, H. VHCF Properties of Nitrided 18Ni Maraging Steel Thin Sheets with Different Co and Ti Content. *Fatigue Fract. Eng. Mater. Struct.* **2015**, *38*, 518–527. [CrossRef]
21. Van Vinh, P.; Van Chinh, N.; Tounsi, A. Static Bending and Buckling Analysis of Bi-Directional Functionally Graded Porous Plates Using an Improved First-Order Shear Deformation Theory and FEM. *Eur. J. Mech.-A/Solids* **2022**, *96*, 104743. [CrossRef]
22. *BS 7608:1993*; BSI Knowledge-Fatigue Design and Assessment of Steel Structures. British Standards Institute: London, UK, 1993.

*materials*

MDPI

*Article*

# Accumulation of Plastic Strain at Notch Root of Steel Specimens Undergoing Asymmetric Fatigue Cycles: Analysis and Simulation

**Faezeh Hatami and Ahmad Varvani-Farahani ***

Department of Mechanical and Industrial Engineering, Toronto Metropolitan University, Toronto, ON M5B 2K3, Canada
* Correspondence: avarvani@torontomu.ca; Tel.: +1-416-979-5000 (ext. 557707)

**Abstract:** The present study evaluates the ratcheting response at notch roots of 1045 steel specimens experiencing uniaxial asymmetric fatigue cycles. Local stress and strain components at the notch root were analytically evaluated through the use of Neuber, Glinka, and Hoffman-Seeger (H-S) rules coupled with the Ahmadzadeh-Varvani (A-V) kinematic hardening model. Backstress promotion through coupled kinematic hardening model with the Hoffman-Seeger, Neuber, and Glinka rules was studied. Relaxation in local stresses on the notched samples as hysteresis loops moved forward with plastic strain accumulation during asymmetric loading cycles was observed. Local ratcheting results were simulated through FE analysis, where the Chaboche model was employed as the materials hardening rule. A consistent response of the ratcheting values was evidenced as predicted, and simulated results were compared with the measured ratcheting data.

**Keywords:** local ratcheting; A-V kinematic hardening model; backstress evolution; neuber; Hoffman-Seeger; Glinka rule; finite element analysis; Chaboche's model

**Citation:** Hatami, F.; Varvani-Farahani, A. Accumulation of Plastic Strain at Notch Root of Steel Specimens Undergoing Asymmetric Fatigue Cycles: Analysis and Simulation. *Materials* **2023**, *16*, 2153. https://doi.org/10.3390/ma16062153

Academic Editors: Grzegorz Lesiuk and Dariusz Rozumek

Received: 28 January 2023
Revised: 25 February 2023
Accepted: 3 March 2023
Published: 7 March 2023



## 1. Introduction

In the presence of stress raisers, load-bearing components are vulnerable to catastrophic failure, especially when they are subjected to asymmetric stress cycles in which the local stress state exceeds the elastic limit resulting in plastic strain accumulation referred to as local ratcheting. The presence of stress raisers intensifies the ratcheting progress during loading cycles. Investigations on how the ratcheting progress at the notch root behaves and how local stresses at the notch root relax out have been the center of attention for researchers [1–4]. Wang et al. [3] investigated steel specimens subjected to asymmetric loading cycles. They proposed an integral approach to define the plastic shakedown rate as the loading cycles proceeded. Hu et al. [4] reported that, as the applied strain increased, the local ratcheting and stress relaxation rate was further promoted at the notch root. Their measured ratcheting data and stress relaxation at the notch roots were reported in consistent agreement with those predicted by means of the Chaboche hardening rule. Rahman et al. [5] performed ratcheting tests on notched 304L steel plates with various notch geometries/shapes. They employed Chaboche, Ohno-Wang, and AbdelKarim-Ohno hardening rules to evaluate the ratcheting response of notched specimens. The ratcheting response and mean stress relaxation of S32750 steel bars were examined in a paper by Lee et al. [6] under the step-loading spectra. They developed a constitutive model to predict the mean stress relaxation and ratcheting, whose accuracy was confirmed through uniaxial loading tests. Strains were measured at the notch root of 1070 steel specimens undergoing axial-torsional loading cycles by Firat [7]. He employed the Chaboche model and Neuber's rule to evaluate progressive plastic strains over cycles. The predicted local ratcheting strain values at the notch root of 1070 steel specimens were found in close agreement with those measured values reported in reference [8]. Kolahsangiani and Shekarian [9,10] examined local ratcheting at the notch root of 1045 steel plate specimens with various circular notch sizes. They discussed the influence of the notch size and stress levels on the local ratcheting magnitude and rate as the stress cycles increased. The coupled A-V hardening model [1] with the Neuber rule [11] was utilized to study the local ratcheting and stress relaxation at the notch vicinity of 1045 steel specimens [12]. The coupled framework governed the progressive plastic strain and stress relaxation at the notch root concurrently. Liu et al. [13] carried out some experimental tests with a high-stress concentration on Z2CND18.12N austenitic stainless steel elbow pipes. The measured local strains through strain gauges mounted on the circumference of pressurized elbow pipes closely agreed with the predicted ratcheting by means of the Chen-Jiao-Kim (CJK) model [14]. Ratcheting progress over loading cycles as well as stress relaxation at the notch roots of steel plates at constant strain ranges, was evaluated by Shekarian et al. [15]. They coupled the kinematic hardening rules of Chaboche and A-V with the Neuber rule to estimate local ratcheting at the notch roots. They found a close agreement between the predicted and measured values at various notch sizes and load levels. In another study, ratcheting at the roots of different elliptical and circular notches in 316 stainless steel specimens [16] was evaluated by means of the A-V and Chaboche hardening models in conjunction with the Neuber rule. Local ratcheting results were discussed on the basis of the choice of hardening models and their related influential parameters.

The present study evaluated the measured local ratcheting response of 1045 steel specimens [17–19] at the notch root employing the A-V hardening rule coupled with the Neuber, Glinka, and H-S models. Local ratcheting strains at the notch roots were calculated through the hardening framework during asymmetric loading cycles. Backstress evolution through the A-V model directly affected the local ratcheting rate and magnitude as different models were employed. The H-S model with equivalent stress terms presented a slightly sharper drop in stress at the notch root as loading cycles were applied. This model, however, resulted in lower ratcheting as coupled with the A-V hardening model. The corresponding backstress terms in this model affected the ratcheting rate and magnitude as the H-S rule was coupled in the framework. Numerically determined local ratcheting strains at notch root by means of the FE analysis fell below the experimental and predicted results. Simulated results were affected by the FE element size and types taken at different distances at the notch root vicinity and their related convergence. The choice of models in assessing the local strain components and their terms/constants was found to affect the rate and magnitude of ratcheting and stress relaxation during stress cycles.

## 2. Modeling and Formulation

### 2.1. Elastic and Plastic Strains

The total strain increment tensor was determined through the summation of elastic and plastic strain increments tensors as:

$$d\bar{\varepsilon} = d\bar{\varepsilon}_e + d\bar{\varepsilon}_p \tag{1}$$

While the elastic strain increment tensor was determined by Hooke's law as:

$$d\bar{\varepsilon}_e = \frac{d\bar{\sigma}}{2G} - \frac{\vartheta}{E}(d\bar{\sigma}.\,\bar{I})\bar{I} \tag{2}$$

where $E$ is the modulus of elasticity, $G$ is the shear modulus, $\vartheta$ is Poisson's ratio, and terms $\bar{I}$ and $\bar{\sigma}$ correspond to unit and stress tensors, respectively.

Based on the associated flow rule, the plastic strain increment tensor can be defined as:

$$d\bar{\varepsilon}_p = \frac{1}{H_p}(d\bar{s}.\,\bar{n})\bar{n} \tag{3}$$

where $H_p$ is the plastic modulus, $d\bar{s}$ is the increment of the deviatoric tensor, and $\bar{n}$ corresponds to the normal vector of the yield surface. The yield criterion represents the onset of yielding where the yield contour separates the elastic domain from the plastic region through:

$$f\left(\bar{s}, \alpha, \sigma_y\right) = \frac{3}{2}(\bar{s} - \bar{\alpha})(\bar{s} - \bar{\alpha}) - \sigma_y^2 \quad (4)$$

In Equation (4), $\bar{\alpha}$ corresponds to the backstress tensor, translating the yield surface within the deviatoric stress space as the loading exceeds an elastic domain.

### 2.2. The Ahmadzadeh-Varvani (A-V) Kinematic Hardening Rule

During plastic deformation, the movement direction of the yield surface in the stress space is governed by the kinematic hardening rule. The A-V nonlinear hardening model [1] was structured to control the evolution of backstress increments over the loading process. The general form of the A-V rule is given as:

$$d\bar{\alpha} = Cd\bar{\varepsilon}_p - \gamma_1\left(\bar{\alpha} - \delta\bar{b}\right)dp \quad (5a)$$

$$d\bar{b} = \gamma_2\left(\bar{\alpha} - \bar{b}\right)dp \quad (5b)$$

The first term of the A-V model corresponds to strain hardening, and the second term presents the dynamic recovery term to accommodate plastic strain accumulation. The internal variable, $\bar{b}$, with a zero initial value, was introduced to the dynamic recovery term of the hardening rule to gradually control backstress $\bar{\alpha}$ over loading cycles. Details of how to determine variables in Equations (5a) and (5b) were given in Ref. [1]. In Equations (5a) and (5b), $dp$ is defined through the dot product of plastic strain increment $d\bar{\varepsilon}_p$ as:

$$dp = \sqrt{d\bar{\varepsilon}_p . d\bar{\varepsilon}_p} \quad (6)$$

Coefficients C and $\gamma_1$ in Equation (5) are defined from uniaxial stress-strain hysteresis loops. Constants $\gamma_2$ and $\delta$ are material-dependent coefficients [17]. For uniaxial loading conditions, $\delta$ is defined as $(\bar{\alpha}/k)^m$ expanding Equation (5a) to:

$$d\bar{\alpha} = Cd\bar{\varepsilon}_p - \gamma_1\left(\bar{\alpha} - (\bar{\alpha}/k)^m\,\bar{b}\right)dp \quad (7)$$

where the coefficient $k$ is defined as $k = C/\gamma_1$. Exponent m is material dependent and stays less than the unity $0 < m < 1.0$.

### 2.3. Local Components of Stress and Strain at Notch Root and Local Ratcheting Strains

#### 2.3.1. Neuber's Rule

Neuber's rule relates the stress and strain concentration factors, $K_\sigma$, and $K_\varepsilon$ to the theoretical stress concentration factor, $K_t$ through:

$$K_\sigma K_\varepsilon = K_t^2 \quad (8)$$

where:

$$K_\sigma = \frac{\sigma}{S} \quad (9)$$

$$K_\varepsilon = \frac{\varepsilon}{e} \quad (10)$$

Substituting Equations (9) and (10) into Equation (8) resulted in:

$$\left(K_t S\right)^2 = E\sigma\varepsilon \quad (11)$$

For the cyclic stress and strain ranges, Equation (11) is rewritten as:

$$\frac{\left(K_t \Delta S\right)^2}{E} = \Delta\sigma\Delta\varepsilon \quad (12)$$

where $S$ is the nominal stress, $e$ is the nominal strain, $E$ is the modulus of elasticity, $\sigma$, and $\varepsilon$ are the local stress and strain at the notch root, respectively. To predict the local stress and strain at the notch root, Neuber's rule [11] was developed for plane stress conditions as [9,10,12,15,16,18]:

$$(\varepsilon_{BL}-\varepsilon_{AL})(\sigma_{BL}-\sigma_{AL})=K_t^2(S_B-S_A)(e_B-e_A)\ \ d\bar{\varepsilon}_p<0 \tag{13}$$

$$(\varepsilon_{CL}-\varepsilon_{BL})(\sigma_{CL}-\sigma_{BL})=K_t^2(S_C-S_B)(e_C-e_B)\ \ d\bar{\varepsilon}_p\geq 0 \tag{14}$$

where subscripts $A$, $B$, $C$ correspond to the loading turning points starting from zero to the maximum load (point $A$), minimum load (point $B$), and maximum load (point $C$), and $K_t$ is the stress concentration factor. The uniaxial nominal strain and stress range are related through the Ramberg-Osgood equation as:

$$\Delta e=\frac{\Delta S}{E}+2\left(\frac{\Delta S}{2K\prime}\right)^{\frac{1}{n\prime}} \tag{15}$$

where $K\prime$ and $n\prime$ correspond to the cyclic hardening coefficient and exponent, respectively.

Substituting Equation (15) into Equations (13) and (14) resulted in Equations (16) and (17) being related to the local strain and stress components as [9,10,12,15,16,18]:

$$(\varepsilon_B-\varepsilon_A)(\sigma_B-\sigma_A)=K_t^2(S_B-S_A)\left(\frac{S_B-S_A}{E}+2\left(\frac{S_B-S_A}{2K\prime}\right)^{\frac{1}{n\prime}}\right),\ d\bar{\varepsilon}_p<0 \tag{16}$$

$$(\varepsilon_C-\varepsilon_B)(\sigma_C-\sigma_B)=K_t^2(S_C-S_B)\left(\frac{S_C-S_B}{E}+2\left(\frac{S_C-S_B}{2K\prime}\right)^{\frac{1}{n\prime}}\right),\ d\bar{\varepsilon}_p\geq 0 \tag{17}$$

For components subjected to uniaxial loading, the local backstress component for loading the half-cycle could be related to local stress at turning points [3]. It can be used for unloading (A→B) and reloading conditions (B→C) through Equations (18)–(20):

$$\alpha_{AL}=\frac{2}{3}(\sigma_{AL}-\sigma_y) \tag{18}$$

$$\alpha_{BL}=\frac{2}{3}(\sigma_{BL}+\sigma_y) \tag{19}$$

$$\alpha_{CL}=\frac{2}{3}(\sigma_{CL}-\sigma_y) \tag{20}$$

2.3.2. Glinka's Rule

Molski and Glinka [20] presented an alternative to Neuber's rule, which was based on the equivalent strain energy density (ESED). This method took the strain energy density at the notch root equal to a condition at which the loaded specimen stayed within the elastic domain. They attributed the stress concentration factor, $K_t$ to the strain energy through:

$$K_t=\frac{\sigma}{S}=\left(\frac{W_\sigma}{W_S}\right)^{1/2} \tag{21}$$

where terms $W_S$ and $W_\sigma$ corresponded to the elastic strain energy per unit volume due to the nominal remote stress $S$ and the strain energy per unit volume due to the local strain and stress at the notch root, respectively. For the notched specimen, the elastic and total strain energy per unit volume were determined by:

$$W_S=\frac{1}{2}S.e=S^2/2E \tag{22}$$

and

$$W_\sigma = \int_0^\varepsilon \sigma(\varepsilon)d\varepsilon = \frac{\sigma^2}{2E} + \frac{\sigma}{n'+1}\left(\frac{\sigma}{K'}\right)^{\frac{1}{n'}} \tag{23}$$

where, through Hooke's law, the elastic strain and stress is related by $\sigma(\varepsilon) = E\varepsilon$. By replacing terms $W_S$ and $W_\sigma$ from Equations (22) and (23) with Equation (21), the stress concentration factor $K_t$ could be rewritten as:

$$K_t = \left\{ \frac{\left( \frac{\sigma^2}{2E} + \frac{\sigma}{n'+1}\left(\frac{\sigma}{K'}\right)^{\frac{1}{n'}} \right)}{\frac{S^2}{2E}} \right\}^{1/2} \tag{24}$$

Through the use of Equation (24), the relationship between the applied and local stress ranges could be determined as:

$$\frac{(K_t\Delta S)^2}{4E} = \frac{\Delta\sigma^2}{4E} + \frac{\Delta\sigma}{n'+1}\left(\frac{\Delta\sigma}{2K'}\right)^{\frac{1}{n'}} \tag{25}$$

To relate the stress and strain terms, the Ramberg-Osgood equation was adapted along with Equation (25) as:

$$\Delta\varepsilon(2E(\Delta\sigma)) + (n'-1)\Delta\sigma^2 = K_t^2(\Delta S)^2(n'+1) \tag{26}$$

Considering subscripts turning points A, B, C over A→B and B→C loading paths, Equation (26) was expended as:

$$(\varepsilon_B - \varepsilon_A)(2E(\sigma_B - \sigma_A)) + (n'-1)(\sigma_B - \sigma_A)^2 = K_t^2(S_B - S_A)^2(n'+1) \;\; d\bar{\varepsilon}_p < 0 \tag{27}$$

$$(\varepsilon_C - \varepsilon_B)(2E(\sigma_C - \sigma_B)) + (n'-1)(\sigma_C - \sigma_B)^2 = K_t^2(S_C - S_B)^2(n'+1) \;\; \bar{\varepsilon}_p \geq 0 \tag{28}$$

### 2.3.3. Hoffman and Seeger (H-S) Approach

Hoffman and Seeger (H-S) [21,22] proposed a method to establish a load-equivalent notch stress and strain relationship. The H-S method consisted of two steps: (i) uniaxial quantities $\sigma$, $\varepsilon$, and $K_t$ were initially replaced by the equivalent quantities ($\sigma_q$, $\varepsilon_q$, and $K_{tq}$) on the basis of the von-Mises yield criterion; (ii) the equivalent values were related to the principal stress and strain components at the notch root. Within the elastic limit, the general form of the H-S model was presented as:

$$\varepsilon_q = \frac{\sigma_q}{E} F\left(\frac{\sigma_{e,q}}{\sigma_q}\right) \tag{29}$$

where $\varepsilon_q$ and $\sigma_q$ respectively correspond to the equivalent strain and stress components at notch root. Function $F\left(\frac{\sigma_{e,q}}{\sigma_q}\right)$ falls between $1 \leq \frac{\sigma_{e,q}}{\sigma_q} < K_p$. Term $K_p$ corresponds to the limit load factor, which is the ratio of the ultimate load $L_p$ to the yield initiation load $L_y$ for elastic-perfectly plastic material.

$$K_p = \frac{L_p}{L_y} \tag{30}$$

Using the von-Mises yield criterion, the theoretical elastic equivalent stress at notch root, $\sigma_{eq}$, can be defined as:

$$\sigma_{eq} = \sigma_{e1}\sqrt{\frac{1}{2}\left[(1-a_e)^2 + (1-b_e)^2 + (a_e - b_e)^2\right]} \tag{31}$$

Subscripts 1, 2, and 3 denote principal stress directions, and subscripts $e$ represent stress levels within the elastic domain. Stress ratios within the elastic domain $a_e$ and $b_e$ are defined as:

$$a_e = \frac{\sigma_{e2}}{\sigma_{e1}} \tag{32}$$

$$b_e = \frac{\sigma_{e3}}{\sigma_{e1}} \tag{33}$$

The equivalent stress concentration $K_{tq}$ is defined as:

$$K_{tq} = \frac{\sigma_{eq}}{S} \tag{34}$$

The relationship between $K_{tq}$ and $K_t$ is given as [21]:

$$K_{tq} = K_t\sqrt{\frac{1}{2}\left[(1-a_e)^2 + (1-b_e)^2 + (a_e - b_e)^2\right]} \tag{35}$$

The H-S approach [21] was developed based on the equivalent strain at notch root through:

$$\varepsilon_q = \frac{K_{tq}^2 S^2}{\sigma_q} \frac{e^*}{S^*} \tag{36}$$

where $\sigma_q$ is the equivalent stress obtained through the von-Mises yield criterion. Terms $S^*$ and $e^*$ are respectively defined as $S^* = (K_{tq}/K_p)S$ and $e^* = (\sigma_y/E)(S^*/\sigma_y)^{\frac{1}{n}}$.

The equivalent applied stress and strain terms are then related to the local components over the unloading (A→B) path and reloading (B→C) path for each stress cycle through Equations (37) and (38), respectively.

$$(\varepsilon_{qB} - \varepsilon_{qA})(\sigma_{qB} - \sigma_{qA}) = \frac{K_{tq}^2(S_B - S_A)}{(S_B^* - S_A^*)}(e_B^* - e_A^*)\ \ d\bar{\varepsilon}_p < 0 \tag{37}$$

$$(\varepsilon_{qC} - \varepsilon_{qB})(\sigma_{qC} - \sigma_{qB}) = \frac{K_{tq}^2(S_C - S_B)}{(S_C^* - S_B^*)}(e_C^* - e_B^*)\ \ d\bar{\varepsilon}_p \geq 0 \tag{38}$$

### 2.4. *Ratcheting Analysis Algorithm*

An algorithm was developed to predict the ratcheting and stress relaxation of notched 1045 steel specimens through the coupled hardening framework. Through the A-V hardening rule, the yield surface evolution was controlled within the plastic domain and backstress term $\left(\bar{a} - \delta\bar{b}\right)$ and dropped up to a steady-state condition as asymmetric loading cycles progressed. To evaluate local ratcheting, local cyclic stress and strain components at the notch root were calculated by coupling Neuber, Glinka, and H-S rules to the hardening framework. The algorithm program enabled an assessment of local ratcheting at constant stress cycles, and stress relaxation was monitored over asymmetric loading cycles at a given constant strain. Backstress $\bar{\alpha}$ and internal variable $\bar{b}$ were to control the increments of plastic strain $d\bar{\varepsilon}_p$. The algorithm to run the ratcheting program through the hardening rule framework was developed through a number of steps:

(i) Applied cyclic stresses to the notched specimens were introduced into the program,
(ii) Through Equations (5a) and (5b), the backstress component $\bar{\alpha}$, internal variable $\bar{b}$, and term $\left(\bar{a} - \delta\bar{b}\right)$ were related to plastic strain increments over the loading progress,
(iii) The plastic strain increment, $d\bar{\varepsilon}_p$, was computed through (i) Equations (16) and (17) based on Neuber's rule, (ii) Equations (27) and (28) based on the Glinka approach, and (iii) Equations (37) and (38) by means of the H-S model.

(iv) The accumulation of the progressive local plastic strain at the notch root, $d\bar{\varepsilon}_p$ was controlled through the A-V hardening model while Neuber, Glinka, and H-S rules were coupled to the hardening framework.
(v) Through Equations (18)–(20), the backstress components were defined during unloading ($A \rightarrow B$)/reloading ($B \rightarrow C$) paths and set as equal to their counterpart increments in Equations (5a)–(7). This enabled us to set relationships between nominal and local stress components in the coupled framework.
(vi) The ratcheting strain was calculated from the average of maximum and minimum local strains over asymmetric loading cycles.

## 3. Testing Conditions and Ratcheting Data

Ratcheting data sets on notched 1045 steel specimens were taken from an earlier article conducted by Varvani and coworkers [18]. Local strain data were measured in the vicinity of the notch roots through the use of strain gauges. Strain gauges were mounted to make an approximate distance of 0.5 mm from the grid circuit edge of the strain gauge to the notch root [19]. Asymmetric cyclic tests were conducted on rectangular specimens with dimensions of 100 × 50 × 3 mm. Specimens with different central notch diameters between 9 mm and 15 mm were cyclically tested with a Zwick/Roell HB 100 servo-hydraulic machine. Figure 1 illustrates a drawing of the specimen with a notch diameter of 15 mm. Experiments were conducted under stress-controlled conditions with a stress ratio of R = 0, an asymmetric loading frequency of 0.5 Hz, and at room temperature. Details of ratcheting tests, including the notch diameter, $D$, stress concentration factor, $K_t$, and nominal stress level $S_m \pm S_a$ applied on notched specimens, are listed in Table 1. Figure 2 presents the measured ratcheting strains at the notch root of 1045 steel specimens undergoing asymmetric loading cycles for different notch diameters and various stress levels.

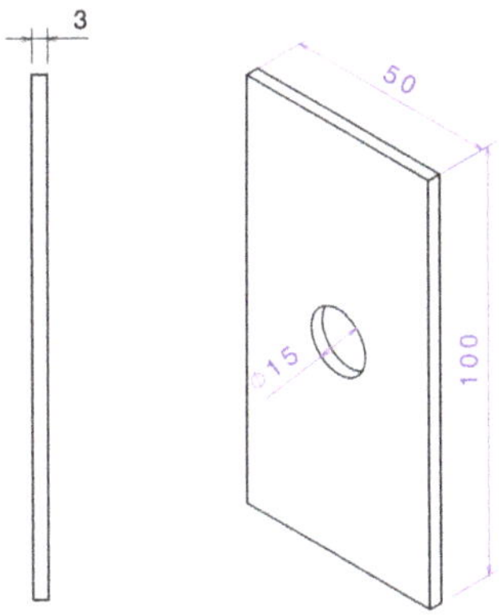


**Figure 1.** Drawing of the specimen with 15 mm notch diameter (dimensions are in mm).

**Table 1.** Loading condition for 1045 steel specimens with different notch diameters [18].

| Test Specimen | Notch Diameter (D) (mm) | $K_t$ | $S_m \pm S_a$(MPa) |
|---|---|---|---|
| S3 | 9 | 2.53 | 105 ± 105 |
| S7 | 9 | 2.53 | 130 ± 130 |
| S11 | 9 | 2.53 | 155 ± 155 |
| S16 | 9 | 2.53 | 203 ± 203 |
| S4 | 15 | 2.36 | 124 ± 124 |
| S8 | 15 | 2.36 | 152 ± 152 |
| S12 | 15 | 2.36 | 181 ± 181 |

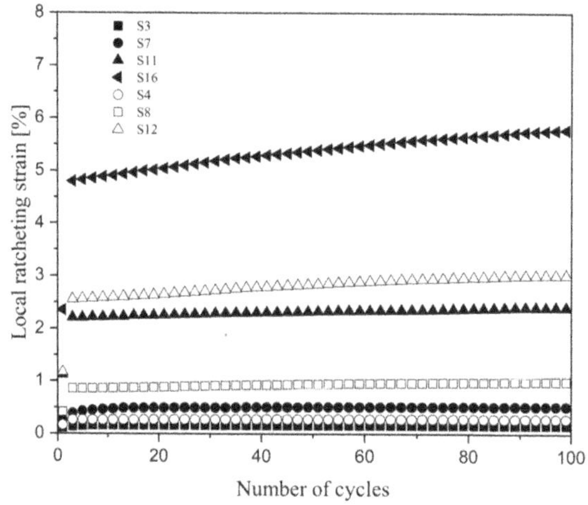


**Figure 2.** Ratcheting data collected at notch roots of 1045 steel specimens with different notch diameters and stress levels [18].

Figure 3 compares the measured stress-strain hysteresis loops and local maximum strain data at the notch root of a typical 1045 steel specimen with the loops and local strains generated through analysis. Different coefficients C, $\gamma_1$, and $\gamma_2$ in Figure 3a–c were implemented through several trials to achieve a consistency condition. Figure 3c presents a set of coefficients C = 50,000 MPa, $\gamma_1$= 350, and $\gamma_2 = 10$ representing a close agreement between the measured and predicted loops, while different values of coefficients in Figure 3a,b resulted in a noticeable difference within the measured loop. Figure 3d plots measured and predicted maximum strain values at the notch root versus loading cycles. The coefficient $\gamma_2 = 10$ resulted in a great agreement between the measured ratcheting data and the predicted curve. The predicted curve position below and above the experimental data for coefficients were $\gamma_2 > 10$ and $\gamma_2 < 10$, respectively.

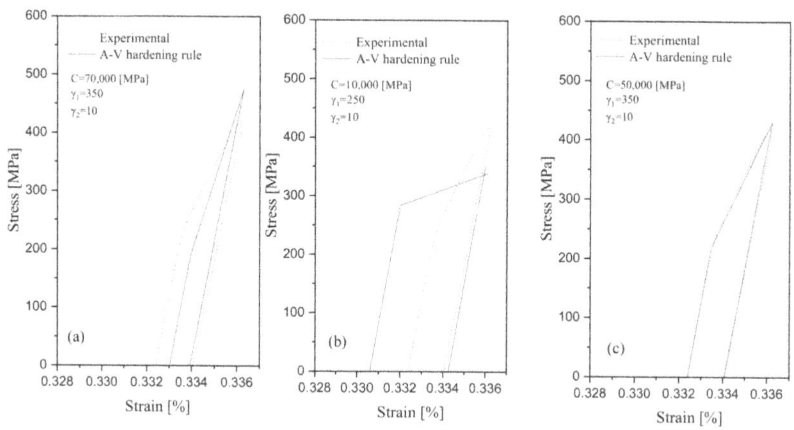


**Figure 3.** *Cont.*

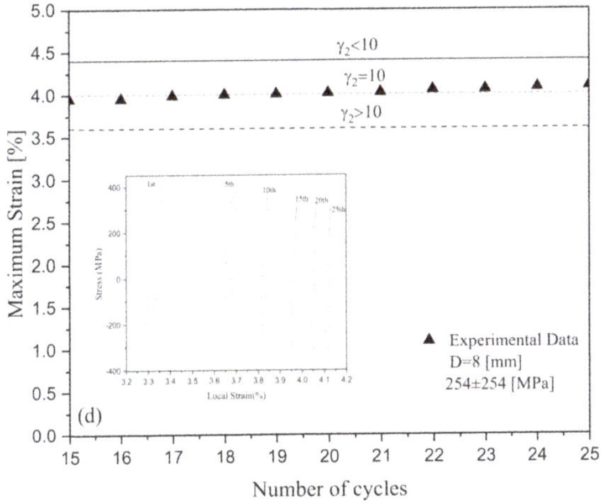


**Figure 3.** Determination of coefficients C, $\gamma_1$, and $\gamma_2$ to achieve consistency condition based on the A-V model. (**a–c**) several trials of coefficients C, $\gamma_1$, and $\gamma_2$ to achieve a consistency condition, and (**d**) measured and predicted maximum strain values at the notch root versus loading cycles.

The measured hysteresis loops in Figure 3d present stress relaxation as the number of stress cycles progressed. A drop in the width of loops with asymmetric loading cycles in this figure verifies the cyclic hardening phenomenon at the notch root of 1045 steel specimen.

## 4. Simulation of Local Ratcheting Strain through Finite Element Analysis

The finite element software ABAQUS version 6.13 [23] was used to simulate the local ratcheting response of steel specimens. Figure 4 shows the meshed specimen undergoing an axial load and its constraints surrounding the circular notch with quadratic elements. Elements were extended in size from the notch root to a distance of 1 mm over the *X*-direction with a mesh size increment of 0.15 mm. The smaller elements were taken at the vicinity of the notch root to achieve a realistic strain/stress comparable with the Neuber, Glinka, and H-S models. The gradual increase in the element size from the notch root enabled them to achieve a better assessment of the strain distribution throughout the modeling process. Translational and rotational axes of the lower end surface of the meshed specimen were constrained (along the *X*- and *Z*-axes) through adapted fixed supports, and the specimen was allowed to take the load along the *Y*-axis. The upper-end surface of the specimen was fixed. The axial load cycles were applied to the lower end of the specimen under a stress-controlled condition with a testing frequency of 0.5 Hz.

The total number of quadratic elements of type C3D8R for the samples with notch diameters of 9 mm and 15 mm were taken, respectively, at 24,306 and 23,748. The former consisted of 29,952 nodes, and the latter possessed 29,670 nodes, respectively. Elements were featured with twenty-four degrees of freedom and with three degrees of freedom per node (eight nodes for each quadratic element). The smallest size of 0.15 mm at the notch root resulted in a consistent convergence as the FE program was run at different applied stress levels and notch sizes. Convergence was consistently achieved for element sizes ranging between 0.15 mm and 0.40 mm, as the program was run for samples during the first hundred loading cycles. For this element, the range size ratcheting at the vicinity of the notch root stayed nearly constant, as presented in Figure 5. In this figure, as elements increased in size beyond 0.40 mm, local ratcheting resulted in decay at different stress levels.

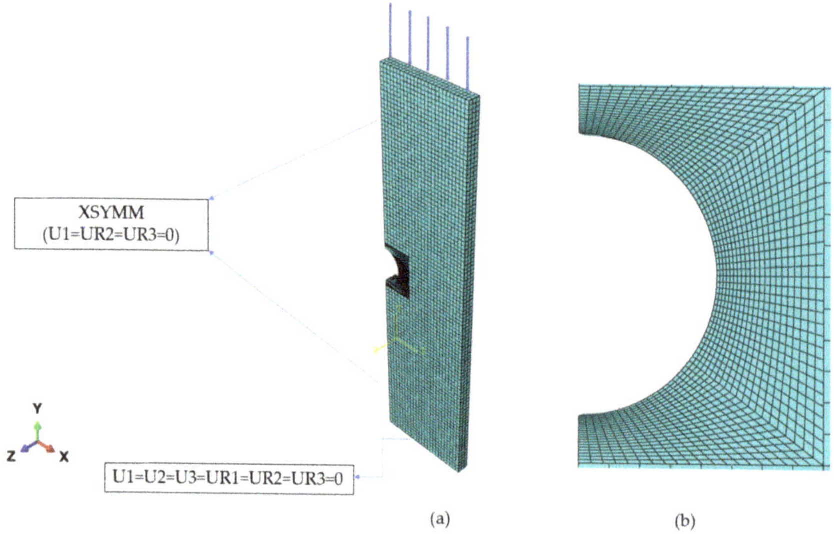


**Figure 4.** (**a**) Meshed specimen loaded axially along *Y*-axis and boundary conditions, (**b**) Meshing at the vicinity of notch root through quadratic elements.

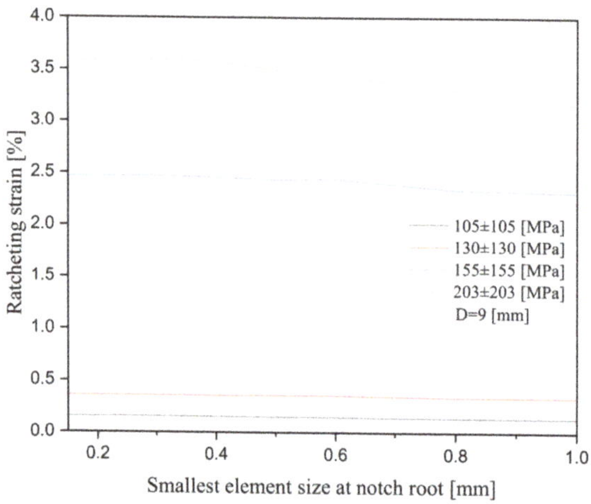


**Figure 5.** Convergence of ratcheting strain at notch root of the 1045 steel specimen (D = 9 mm) versus the quadratic mesh size at different stress levels for a given hysteresis loop.

The simulation of ratcheting at the notch root was conducted on the basis of the elastic-plastic materials kinematic hardening model of Chaboche [24]. Based on Chaboche's non-linear model, the yield surface was translated in the deviatoric stress space as the materials were deformed beyond the elastic limit. The yield surface translation was described based on Chaboche's postulation as backstress increments were integrated through:

$$d\overline{\alpha} = \sum_{i=1}^{3} d\overline{\alpha}_i, \;\; d\overline{\alpha}_i = \frac{2}{3} C_i d\overline{\varepsilon}_p - {\gamma_i}' \overline{\alpha}_i dp \tag{39}$$

Components of backstress $\alpha_i$ during the unloading and reloading paths were defined as [24]:

$$\alpha_i = \frac{2C_i}{3\gamma_i{'}} + \left(\alpha_{i0} - \frac{2C_i}{3\gamma_i{'}}\right) \exp\left[-\gamma_i{'}(\varepsilon_p - \varepsilon_{p0})\right] \; d\varepsilon_p \geq 0 \tag{40a}$$

$$\alpha_i = -\frac{2C_i}{3\gamma_i{'}} + \left(\alpha_{i0} + \frac{2C_i}{3\gamma_i{'}}\right) \exp\left[\gamma_i{'}(\varepsilon_p - \varepsilon_{p0})\right] \; d\varepsilon_p < 0 \tag{40b}$$

where $\varepsilon_{p0}$ represents the initial plastic strain and $\alpha_{i0}$ corresponds to the initial backstress. Coefficients $C_1$, $C_2$, $C_3$ and $\gamma'_1$, $\gamma'_2$, $\gamma'_3$ are Chaboche's materials constants. These coefficients for the 1045 steel alloy were determined from a stress-strain hysteresis loop that was generated based on a strain-controlled test of $\pm 0.8\%$. Chaboche parameters were obtained by simulating the half, or the lower half of the stabilized hysteresis curve, from the strain-controlled test [24]. The parameter $C_1$ was obtained from the slope of the initial part of the stabilized hysteresis curve with a high plastic modulus at the yield point and the parameter $C_3$ was determined from the linear part of the stabilized hysteresis curve with a high strain range. The coefficient $\gamma'_1$ should be large enough to stabilize the first hardening parameter of Chaboche's rule. Figure 6 presents an experimentally obtained stress-strain hysteresis loop for the 1045 steel alloy. This figure presents three sets of coefficients $C_{1-3}$ and $\gamma'_{1-3}$ and their corresponding loops simulated through FE analysis. These coefficients are chosen to achieve a close agreement between the experimental and simulated hysteresis loops. Figure 6c shows a close agreement of experimental and simulated loops for the 1045 steel alloy for coefficients $C_{1-3}$ = 75,000, 40,000, and 2500 MPa and $\gamma'_{1-3}$ = 2200, 215, and 1. Figure 6a,b shows deviations and changes in the simulated loops as different sets of $C_{1-3}$ and $\gamma'_{1-3}$ were taken.

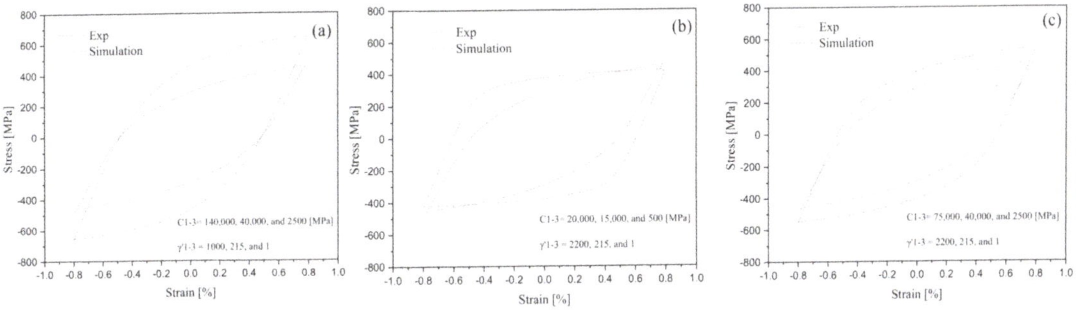


**Figure 6.** Coefficients $C_{1-3}$ and $\gamma'_{1-3}$ achieved a close agreement between the stress-strain hysteresis loop obtained from a test conducted under the strain-controlled condition and the one simulated through FE analysis [25]. (**a–c**) the strain-based hysteresis loops for different sets of $C_{1-3}$ and $\gamma'_{1-3}$.

## 5. Results and Discussion

The local ratcheting and stress relaxation at the notch root of steel specimens were evaluated through the A-V hardening framework. Local stress and strain components were coupled with the framework through the use of different model choices of Neuber, Glinka, and H-S. Ratcheting at the notch roots was also simulated by FE analysis where Chaboche's hardening rule was employed.

### 5.1. *Local Ratcheting Prediction through the A-V Hardening Rule*

#### 5.1.1. Estimation of Local Strain/Stress at Notch Roots through Different Models

To better estimate the local stress and strain terms at the notch root, different choice models of Neuber, Glinka, and H-S were examined. Steel specimens with notch diameters of 9 and 15 mm were tested under nominal stress levels of 155 ± 155 MPa and 203 ± 203 MPa loading conditions. These tests enabled us to evaluate the stress and strain components at the notch root and compare the employed models for their strain energies at applied nominal stresses and at different notch diameters. Figure 7 presents the state of nominal and local stress and strain components for these models at different applied stress levels and notch sizes. In this figure, the pseudo-elastic lines were obtained by applying the constraints of each model, and stress-strain curves were developed based on the Ramberg-Osgood equation. In Figure 7a, the H-S model presented a noticeable increase in the stress and strain components on a specimen with a notch size of 9 mm as the stress level changed from $155 \pm 155$ MPa to $203 \pm 203$ MPa. At the given nominal stress level of $155 \pm 155$ MPa in Figure 7b, the H-S model depicted smaller progress in the local stress and strain components as the notch size of the specimens changed from 9 to 15 mm. A change in the nominal stress level and notch diameter directly affected the extent of stress-strain area based on Glinka and Neuber models. In Figure 7c,d, at higher applied stress levels of $203 \pm 203$ MPa, the Glinka model resulted in a greater area underneath the stress—strain curve while, for the constant stress level, the specimen with a notch size of 15 mm caused a smaller increase in the local stress/strain data on the curve. Neuber's rule, however, involved a greater amount of energy from the product of stress and strain obtained from the rectangular area in Figure 7e,f. In the Neuber and Glinka rules, the total strain energy density at the notch root was taken as equal to the total pseudo strain energy density assuming that the specimen did not exceed the elastic domain even beyond its yield point.

#### 5.1.2. Backstress Evolution during Loading

Over the loading paths, backstress evolution was controlled through the A-V kinematic hardening model. Backstress $\overline{\alpha}$ and the internal variable $\overline{b}$ controlled the plastic strain increment $d\varepsilon_p$ and its accumulation during asymmetric loading cycles. The magnitude of backstress $\overline{\alpha}$ gradually stabilized over loading cycles in a nonlinear form through the term $(\overline{\alpha} - \delta\overline{b})$ in the dynamic recovery of the A-V model. This term in the A-V model was analogous to the integration of backstress increments $d\overline{\alpha} = \sum_{i=1}^{3} d\overline{\alpha}_i$, as proposed earlier by Chaboche [24]. The plastic strain accumulation was attributed to the cross-slip, and as the stress cycles proceeded, the accumulation of dislocations and their interactions led to a decrease in the ratcheting strain rate [26]. Figure 8 shows the evolution of the backstress term $(\overline{\alpha} - \delta\overline{b})$ over the first thirty loading cycles on a typical 1045 steel specimen through a choice of different models. The decay in the backstress term during stress cycles was more pronounced as the Neuber and Glinka rules were coupled with the A-V model compared to that of the H-S model. This figure shows a sudden drop in term $(\overline{\alpha} - \delta\overline{b})$ over the first few cycles. Following the initial loading cycles, a steady state was achieved. The smaller difference in terms $(\overline{\alpha} - \delta\overline{b})$ between the Neuber and Glinka rules was associated in relation to the nominal and local stresses through Equation (13). A larger product of the local stress and strain components in Figure 7e,f resulted in a larger nominal stress component and a small drop in term $(\overline{\alpha} - \delta\overline{b})$ over the loading cycles. The lowest backstress term in Figure 8 was attributed to the equivalent stress components, as defined in the H-S model. In the presence of the H-S model, the backstress term achieved its stability after the first seven cycles, while backstress curves that were generated through the use of Neuber and Glinka models required an even smaller number of cycles to gain a steady condition.

### *5.2. Predicted and Simulated Local Ratcheting Curves*

The coupled kinematic hardening framework was employed to assess the ratcheting response of notched 1045 steel specimens undergoing asymmetric stress cycles. The predicted and experimental ratcheting results at various stress levels and notch sizes were plotted in Figure 9. The predicted ratcheting curves in this figure show a consistent response compared with the experimental data. Predicted local ratcheting over the first few cycles showed an abrupt increase, and shortly after, as the number of cycles increased, the ratcheting rate dropped, and the slope of the ratcheting progress stayed nearly constant. Considering Glinka, Neuber, and H-S models, the choice of the H-S model resulted in lower ratcheting curves, and those curves were predicted on the basis of the hardening framework coupled with the Glinka model, which possessed the highest ratcheting values. The H-S model has, however, shown closer agreements with the experimental data at different stress levels and notch sizes over the first 20–40 cycles. Ratcheting curves predicted by the coupled framework of the A-V hardening rule and Neuber model closely agreed with the measured values of ratcheting strains over the loading cycles. Lower local stress and strain at the notch root were calculated based on the H-S model and suppressed the predicted ratcheting curves by the hardening framework. Higher strain energy was achieved through the Neuber and Glinka models, which increased the predicted local ratcheting as local stress and strain values at the notch root increased. Figure 9 shows the simulated ratcheting results through the use of FE analysis for 1045 steel specimens with different notch sizes undergoing different stress levels, which fell below the experimental and predicted ratcheting curves. Simulated curves were affected by the FE element size taken at different distances from the notch root to achieve a consistent convergence. The simulated curves correspond to lower local ratcheting values with a noticeable difference between the measured ratcheting data and those of the predicted curves.

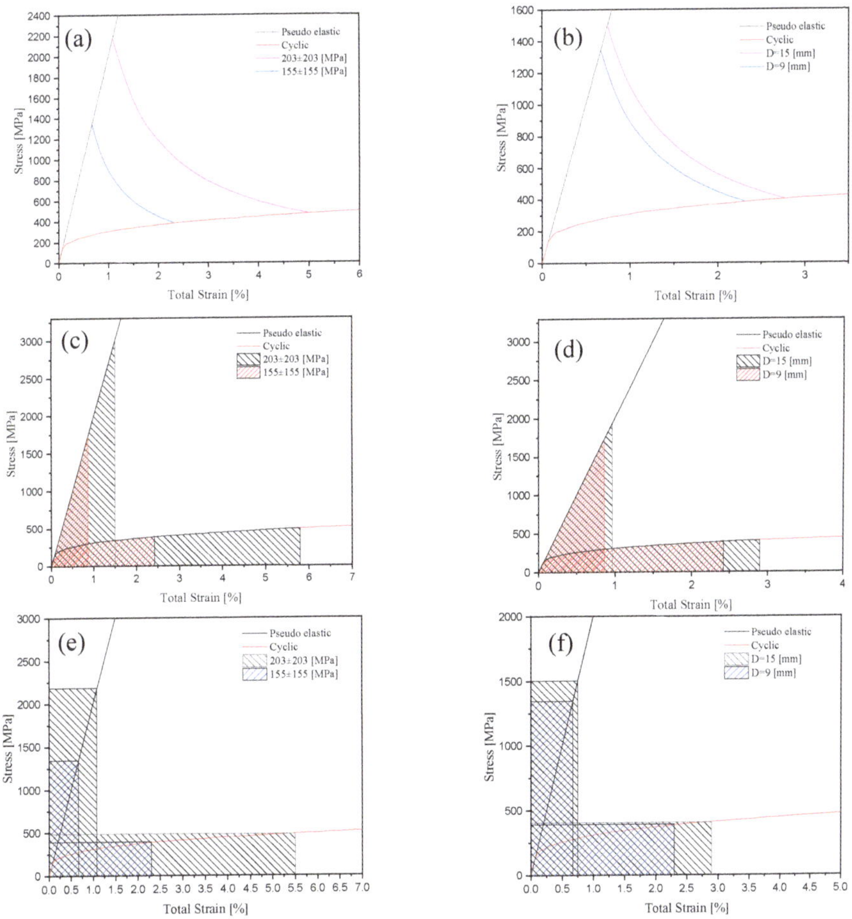


**Figure 7.** (**a**,**c**,**e**) Cyclic and pseudo–elastic curves for steel specimens with D = 9 mm notch diameter subjected to different loading values using H-S, Glinka and Neuber models. (**b**,**d**,**f**) cyclic and pseudo–elastic curves for a steel specimen undergoing 155 ± 155 MPa for notched specimens with D = 9 and D = 15 mm notch diameters using the H-S, Glinka, and Neuber models.

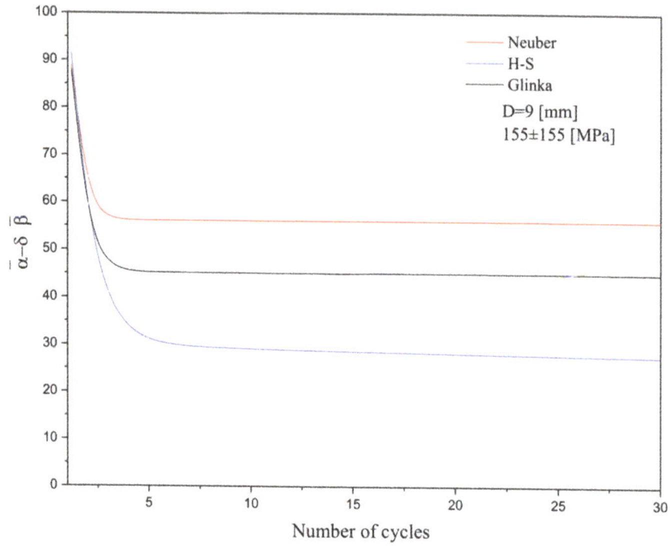


**Figure 8.** Backstress evolution over the first 30 loading cycles based on the Neuber, H-S, and Glinka rule coupled with the A-V kinematic hardening rule for a steel specimen and an 8 mm diameter under 155 ± 155 MPa.

The test samples with the same diameter consistently showed that an increase in the applied stress led to a higher ratcheting strain. At the same applied stress level, an increase in the diameter of the specimen from D = 9 mm to 15 mm decreased the ratcheting strain noticeably. While the ratcheting strain was promoted during asymmetric stress cycles in Figure 9, the cyclic stress levels gradually dropped to lower levels revealing stress relaxation at the notch root of the steel specimens. Stress relaxation over the stress cycles at the notch root of specimens was found at slightly different rates as the A-V hardening model was coupled with different models. Hysteresis loops of a typical 1045 steel specimen with a notch diameter of D = 9 mm were tested at 203±203MPa and showed that the H-S model resulted in the widening of loops while the Glinka model lowered the plastic strain range in the progressing loops. Loops generated by the hardening rule coupled with the Neuber model possessed an intermediate width. These hysteresis loops are presented in Figure 10. It is intended to keep the stress and strain axes in this figure within the same scale for a better comparison of these models and their generated loops. The peaks of progressing loops were connected as the number of cycles proceeded in Figure 11, resulting in a decreasing trend with nearly the same pace for all models. The horizontal axis in these figures was normalized with the strain at the 100th cycle to be able to present stress relaxation when three different models were compared. The ratcheting progress was controlled through the hardening framework at constant stress cycles, while stress relaxation at the notch root was controlled at a constant local strain leading to progressive loops experiencing a drop in the stress magnitude. Predicted curves in Figure 11 show relatively lower values of stresses and noticeable rates of relaxation compared with the simulated curve through the FE analysis. In this figure, the simulated stress relation for cycles beyond the first 20 cycles is presented for a consistent comparison with the predicted curves.

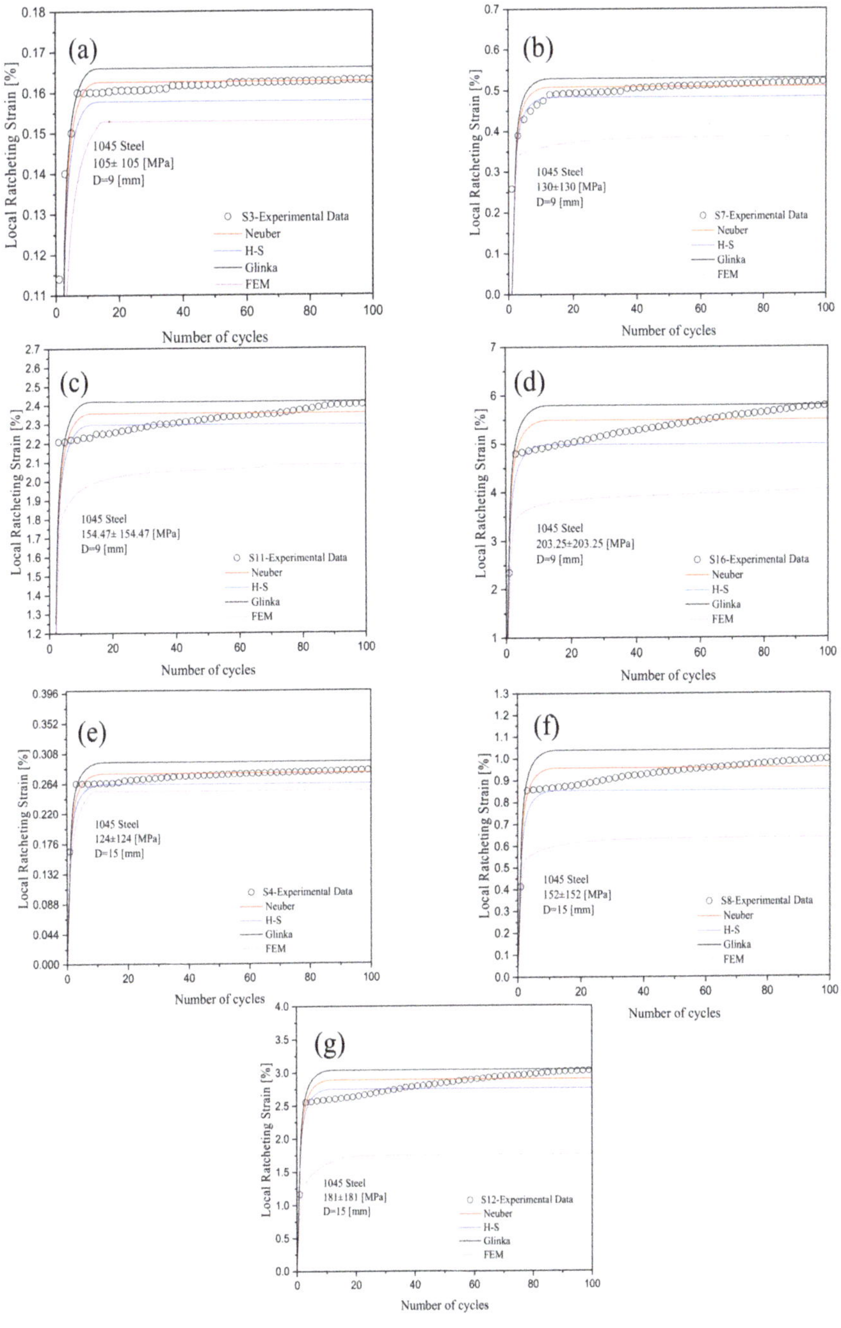


**Figure 9.** Predicted ratcheting curves using three different models coupled with the A-V hardening rule versus experimental ratcheting data for 1045 steel notched specimens undergoing various stress levels of (**a**) 105 ± 105 MPa with D = 9 mm, (**b**) 130 ± 130 MPa with D = 9 mm, (**c**) 155 ± 155 MPa with D = 9 mm, (**d**) 203 ± 203 MPa with D = 9 mm, (**e**) 124 ± 124 MPa with D = 15 mm, (**f**) 152 ± 152 MPa with D = 15 mm, (**g**) 181 ± 181 MPa with D = 15 mm.

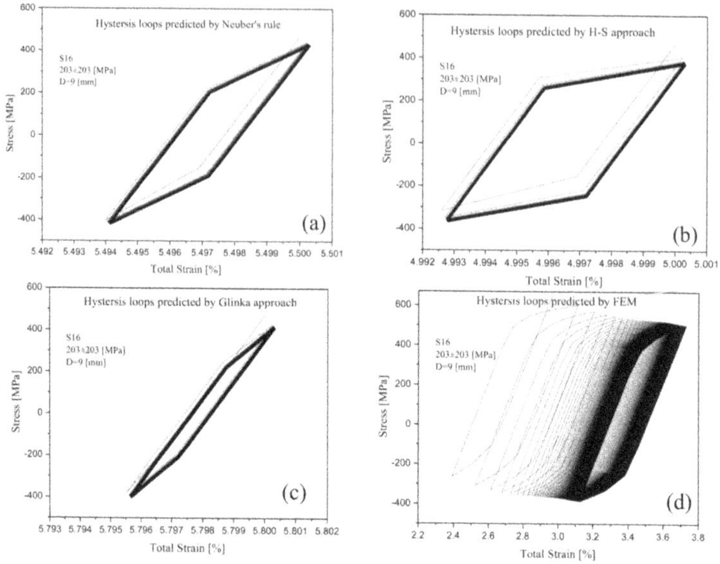


**Figure 10.** (**a**–**c**) Predicted stress-strain hysteresis loops based on Neuber's rule, H-S rule, and Glinka's approach coupled with the A-V model, respectively, and (**d**) simulated stress-strain hysteresis loops by FE analysis on the basis of Chaboche's model.

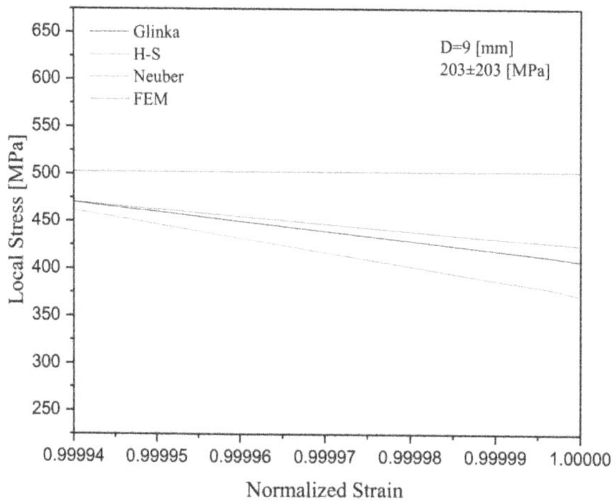


**Figure 11.** Stress relaxation for a steel specimen with a 9 mm notch diameter subjected to 203 ± 203 MPa, using three different rules coupled with the A-V kinematic hardening rule and through FE analysis.

### 5.3. Discussion

Ratcheting strain progress with asymmetric loading cycles took place over the transient and steady stages. Over stage II, ratcheting was found to be associated with the coefficient $\delta$ in the dynamic recovery of the A-V hardening rule. This coefficient was defined to be related to the backstress evolution and materials constants, C, $\gamma_1$, and $m$ through $(\overline{\alpha}/k)^m$. The local ratcheting rates (slopes) were predicted based on the coupled framework with different models and were found to be relatively smaller than that of the experimental data for stage II.

Various cyclic tests were conducted to examine the local ratcheting response of 1045 steel specimens at the vicinity of notch roots. While the present authors studied a number of influential parameters on the local ratcheting in the presence of the notch in steel plates, including stress levels and notch sizes, the open literature lacks a pertinent volume of experimental and theoretical research on local ratcheting phenomenon. More investigation is required to fully address the local ratcheting and stress relaxation at the notch root. Challenges in research include the complexity of cyclic tests to examine/detect the notch root plasticity and accurately measure local strains, as well as the lack of hardening frameworks/theories to sufficiently address the plastic flow in the vicinity of notch roots. The authors believe that more experimental investigation and analysis are required to fully understand notch root plasticity and its progress during asymmetric loading cycles. To measure progressive plastic deformation at the notch root and at various distances from the notch root, the use of strain gauges along longitudinal and lateral directions are inevitable. The author's further plan is to control the local ratcheting progress through technical/mechanical processes, including cold-pressing the notch root and localized heat-treatment processes. While such processes are expected to improve mechanical properties at the notch edges, they will also noticeably lessen the ratcheting magnitude and rate. The coupled hardening model, along with the Neuber and Glinka rules, will be employed to assess the ratcheting of cold-worked notches. Analytical and numerical approaches will be used to evaluate the choice of hardening rules and to encounter more variables such as time dependency, stress rate, testing frequency, and temperature in the ratcheting assessment program.

## 6. Conclusions

Local ratcheting was evaluated at the notch root of 1045 steel specimens by means of the A-V hardening rule coupled with the Neuber, Glinka, and H-S models. The evolution of backstress was governed by the coupled hardening framework. The local ratcheting rate and magnitude and stress relaxation at the notch root of steel specimens was assessed at different stress levels and notch sizes through various coupled models and FE analysis. The hardening rule algorithm was developed to assess local ratcheting coupled with different model choices to assess local strain and stress at the notch root. Predicted ratcheting curves through the coupled hardening framework with the Glinka model shifted above the measured ratcheting data, and those evaluated by means of H-S fell below experimental data. The Neuber model, however, closely agreed with the experimental ratcheting data at different stress levels and specimen notch sizes. Local ratcheting simulated through FE analysis fell below the experimental data and predicted curves. The choice of Neuber, Glinka, and H-S models in assessing local strain components and their terms/constants was found to affect the rate and magnitude of predicted ratcheting and stress relaxation by means of the coupled hardening framework.

**Author Contributions:** Conceptualization, A.V.-F.; methodology, A.V.-F.; software, A.V.-F., F.H.; validation, A.V.-F., F.H.; formal analysis, A.V.-F., F.H.; investigation, A.V.-F., F.H.; resources, A.V.-F., F.H.; data curation, A.V.-F., F.H.; writing—original draft preparation, A.V.-F., F.H.; writing—review and editing, F.H.; visualization, A.V.-F.; supervision, A.V.-F.; project administration, A.V.-F.; funding acquisition, A.V.-F. All authors have read and agreed to the published version of the manuscript.

**Funding:** This research was funded by the Natural Sciences and Engineering Research Council (NSERC) of Canada, grant number RGPIN-2021-03047 and the APC was waived by the journal.

**Institutional Review Board Statement:** Not applicable.

**Informed Consent Statement:** Not applicable.

**Data Availability Statement:** Experimental data can be available upon request of researchers through communication with corresponding author.

**Acknowledgments:** Special thanks go to K. Kolasangiani (the former graduate student) for conducting ratcheting tests in notched 1045 steel samples during his PhD program in the group.

**Conflicts of Interest:** The authors declare no conflict of interest. The funders had no role in the design of the study; in the collection, analyses, or interpretation of data; in the writing of the manuscript; or in the decision to publish the results.

## Nomenclature

| | |
|---|---|
| $d\bar{\varepsilon}$ | Total strain increment tensor |
| $d\bar{\varepsilon}_e$ | Elastic strain increment tensor |
| $d\bar{\varepsilon}_p$ | Plastic strain increment tensor |
| $E$ | Modulus of elasticity |
| $H_p$ | Plastic modulus |
| $\bar{\alpha}$ | Backstress tensor |
| $\bar{\sigma}$ | Stress tensor |
| $\bar{I}$ | Unit tensor |
| $\vartheta$ | Poisson's ratio |
| $G$ | Shear modulus |
| $\bar{s}$ | Deviatoric stress tensor |
| $\sigma_y$ | Yield strength |
| $D$ | Circular notch diameter |
| $\gamma_1, \gamma_2, C, \delta$ | Coefficients of the A-V model |
| $C_{1-3}, \gamma'_{1-3}$ | Chaboche materials coefficients |
| $K_t$ | Stress concentration factor |
| $\bar{b}$ | Internal variable of the A-V model tensor |
| $n', K\prime$ | Ramberg-Osgood coefficients |
| $S, e$ | Nominal stress and strain |
| $R$ | Stress ratio |
| $K_\sigma, K_\varepsilon$ | Stress and strain concentration factors |
| $\sigma, \varepsilon$ | Uniaxial local stress and strain at the notch root |
| $W_\sigma$ | Strain energy per unit volume at the notch root |
| $W_S,$ | Elastic strain energy per unit volume due to the nominal remote stress S |
| $\sigma_{eq}$ | Theoretical elastic equivalent stress at notch root |
| $\varepsilon_q$ | Equivalent strain at notch root |
| $\sigma_q$ | Equivalent stress at notch root |
| $K_p$ | Limit load factor |
| $a_e, b_e$ | Stress ratios within the elastic domain |

## References

1. Ahmadzadeh, G.R.; Varvani-Farahani, A. Ratcheting assessment of materials based on the modified Armstrong–Frederick hardening rule at various uniaxial stress levels. *Fatigue Fract. Eng. Mater. Struct.* **2013**, *36*, 1232–1245. [CrossRef]
2. Leis, B.N.; Topper, T.H. Cyclic deformation and fatigue analysis for notched component. *Nucl. Eng. Des.* **1974**, *29*, 370–383.
3. Wang, C.H.; Rose, L.R.F. Transient and steady-state deformation at notch root under cyclic loading. *Mech. Mater.* **1998**, *30*, 229–241. [CrossRef]
4. Hu, W.; Wang, C.H.; Barter, S. *Analysis of Cyclic Mean Stress Relaxation and Strain Ratchetting Behavior of Aluminum 7050*; Technical Report AR 010-989; DSTO Aeronautical and Maritime Research Laboratory: Fishermens Bend, VI, Australia, 1999.
5. Rahman, S.M.; Hassan, T. Advanced cyclic plasticity models in simulating ratcheting responses of straight and elbow piping components, and notched plates. *ASME Press. Vessel. Pip. Conf.* **2005**, *4188*, 421–427.
6. Lee, C.H.; Van Do, V.N.; Chang, K.H. Analysis of uniaxial ratcheting behavior and cyclic mean stress relaxation of a duplex stainless steel. *Int. J. Plast.* **2014**, *62*, 17–33. [CrossRef]
7. Firat, M. A notch strain calculation of a notched specimen under axial-torsion loadings. *Mater. Des.* **2011**, *32*, 3876–3882. [CrossRef]
8. Barkey, M.E. Calculation of Notch Strains under Multiaxial Nominal Loading. Ph.D. Thesis, Department of Theoretical and Applied Mechanics, University of Illinois, Urbana-Champaign, IL, USA, 1993.
9. Kolasangiani, K.; Farhangdoost, K.; Shariati, M.; Varvani-Farahani, A. Ratcheting assessment of notched steel samples subjected to asymmetric loading cycles through coupled kinematic hardening-Neuber rules. *Int. J. Mech. Sci.* **2018**, *144*, 24–32. [CrossRef]
10. Shekarian, A.; Varvani-Farahani, A. Ratcheting prediction at the notch root of steel samples over asymmetric loading cycles. *J. Eng. Mater. Technol. Trans.* **2020**, *142*, 1–23. [CrossRef]
11. Neuber, H. Theory of stress concentration for shear-strained prismatical bodies with arbitrary nonlinear stress-strain law. *J. Appl. Mech.* **1961**, *28*, 544–550. [CrossRef]
12. Shekarian, A.; Varvani-Farahani, A. Concurrent ratcheting and stress relaxation at the notch root of steel samples undergoing asymmetric tensile loading cycles. *Fatigue Fract. Eng. Mater. Struct.* **2019**, *42*, 1402–1413. [CrossRef]

13. Liu, C.; Shi, S.; Cai, Y.; Chen, X. Ratcheting behavior of pressurized-bending elbow pipe after thermal aging. *Int. J. Press. Vessel. Pip.* **2019**, *169*, 160–169. [CrossRef]
14. Chen, X.; Jiao, R.; Kim, K.S. On the Ohno-Wang kinematic hardening rules for multiaxial ratcheting modeling of medium carbon steel. *Int. J. Plast.* **2005**, *21*, 161–184. [CrossRef]
15. Shekarian, A.; Varvani-Farahani, A. Ratcheting behavior of notched stainless-steel samples subjected to asymmetric loading cycles. *J. Iron Steel Res. Int.* **2021**, *28*, 86–97. [CrossRef]
16. Shekarian, A.; Varvani-Farahani, A. Ratcheting response of SS316 steel samples with different notch shapes under various loading spectra. *J. Mater. Eng. Perform.* **2021**, *30*, 3524–3535. [CrossRef]
17. Varvani-Farahani, A. A comparative study in descriptions of coupled kinematic hardening rules and ratcheting assessment over asymmetric stress cycles. *Fatigue Fract. Eng. Mater. Struct.* **2017**, *40*, 882–893.
18. Kolasangiani, K.; Farhangdoost, K.; Shariati, M.; Varvani-Farahani, A. Ratcheting progress at notch root of 1045 steel samples over asymmetric loading cycles: Experiments and analyses. *Fatigue Fract. Eng. Mater. Struct.* **2018**, *41*, 1870–1883. [CrossRef]
19. Kolasangiani, K.; Shariati, M.; Varvani-Farahani, A. Ratcheting examination of 1045 notched steel plates under Low-High and High-Low sequences. *J. Test. Eval.* **2020**, *49*, 20190728. [CrossRef]
20. Molski, K.; Glinka, G. A method of elastic-plastic stress and strain calculation at a notch root. *Mater. Sci. Eng.* **1981**, *50*, 93–100. [CrossRef]
21. Hoffmann, M.; Seeger, T. A generalized method for estimating multiaxial elastic–plastic notch stresses and strains. Part I: Theory. *J. Eng. Mater. Technol.* **1985**, *107*, 250–254. [CrossRef]
22. Hoffmann, M.; Seeger, T. Stress–strain analysis and life predictions of a notched shaft under multiaxial loading. *Multiaxial Fatigue Anal. Exp.* **1989**, *AE-14*, 81–101.
23. *ABAQUS User's Manual*; Version 6.13 User's Manual; Hibbitt, Karlsson and Sorensen, Dassault Systèmes Simulia Corp.: Providence, RI, USA, 2016.
24. Chaboche, J. On some modifications of kinematic hardening to improve the description of ratcheting effects. *Int. J. Plast.* **1991**, *7*, 661–678. [CrossRef]
25. Kolasangiani, K. Analysis of Local and Global Ratcheting Behavior of AISI 1045 Steel Specimens with Cutout. Ph.D. Thesis, University of Mashhad, Mashhad, Iran, 2018.
26. Gaudin, C.; Feaugas, X. Cyclic creep process in AISI 316L stainless steel in terms of dislocation patterns and internal stresses. *Acta Mater.* **2004**, *52*, 3097–3110. [CrossRef]

*materials*

MDPI

*Article*

# High-Cycle Fatigue Behaviour of the Aluminium Alloy 5083-H111

**Branko Nečemer, Franc Zupanič, Tomaž Vuherer and Srečko Glodež ***

Faculty of Mechanical Engineering, University of Maribor, Smetanova 17, 2000 Maribor, Slovenia
* Correspondence: srecko.glodez@um.si

**Abstract:** This study presents a comprehensive experimental investigation of the high-cycle fatigue (HCF) behaviour of the ductile aluminium alloy AA 5083-H111. The analysed specimens were fabricated in the rolling direction (RD) and transverse direction (TD). The HCF tests were performed in a load control (load ratio $R = 0.1$) at different loading levels under the loading frequency of 66 Hz up to the final failure of the specimen. The experimental results have shown that the S–N curves of the analysed Al-alloy consist of two linear curves with different slopes. Furthermore, RD-specimens demonstrated longer fatigue life if compared to TD-specimens. This difference was about 25% at the amplitude stress 65 MPa, where the average fatigue lives 276,551 cycles for RD-specimens, and 206,727 cycles for TD-specimens were obtained. Similar behaviour was also found for the lower amplitude stresses and fatigue lives between $10^6$ and $10^8$ cycles. The difference can be caused by large $Al_6$(Mn,Fe) particles which are elongated in the rolling direction and cause higher stress concentrations in the case of TD-specimens. The micrography of the fractured surfaces has shown that the fracture characteristics were typical for the ductile materials and were similar for both specimen orientations.

**Keywords:** aluminium alloys; rolling direction; high-cycle fatigue; fracture analysis

## 1. Introduction

Aluminium alloy series 5xxx is a group of alloys based on aluminium–magnesium (Al-Mg) compositions [1]. These alloys are known for their high strength-to-weight ratio, excellent corrosion resistance, and good formability [2–4]. Based on these characteristics and their mechanical properties, these alloys are used widely in the automotive and aerospace industries, transportation, and other applications where lightweight and corrosion-resistant materials are required [5–9].

In the past, numerous studies have been conducted to understand the fatigue behaviour of AA 5xxx alloys [10–14]. Fatigue is an essential factor in determining the behaviour of mechanical parts under variable loads and is the primary cause of 80–90% of engineering failures. In applications that use aluminium alloys frequently, it is necessary to understand the fatigue performance of the components and the effects of operating parameters on the fatigue behaviour of the analysed constructional components. A fatigue assessment can be carried out using the stress–life (S–N) or the strain–life approach, depending on whether stresses in the critical cross-sections of the analysed component are in the elastic or plastic areas [15,16]. In the case that only elastic stresses occur, the S–N approach is usually used to obtain the fatigue life up to the final failure of the component. Establishing comprehensive databases, including S–N information, is essential for evaluating the fatigue characteristics of components under different operating conditions accurately. The fatigue life of Al-alloys is influenced not only by material characteristics but also by the characteristics of the specimens, such as microcavities formed during production, surface defects, hot or cold deformation, and changes in grain structure [17–19]. The tensile strength and fatigue life of Al-alloys are affected slightly by the rolling direction

**Citation:** Nečemer, B.; Zupanič, F.; Vuherer, T.; Glodež, S. High-Cycle Fatigue Behaviour of the Aluminium Alloy 5083-H111. *Materials* **2023**, *16*, 2674. https://doi.org/10.3390/ma16072674

Academic Editor: Qianhua Kan

Received: 30 January 2023
Revised: 16 March 2023
Accepted: 22 March 2023
Published: 28 March 2023

at room temperature, but the effects become more significant at higher temperatures due to the expansion of the grain structure [20,21]. Hockauf et al. [22] investigated the effect of different precipitate morphologies on low-cycle fatigue (LCF) and fatigue crack growth in an ultrafine-grained (UFG) age-hardening aluminium alloy. Their experimental results showed that the newly formed, coherent precipitates in the thermally recovered condition contribute to more planar slip behaviour, slip localisation and early failure during LCF loading. In [23], the researchers investigated the high-cycle fatigue behaviour of an Al-Si-Cu-Mn aluminium alloy at room temperature and 350 °C and evaluated the effect of different second phases on its fatigue behaviour. The alloy demonstrated excellent fatigue resistance at both temperatures, with a high-cycle fatigue strength of 125.0 MPa at room temperature and 47.5 MPa at 350 °C. The study found that fatigue cracks in the alloy at room temperature originated from casting defects, while at 350 °C, they nucleated from the primary Al-Mn-Si phase on the specimen's surface. The study also demonstrated that the modulus difference between the Al-Mn-Si phase and the alpha aluminium phase was higher at 350 °C than at room temperature, leading to more crack initiation and propagation along the phase interface at 350 °C. Khisheh et al. [24] investigated the influences of surface roughness and heat treatment on the high-cycle bending fatigue properties of A380 aluminium alloy under stress-controlled cyclic loading. The results showed that the heat treatment increased the high-cycle bending fatigue lifetime by 26% (for the highest stress level) and 85% (for the lowest stress level), respectively. The study also found that samples with low roughness had a longer fatigue life than those with high roughness. In the research presented by Gao et al. [25], the authors investigated the effect of surface mechanical attrition treatment (SMAT) on the fatigue performance of a 7075-T6 aluminium alloy in both high-cycle fatigue (HCF) and very-high-cycle fatigue (VHCF) regimes. The results showed that SMAT could improve fatigue strength in the HCF regime but decrease it in the VHCF regime. Meng et al. [26] investigated the vibration fatigue improvement of 2024-T351 aluminium alloy by ultrasonic-assisted laser shock peening (ULP). The results showed that the ULP increases the dislocation density significantly and refines the grains, leading to increased compressive residual stress and microhardness and inhibited crack initiation and propagation, resulting in a significant increase in vibration fatigue life. Xing et al. [27] studied the transition fatigue failure of weld toe cracking and weld root cracking in aluminium fillet welds. Their experimental results showed that the weld root cracking has a lower fatigue life and wider scatter band than the weld toe cracking. The researchers also proposed the weld sizing criterion for avoiding weld root cracking in fillet-welded aluminium connections. Sakin et al. [28] investigated the bending fatigue lives of AA 1100 and AA 1050 aluminium sheets experimentally under both high-cycle fatigue (HCF) and low-cycle fatigue (LCF) conditions. The specimens were tested along four different directions, including the rolling direction (RD) and transverse direction (TD). The results showed that the longest fatigue lives in the LCF region were observed in the AA1050 (RD) specimens, while the AA 1100 (RD) specimens had the longest fatigue lives in the HCF region. In the research works presented in [29–32], the researchers investigated the fatigue behaviour of cellular structures made of different aluminium sheets. The specimens were fabricated in a rolling direction using water jet cutting technology.

Material testing provides valuable information on the material's mechanical properties, such as its stress–strain response, deformation, fatigue life, and fracture behaviour. This information is crucial for optimising the design and manufacture of engineering structures and components made from this material and assessing their performance and durability in various applications. The proposed study is the continuation of the author's previous work [33], where the LCF behaviour of the aluminium alloy 5083-H111 was investigated. The obtained experimental results (cyclic S–N curve, LCF-fatigue parameters) were then used in work [34], where Nečemer et al. analysed the LCF behaviour of auxetic cellular structures. As presented by Lehmus et al. [35], cellular structures represent a unique opportunity for adoption in lightweight design due to their favourable characteristics regarding sound isolation, damping, energy absorption, etc. Using the advanced additive manufacturing (AM) technologies, abrasive water jet (AWJ) cutting technology, etc., different types of cellular structures (see two examples in Figure 1) can be manufactured for specific mechanical properties and other characteristics useful for different engineering applications. When analysing the fatigue behaviour of cellular structures made of aluminium alloy AA 5083-H111, the effect of rolling direction on the fatigue life may be significant, especially in the high-cycle fatigue (HCF) area. For that reason, the proposed research considers the influence of the rolling direction on the fatigue life and fracture behaviour of aluminium alloy AA 5083-H111 in the HCF regime. Additionally, the obtained results could help engineers make the appropriate decisions about the use and performance of this alloy in various engineering applications.

**Figure 1.** Two examples of cellular structures: (**a**) chiral structure; (**b**) re-entrant structure.

## 2. Materials and Methods

### *2.1. Material and Specimen Geometry*

In the experimental investigation, the specimens were made from the aluminium alloy AA 5083-H111, which is a commonly used aluminium alloy known for its excellent corrosion resistance and high strength-to-weight ratio. The chemical composition of this alloy is presented in Table 1.

**Table 1.** The chemical composition of AA 5083-H111.

| Al [wt %] | Mg [wt %] | Mn [wt %] | Si [wt %] | Fe [wt %] | Zn [wt %] | Cr [wt %] | Ti [wt %] | Cu [wt %] |
|---|---|---|---|---|---|---|---|---|
| 92.55 | 4.9 | 1 | 0.4 | 0.4 | 0.25 | 0.25 | 0.15 | 0.1 |

The flat tensile specimen shown in Figure 2a, taken from one batch of material, was used in the uniaxial quasi-static tensile tests and high-cycle fatigue (HCF) tests. All the specimens were fabricated using the abrasive water jet (AWJ) cutting technology from a 4 mm thick sheet. The specimens were fabricated in the rolling direction (RD) and transverse direction (TD); see Figure 2b.

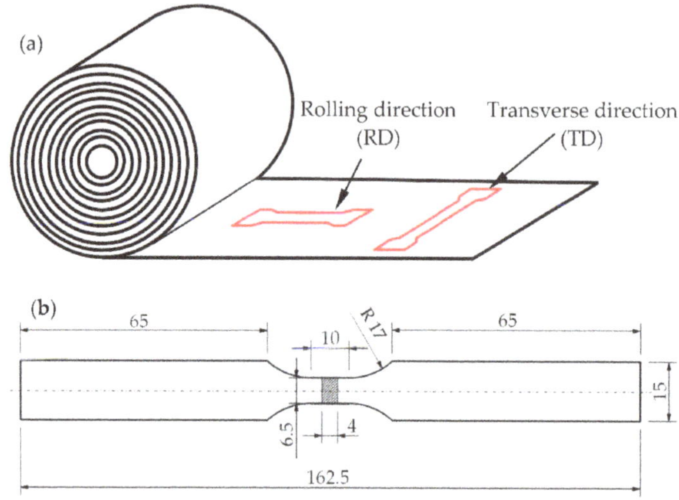


**Figure 2.** (**a**) Specimen's orientations. (**b**) Geometry and dimensions of the specimen.

Figure 3 shows the backscattered electron images of the investigated alloy in its as-received condition for both orientations (in the rolling and transverse directions). The alloy is composed of an Al-rich solid solution, alpha-Al, dark $Mg_2Si$ particles, large bright iron-rich $Al_6$(Fe, Mn) particles, and smaller plate-like Mn-rich $Al_6$(Fe, Mn) particles. The particles were identified using EDS. A detailed investigation of particles was carried out in our previous publication [33], and the results were consistent with the findings of Liu et al. [13]. The microstructures in both sections are very similar; however, the dimensions of Fe-rich $Al_6$(Fe, Mn) particles are longer in the direction of rolling.

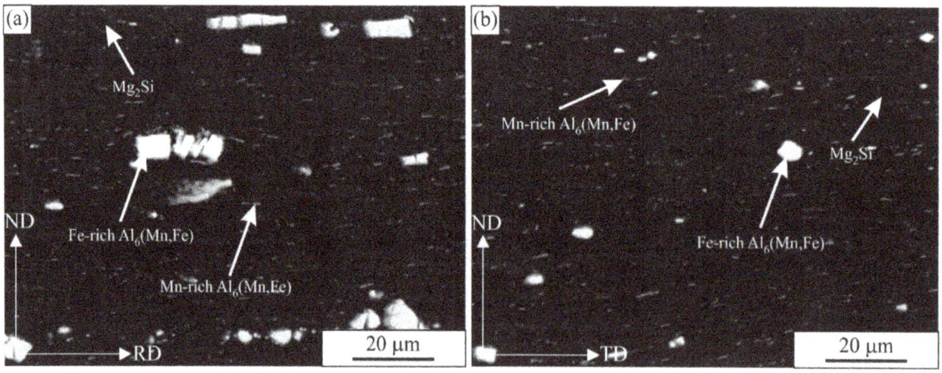


**Figure 3.** The backscattered electron images of the AA 5083-H111 alloy. (**a**) In the longitudinal direction. (**b**) In the transverse direction: RD-rolling direction, ND-normal direction, TD-transverse direction.

## 2.2. Experimental Methods

In the scope of the experimental investigation, the quasi-static tensile tests were performed, as well as the HCF tests. All the tests were carried out on the electrodynamic pulsating testing machine Vibrophore 100 ZwickRoell.

The quasi-static tensile tests were performed in a force control at an ambient temperature of 22 °C. The force was monitored with a 100 kN tensile–compressive load cell mounted on the testing machine, while the strain was controlled on the narrowed part of the specimen with an axial mechanical extensometer ZwickRoell DigiClip 40. Four quasi-static tensile tests were performed for each specimen orientation.

The HCF tests were performed in load control at different load levels at the same ambient temperature of 22 °C. The loading frequency was fixed to 66 Hz considering the sinusoidal wave shape with the constant load ratio $R = 0.1$. In the experimental testing, eight different values of maximum force $F_{max}$ were selected for each specimen orientation. The first stress level (maximum force $F_{max}$) was defined at approximately 50% of the yield stress of the material, and each next stress level was reduced by 3 %. At least two or three specimens were tested for each stress level, up to the final failure of the specimen. The run-out condition (without failure) was set to $10^8$ cycles.

The metallographic samples were prepared by grinding with SiC paper and polished using diamond paste. They were examined using the scanning electron microscope (SEM) Quanta 3D (FEI, Eindhoven, the Netherlands). The fractography study was carried out using a light microscope, an Olympus EP 50, and the aforementioned SEM. The micro-chemical analysis of the particles in a scanning electron microscope Sirion 400 NC (FEI, Eindhoven, The Netherlands) equipped with an energy-dispersive spectrometer (Oxford Analytical, Bicester, UK).

## 3. Results and Discussion

### 3.1. *Quasi-Static Tensile Tests*

Figure 4 shows the engineering stress–strain curves for both specimens' layouts. The average mechanical properties of the analysed aluminium alloy AA 5083-H111 are summarised in Table 2. The Young's modulus for the rolling direction (RD) was evaluated at around 70.8 GPa, while, for the transverse direction (TD), it was around 71.2 GPa. Furthermore, the yield stress and ultimate tensile strength were found to be higher for the RD if compared to the TD. Finally, the strain at fracture was found to be almost the same for both specimens' orientations.

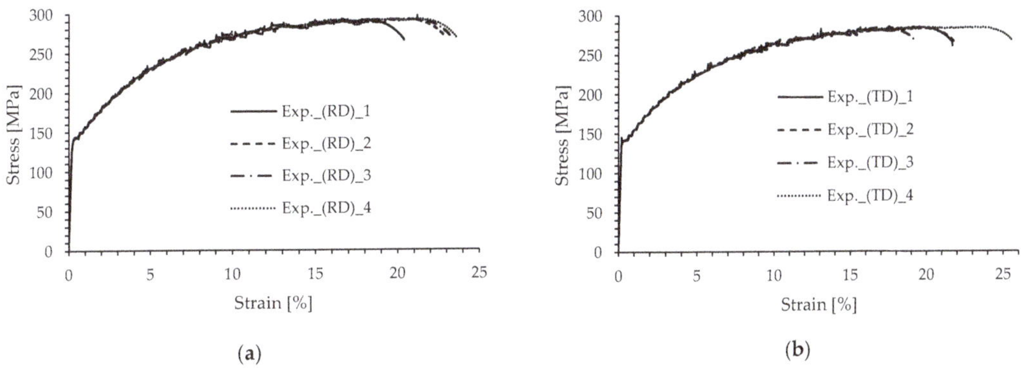


**Figure 4.** Engineering stress–strain curves of the AA 5083-H111 alloy. (**a**) Rolling direction (RD). (**b**) Transverse direction (TD).

**Table 2.** Average mechanical properties of AA 5083-H111.

| **Young's Modulus** *E* **[GPa]** | | **Yield Stress** $\sigma_y$ **[MPa]** | | **UTS** $\sigma_{UTS}$ **[MPa]** | | **Elongation** *A* **[%]** | |
|---|---|---|---|---|---|---|---|
| **(RD)** | **(TD)** | **(RD)** | **(TD)** | **(RD)** | **(TD)** | **(RD)** | **(TD)** |
| 70.8 | 71.2 | 143.2 | 141.9 | 293.9 | 284.2 | 22.2 | 21.9 |

(RD)—rolling direction, (TD)—transverse direction.

Based on the experimental results presented in Figure 4 and Table 2, it can be concluded that the mechanical properties (Young's modulus, yield stress, ultimate tensile strength, elongation at fracture) were quite similar for both specimens. However, the specimens manufactured in the longitudinal/rolling direction demonstrated slightly better properties (except for Young's modulus, $E$) if compared to the specimens manufactured perpendicular to the rolling direction (transverse direction).

Figure 5 shows the fractured surfaces of the tensile specimens for both directions. The elongation at fracture was about 18 %, while the contraction was very low. The fracture surface was inclined about 45 ° in relation to the loading direction. The fracture shows a ductile character with many dimples. The diameters of the larger pores were 20 to 30 µm, and the smaller ones were a few micrometres. Inside the larger voids, coarse intermetallic particles were present. Inside the smaller pores, some particles were found occasionally. Thus, the pores started to form at the interfaces between the matrix and intermetallic particles. In addition, some of the larger particles also fractured during deformation. The pore sizes were several times larger than the particle sizes.

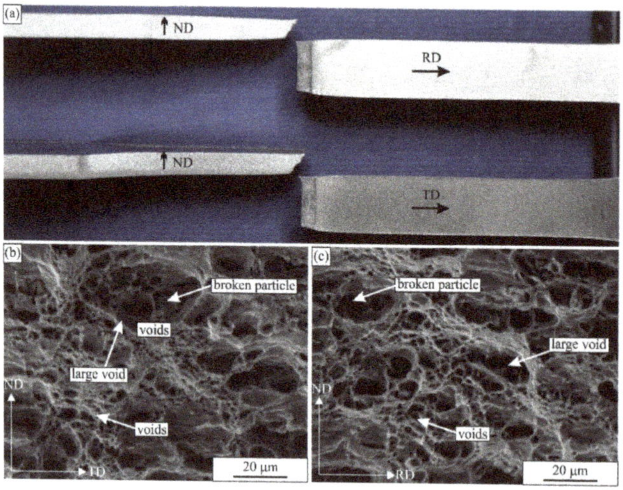


**Figure 5.** The fractured tensile specimens. (**a**) The photography of the samples. The secondary electron micrographs of the fractured surfaces of the tensile specimens: (**b**) in the RD and (**c**) in the TD.

### 3.2. *High-Cycle Fatigue Tests*

The high-cycle fatigue (HCF) tests were performed in a force control under pulsating loading (load ratio $R$ = 0.1) at a constant frequency of 66 Hz. Eight loading levels were selected, to obtain the S–N curves for both specimens' layouts. The S–N curves presented in Figures 6 and 7 were defined by plotting the stress amplitude $\sigma_a$ versus the number of cycles to failure $N_f$. Some scatter of the experimental results presented in Figures 6 and 7 is evident for both specimens' layouts. However, the scatter is greater for the transverse direction. Furthermore, the S–N curves are "bilinear" and consist of two linear curves with different slopes, which intersect at the knee point $N_k$ [36]. The fatigue behaviour below and upper the knee point $N_k$ can be expressed as follows:

$$\sigma_a = \sigma_{a,k} \cdot \left( \frac{N_k}{N_f} \right)^{\frac{1}{k}} \text{ below the knee point} \quad (1)$$

$$\sigma_a = \sigma_{a,k} \cdot \left(\frac{N_k}{N_f}\right)^{\frac{1}{k^*}} \text{ above the knee point} \quad (2)$$

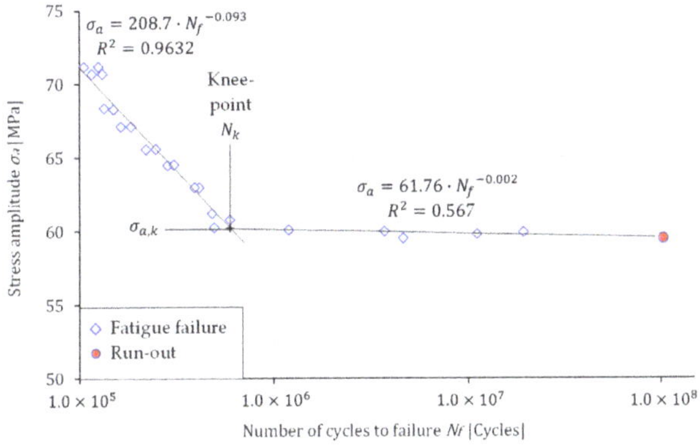


**Figure 6.** The S–N curve for the rolling direction.

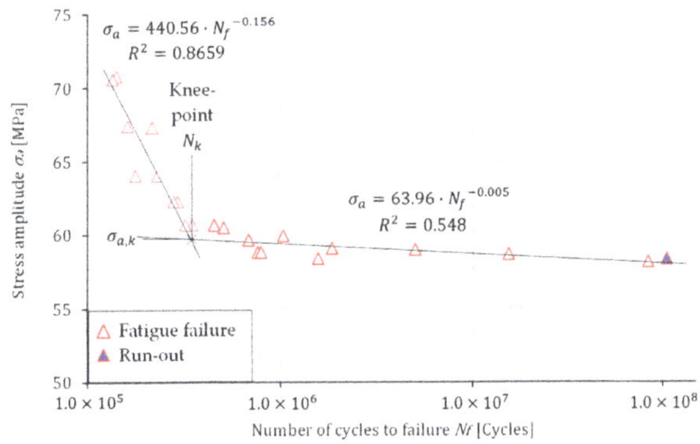


**Figure 7.** The S–N curve for the transverse direction.

In Equations (1) and (2), the $k$ represents the slope before the knee point, $\sigma_{a,k}$ is the stress amplitude at the knee point, $N_k$ is the number of cycles at the knee point, $k^*$ is the slope after the knee point, and $\sigma_a$ is the amplitude stress. The material parameters related to the Equations (1) and (2) are summarised in Table 3.

**Table 3.** Fatigue parameters of AA 5083-H111 related to the bilinear S–N curve.

| Number of Cycles at the Knee Point $N_k$ [Cycles] | | Stress Amplitude at the Knee Point $\sigma_{a,k}$ [MPa] | | Slope before the Knee Point $k$ [/] | | Slope after the Knee Point $K^*$ [/] | |
|---|---|---|---|---|---|---|---|
| (RD) | (TD) | (RD) | (TD) | (RD) | (TD) | (RD) | (TD) |
| 596,500 | 348,500 | 60.2 | 59.7 | 10.02 | 6.14 | 389.04 | 199.5 |

(RD)—rolling direction; (TD)—transverse direction.

Figure 8 shows the comparison of S–N curves for both specimen layouts. It is clear that the amplitude stress of the knee point, $\sigma_{a,k}$, is almost the same for both specimen layouts. However, the number of cycles of the knee point is higher for the specimens oriented in the rolling direction, which is then reflected in the slopes $k$ and $k^*$. When analysing the fatigue live for both specimen orientations, it is evident from Figures 6–8 that RD-specimens demonstrate longer fatigue live if compared to the TD-specimens. This difference is about 25 % at the amplitude stress 65 MPa, where fatigue lives 276,551 cycles for RD-specimens, and 206,727 cycles for TD-specimens were obtained. Similar behaviour may also be found for the lower amplitude stresses and fatigue lives between $10^6$ and $10^8$ cycles. The difference can be caused by large $Al_6(Mn,Fe)$ particles which are elongated in the rolling direction and cause higher stress concentrations in the case of TD-specimens.

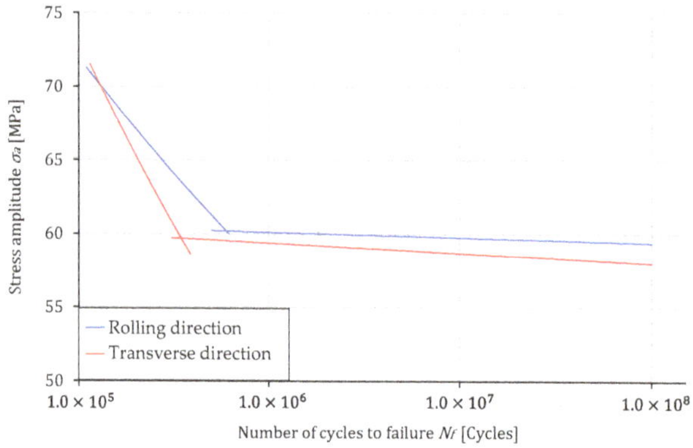


**Figure 8.** The comparison of S–N curve for both specimen's directions.

Figure 9 shows the fractured surfaces of the selected fatigue specimens. The cracks are typically initiated at specimen edges and then propagated in the directions indicated by the arrows. As can be seen, they propagated in the direction parallel to the original sheet surface. In our case, the cracks started at a corner at the top and then propagated towards the bottom.

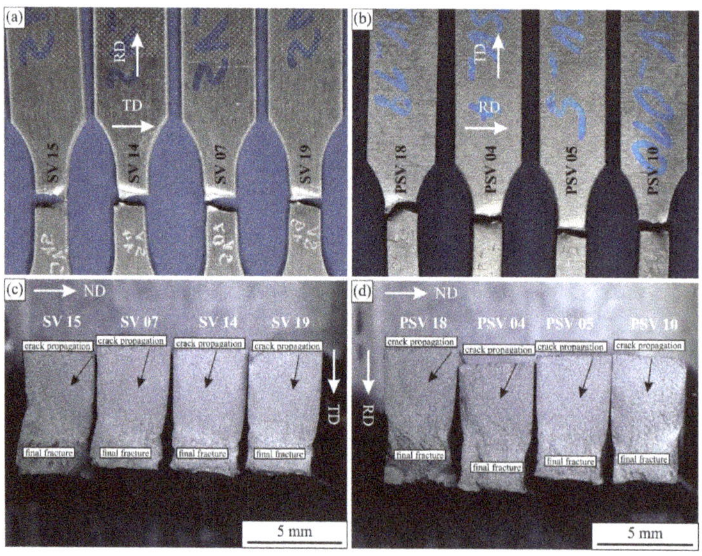


**Figure 9.** The light micrographs of the fractured surfaces of the selected fatigue samples: (**a**,**c**) longitudinal layout and (**b**,**d**) transverse layout (ND—normal direction, RD—rolling direction, TD—transverse direction).

The initial parts of the cross-sections were relatively flat and approximately normal to the loading direction. At some distances, the fracture surface became rougher, and this coincided with the distortion of the specimen in the bottom part caused by the plastic deformation. The initially rectangular shape became distorted, and the distortion was higher at higher stress levels for the samples on the left side of the images.

Figures 10 and 11 show the scanning electron micrographs of the fracture surfaces of the selected specimens at three different positions. The fracture characteristics were similar irrespective of the sample orientation (longitudinal or transverse layout) and the number of cycles to fracture. At the beginning of the cracks, the surface was rough on the micrometre scale. The crack followed the glide planes in the grains and formed rather flat facets [37]. Striations were present on the facets. The distances between them were less than one micrometre, typically between 0.4 and 0.7 μm (see Figure 12a).

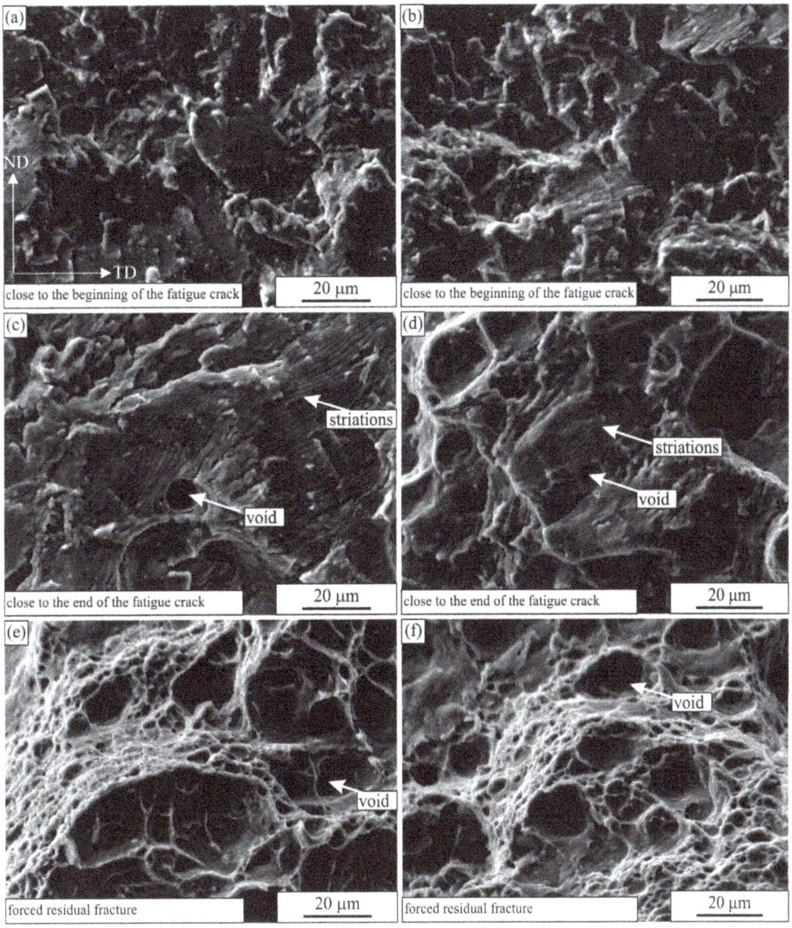


**Figure 10.** The secondary electron micrographs of the fractured surfaces of the fatigue samples tested in the longitudinal layout: (**a**,**c**,**e**) SV 15 (131,581 cycles to failure), (**b**,**d**,**f**) SV 19 (19,392,180 cycles to failure). The orientation of all images is the same. The RD direction is perpendicular to the images.

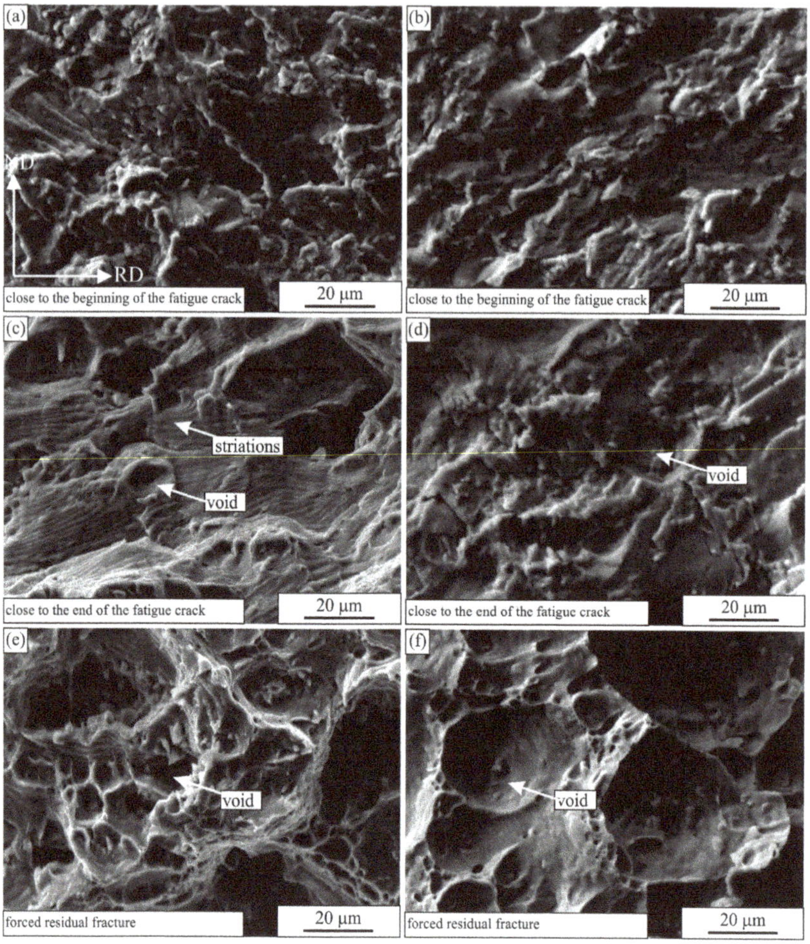


**Figure 11.** The secondary electron micrographs of the fractured surfaces of the fatigue samples tested in the transverse layout: (**a**,**c**,**e**) PSV 18 (135,165 cycles to failure), (**b**,**d**,**f**) SV 10 (15,500,000 cycles to failure). The orientation of all images is the same. The TD direction is perpendicular to the images.

In the last stages of the crack propagation, the surface was rather rough, and it was not perpendicular to the loading direction. The distances between striations increased and were typically between 2 and 4 μm. In these regions, not only striations were present, but also some voids. The voids formed at larger inclusions in the microstructure. They formed in the regions in front of the fatigue crack as the stresses surpassed the yield stress of the material, and they grew until the main fatigue crack overtook them. The final forced residual fracture was almost the same as the tensile fracture (see Figure 3) and consisted of voids of different sizes.

Figure 12 shows some fractured surfaces at a much higher magnification and resolution. The distances between the striations—the striation wavelengths—were measured on the basis of such micrographs. They are presented in Figure 13. The striations were not visible up to distances below 1 mm from the crack initiation site (Figure 12a), which was typically at the specimen corner. This is Region A of crack propagation, which is highly sensitive to microstructure characteristics [38]. Usually, a crack with a length 1–2 mm is considered a small fatigue crack. One can observe grain boundaries and facets in the crystal grains, which typically glide planes. The crack propagation rate is often in the order of nm, and the specimen survives in this region for most of its lifetime [39]. Striations become visible in range B of crack propagation [38]. The striations become stable, and in the first part, their wavelength is in the order of 100 nm; we measured 200 nm at a distance of 1.4 mm from the crack initiation site (Figure 12b).

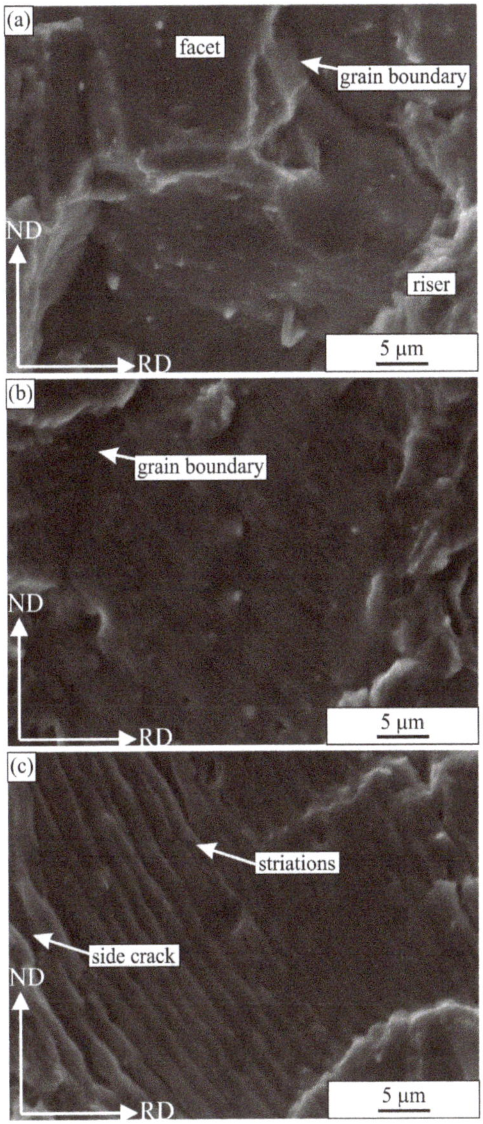


**Figure 12.** Secondary electron micrographs of the fractured surfaces of the fatigue sample PSV 14 (765,121 cycles to failure): (**a**) 0.7 mm from the fatigue crack initiation (Stage Regime A), (**b**) 1.4 mm from the fatigue crack initiation (start of Regime B), (**c**) 3.5 mm from the fatigue crack initiation (near the end of Regime B).

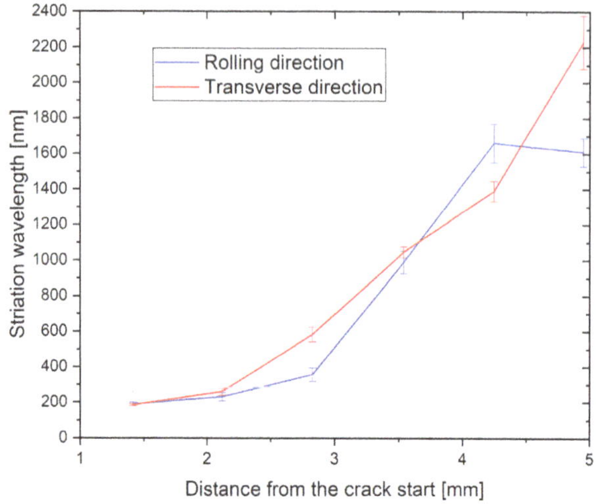


**Figure 13.** Striation wavelength in dependence from the crack start (RD—sample, TD—sample).

With the growing fatigue crack, the stress intensity factor grows, and the striation wavelength increases in micrometre size. It should be stressed that with the wavelength of 1 μm, the fatigue crack will proceed 1 mm after 1000 cycles and leads to a quick fracture of the specimen. This wavelength was achieved approximately 3.5 mm from the crack initiation (Figure 12c). In Regime B, the crack propagation is less sensitive to microstructure. During the main part of the crack path, the distances between the striations were larger for the transverse specimen, which can also explain the shorter lifetime for this specimen. The last two measurements were close to the crack ends, where the surface resembled those to that shown in Figure 10c,d and Figure 11c,d, where the scatter was rather large. At the transition from Regime B to Regime C, the crack propagation becomes more microstructure sensitive. In the case of this alloy, fatigue cracks can form even in the front of the main cracks around the largest inclusions, as was discussed before.

## 4. Conclusions

The comprehensive experimental investigation of the high-cycle fatigue (HCF) behaviour of the ductile aluminium alloy AA 5083-H111 was presented in this study. Based on the metallographic investigation of the material, quasi-static tensile tests, high-cycle fatigue tests, and fractography of the fractured surfaces of the quasi-static and fatigue specimens, the following conclusions can be made:

- The analysed AA 5083-H111 alloy is composed of an Al-rich solid solution, alpha-Al, dark $Mg_2Si$ particles, large bright iron-rich $Al_6$(Fe, Mn) particles, and smaller, plate-like Mn-rich $Al_6$(Fe, Mn) particles. The microstructures in both the longitudinal/rolling (RD) and transversal (TD) directions were found to be very similar. However, the dimensions of the Fe-rich $Al_6$(Fe, Mn) particles were longer in the longitudinal/rolling direction.
- The mechanical properties (yield stress, ultimate tensile strength, elongation at fracture) were quite similar for both specimens. However, the specimens manufactured in the longitudinal/rolling direction demonstrated slightly better properties (except for Young's modulus, $E$) if compared to the specimens manufactured perpendicular to the rolling direction (transverse direction).
- The experimental results have shown that the S–N curves of the analysed Al-alloy consist of two linear curves with different slopes, which intersect at the knee point $N_k$. The corresponding amplitude stress at the knee point, $\sigma_{a,k}$, was found to be almost the same for both specimen layouts, while the number of cycles at the knee point,

$N_k$, was found to be higher for the specimens oriented longitudinally to the rolling direction. The difference can be caused by large, in the rolling direction elongated $Al_6$(Mn, Fe) particles, which cause higher stress concentrations when tested in the TD. Furthermore, the main part of the larger particles has a cuboidal shape, with a larger axis approximately parallel to the rolling direction. Thus, in the RD orientation, the larger axis of particles lay in the direction of the load and in the TD orientation perpendicular to the load. It could be expected that at TD orientation, higher stress concentrations occurred at the particle–matrix interface and that this leads to slightly worse fatigue resistance in the TD direction.

- The micrography of the fractured surfaces of the fatigue specimens has shown that the fracture characteristics are similar for both specimen orientations (longitudinal or transversal). The fracture surface has a typical appearance for the ductile material, characterised by striations during propagation of the fatigue crack and final ductile fracture. The distance between striations increased from the crack beginning (less than 0.5 micrometres) to the crack end (more than 3 micrometres).
- In the proposed research work, we analysed only two specimen orientations: (i) in the rolling direction and (ii) transverse to the rolling direction. The third specimen orientation (45° in regard to the rolling direction) could be investigated in our further research work. Furthermore, further research work should consider the higher number of experiments, especially in the long-life fatigue area (more than $10^7$ loading cycles). In this case, a comprehensive statistical evaluation could be performed to obtain more qualitative results regarding the fatigue behaviour of the analysed aluminium alloy.

**Author Contributions:** Conceptualisation, S.G. and B.N.; methodology, S.G.; validation, S.G.; investigation, B.N., T.V., and F.Z.; writing—original draft preparation, B.N.; writing—review and editing, S.G., F.Z., and T.V.; visualisation, S.G.; supervision, S.G. and F.Z.; project administration, S.G.; All authors have read and agreed to the published version of the manuscript.

**Funding:** This research was funded by the Slovenian Research Agency (ARRS), Research Core Funding No. P2-0063.

**Data Availability Statement:** The data presented in this study are available on request from the corresponding author.

**Acknowledgments:** The authors acknowledge the financial support of the Research Core Funding's (No. P2-0063 and No. P2-0120) from the Slovenian Research Agency. The authors also acknowledge for the use of research equipment "high-frequency pulsator" within the project "Upgrading national research infrastructures—RIUM", which was co-financed by the Republic of Slovenia, the Ministry of Education, Science and Sport and the European Union from the European Regional Development.

**Conflicts of Interest:** The authors declare no conflict of interest.

## References

1. Hirsch, J.; Al-Samman, T. Superior Light Metals by Texture Engineering: Optimised Aluminum and Magnesium Alloys for Automotive Applications. *Acta Mater.* **2013**, *61*, 818–843. [CrossRef]
2. Liu, Y.; Sun, Y.; Zhang, L.; Zhao, Y.; Wang, J.; Liu, C. Microstructure and Mechanical Properties of Al-5Mg-0.8Mn Alloys with Various Contents of Fe and Si Cast under Near-Rapid Cooling. *Metals* **2017**, *7*, 428. [CrossRef]
3. Jin, H. Optimization of Aluminum Alloy AA5083 for Superplastic and Quick Plastic Forming. *Metall. Mater. Trans. A* **2019**, *50*, 3868–3890. [CrossRef]
4. Ilman, M.N.; Triwibowo, N.A.; Wahyudianto, A.; Muslih, M.R. Environmentally Assisted Fatigue Behaviour of Stress Relieved Metal Inert Gas (MIG) AA5083 Welds in 3.5% NaCl Solution. *Int. J. Fatigue* **2017**, *100*, 285–295. [CrossRef]
5. Billy, R.G.; Müller, D.B. Aluminium Use in Passenger Cars Poses Systemic Challenges for Recycling and GHG Emissions. *Resour. Conserv. Recycl.* **2023**, *190*, 106827. [CrossRef]
6. Tianyu, Z.; Teng, Z.; Yuting, H.; Xianghong, F.; Yaping, B. Probabilistic Model of the Fatigue Life of Epoxy-Coated Aluminum Alloys Considering Atmospheric Exposure. *Int. J. Fatigue* **2022**, *162*, 106899. [CrossRef]
7. Raabe, D.; Ponge, D.; Uggowitzer, P.J.; Roscher, M.; Paolantonio, M.; Liu, C.; Antrekowitsch, H.; Kozeschnik, E.; Seidmann, D.; Gault, B.; et al. Making Sustainable Aluminum by Recycling Scrap: The Science of "Dirty" Alloys. *Prog. Mater. Sci.* **2022**, *128*, 100947. [CrossRef]

8. Hajjioui, E.A.; Bouchaâla, K.; Faqir, M.; Essadiqi, E. A Review of Manufacturing Processes, Mechanical Properties and Precipitations for Aluminum Lithium Alloys Used in Aeronautic Applications. *Heliyon* **2022**, *9*, e12565. [CrossRef]
9. Manuel, N.; Beltrão, D.; Galvão, I.; Leal, R.M.; Costa, J.D.; Loureiro, A. Influence of Tool Geometry and Process Parameters on Torque, Temperature, and Quality of Friction Stir Welds in Dissimilar Al Alloys. *Materials* **2021**, *14*, 6020. [CrossRef]
10. Xu, S.; Chen, J.; Shen, W.; Hou, R.; Wu, Y. Fatigue Strength Evaluation of 5059 Aluminum Alloy Welded Joints Considering Welding Deformation and Residual Stress. *Int. J. Fatigue* **2022**, *162*, 106988. [CrossRef]
11. Bay, R.M.; Shrock, D.J.; Akman, A.M.; Bland, L.G.; Thodla, R.; Lockem, J.S. The effect of sensitization and fatigue loading frequency on corrosion fatigue of AA5083-H131. *Int. J. Fatigue* **2019**, *124*, 1–9. [CrossRef]
12. Sidhom, N.; Moussa, N.B.; Janeb, S.; Braham, C.; Sidhom, H. Potential Fatigue Strength Improvement of AA 5083-H111 Notched Parts by Wire Brush Hammering: Experimental Analysis and Numerical Simulation. *Mater. Des.* **2014**, *64*, 503–519. [CrossRef]
13. Liu, Y.; Luo, L.; Han, C.; Ou, L.; Wang, J.; Liu, C. Effect of Fe, Si and Cooling Rate on the Formation of Fe- and Mn-Rich Intermetallics in Al-5Mg-0.8Mn Alloy. *J. Mater. Sci. Technol.* **2016**, *32*, 305–312. [CrossRef]
14. Ding, S.; Zhang, J.; Khan, S.A.; Yanagimoto, J. Static Recovery of A5083 Aluminum Alloy after a Small Deformation through Various Measuring Approaches. *J. Mater. Sci. Technol.* **2022**, *104*, 202–213. [CrossRef]
15. Dowling, N.E. *Mechanical Behavior of Materials*; Cambridge University Press: Cambridge, UK, 2000; Volume 15, ISBN 978-0-13-139506-0.
16. Klemenc, J.; Šeruga, D.; Nagode, A.; Nagode, M. Comprehensive Modelling of the Hysteresis Loops and Strain–Energy Density for Low-Cycle Fatigue-Life Predictions of the AZ31 Magnesium Alloy. *Materials* **2019**, *12*, 3692. [CrossRef] [PubMed]
17. Tomažinčič, D.; Nečemer, B.; Vesenjak, M.; Klemenc, J. Low-Cycle Fatigue Life of Thin-Plate Auxetic Cellular Structures Made from Aluminium Alloy 7075-T651. *Fatigue Fract. Eng. Mater. Struct.* **2019**, *42*, 1022–1036. [CrossRef]
18. Sedighi, M.; Monshi, P.; Joudaki, J. Investigation of Mechanical Properties and Fatigue Life of ECARed AA5083 Aluminium Alloy. *Fatigue Fract. Eng. Mater. Struct.* **2017**, *40*, 412–422. [CrossRef]
19. Ma, M.; Zhang, J.; Yi, D.; Wang, B. Investigation of High-Cycle Fatigue and Fatigue Crack Propagation Characteristic in 5083-O Aluminum Alloy. *Int. J. Fatigue* **2019**, *126*, 357–368. [CrossRef]
20. Meng, L.; Goyal, A.; Doquet, V.; Ranc, N.; Couzinié, J.-P. Ultrafine versus Coarse Grained Al 5083 Alloys: From Low-Cycle to Very-High-Cycle Fatigue. *Int. J. Fatigue* **2019**, *121*, 84–97. [CrossRef]
21. Ben Ali, N.; Estevez, R.; Tanguy, D. Heterogeneity of Grain Boundaries in 5xxx and 7xxx Aluminum Alloys and Its Influence on Intergranular Toughness. *Eng. Fract. Mech.* **2013**, *97*, 1–11. [CrossRef]
22. Hockauf, K.; Wagner, M.F.-X.; Halle, T.; Niendorf, T.; Hockauf, M.; Lampke, T. Influence of Precipitates on Low-Cycle Fatigue and Crack Growth Behavior in an Ultrafine-Grained Aluminum Alloy. *Acta Mater.* **2014**, *80*, 250–263. [CrossRef]
23. Wu, Y.; Liao, H.; Tang, Y. Enhanced High-Cycle Fatigue Strength of Al–12Si–4Cu-1.2Mn-T6 Cast Aluminum Alloy at Room Temperature and 350 C. *Mater. Sci. Eng. A* **2021**, *825*, 141917. [CrossRef]
24. Khisheh, S.; Khalili, K.; Azadi, M.; Hendouabadi, V.Z. Influences of Roughness and Heat Treatment on High-Cycle Bending Fatigue Properties of A380 Aluminum Alloy under Stress-Controlled Cyclic Loading. *Mater. Chem. Phys.* **2021**, *264*, 124475. [CrossRef]
25. Gao, T.; Sun, Z.; Xue, H.; Retraint, D. Effect of Surface Mechanical Attrition Treatment on High Cycle and Very High Cycle Fatigue of a 7075-T6 Aluminium Alloy. *Int. J. Fatigue* **2020**, *139*, 105798. [CrossRef]
26. Meng, X.; Leng, X.; Shan, C.; Zhou, L.; Zhou, J.; Huang, S.; Lu, J. Vibration Fatigue Performance Improvement in 2024-T351 Aluminum Alloy by Ultrasonic-Assisted Laser Shock Peening. *Int. J. Fatigue* **2023**, *168*, 107471. [CrossRef]
27. Xing, S.; Pei, X.; Mei, J.; Dong, P.; Su, S.; Zhen, C. Weld Toe versus Root Fatigue Failure Mode and Governing Parameters: A Study of Aluminum Alloy Load-Carrying Fillet Joints. *Mar. Struct.* **2023**, *88*, 103344. [CrossRef]
28. Sakin, R. Investigation of Bending Fatigue-Life of Aluminum Sheets Based on Rolling Direction. *Alex. Eng. J.* **2018**, *57*, 35–47. [CrossRef]
29. Nečemer, B.; Klemenc, J.; Zupanič, F.; Glodež, S. Modelling and Predicting of the LCF-Behaviour of Aluminium Auxetic Structures. *Int. J. Fatigue* **2022**, *156*, 106673. [CrossRef]
30. Kramberger, J.; Nečemer, B.; Glodež, S. Assessing the Cracking Behavior of Auxetic Cellular Structures by Using Both a Numerical and an Experimental Approach. *Theor. Appl. Fract. Mech.* **2019**, *101*, 17–24. [CrossRef]
31. Nečemer, B.; Kramberger, J.; Vuherer, T.; Glodež, S. Fatigue Crack Initiation and Propagation in Re-Entrant Auxetic Cellular Structures. *Int. J. Fatigue* **2019**, *126*, 241–247. [CrossRef]
32. Nečemer, B.; Vuherer, T.; Glodež, S.; Kramberger, J. Fatigue Behaviour of Re-Entrant Auxetic Structures Made of the Aluminium Alloy AA7075-T651. *Thin-Walled Struct.* **2022**, *180*, 109917. [CrossRef]
33. Nečemer, B.; Zupanič, F.; Gabriel, D.; Tarquino, E.A.; Šraml, M.; Glodež, S. Low Cycle Fatigue Behaviour of Ductile Aluminium Alloys Using the Inelastic Energy Approach. *Mater. Sci. Eng. A* **2021**, *800*, 140385. [CrossRef]
34. Nečemer, B.; Klemenc, J.; Glodež, S. The Computational LCF-Analyses of Chiral and Re-Entrant Auxetic Structure Using the Direct Cyclic Algorithm. *Mater. Sci. Eng. A* **2020**, *789*, 139618. [CrossRef]
35. Lehmhus, D.; Vesenjak, M.; de Schampheleire, S.; Fiedler, T. From Stochastic Foam to Designed Structure: Balancing Cost and Performance of Cellular Metals. *Materials* **2017**, *10*, 922. [CrossRef] [PubMed]
36. Yen, B.T.; Hodgson, I.C.; Zhou, Y.E.; Crudele, B.B. Bilinear S-N Curves and Equivalent Stress Ranges for Fatigue Life Estimation. *J. Bridg. Eng.* **2013**, *18*, 26–30. [CrossRef]

37. Tanaka, K.; Akiniwa, Y. A Model for Propagation of Small Fatigue Crack Interacting with Grain Boundary. *J. Soc. Mater. Sci. Jpn.* **1985**, *34*, 1301–1307. [CrossRef]
38. Tanaka, K. Fatigue Crack Propagation. In *Comprehensive Structural Integrity*; Milne, I., Ritchie, R.O., Karihaloo, B., Eds.; Pergamon: Oxford, UK, 2003.
39. Pineau, A.; McDowell, D.L.; Busso, E.P.; Antolovich, S.D. Failure of Metals II: Fatigue. *Acta Mater.* **2016**, *107*, 484–507. [CrossRef]

*materials*

MDPI

*Article*

# Material's Strength Analysis of the Coupling Node of Axle of the Truck Trailer

**Živilė Decker [1], Jurijus Tretjakovas [2], Kazimierz Drozd [3], Vitalijus Rudzinskas [1], Mariusz Walczak [3,*], Artūras Kilikevičius [4], Jonas Matijosius [4] and Iryna Boretska [5]**

1 Department of Mechanical and Material Engineering, Faculty of Mechanics, Vilnius Gediminas Technical University, J. Basanavičiaus g. 28, LT-03224 Vilnius, Lithuania; zivile.decker@vilniustech.lt (Ž.D.); vitalijus.rudzinskas@vilniustech.lt (V.R.)
2 Department of Applied Mechanics, Faculty of Civil Engineering, Vilnius Gediminas Technical University, Saulėtekio al. 11, LT-10223 Vilnius, Lithuania; jurijus.tretjakovas@vilniustech.lt
3 Faculty of Mechanical Engineering, Lublin University of Technology, Nadbystrzycka 36, 20-618 Lublin, Poland; k.drozd@pollub.pl
4 Mechanical Science Institute, Vilnius Gediminas Technical University, J. Basanavičiaus g. 28, LT-03224 Vilnius, Lithuania; arturas.kilikevicius@vilniustech.lt (A.K.); jonas.matijosius@vilniustech.lt (J.M.)
5 Department of Information Technologies, Ukrainian National Forestry University, 103, Gen. Chuprynky, 79057 Lviv, Ukraine; boretska@nltu.edu.ua
* Correspondence: m.walczak@pollub.pl

**Abstract:** Road transport plays an important role in the transport of goods and people and is important for the national economy. Damage usually excludes the means of transport from operation, which causes disruption of supply chains. One such damage is the failure of the suspension system of the vehicle or trailer, which usually occurs when the vehicle is heavily loaded. Such a defective system has been analyzed in this publication. Mathematical apparatus and finite element method (FEM) numerical simulations were used. A dangerous axle cross-section in terms of load was indicated and the maximum stresses in this area were calculated for two types of roads. On highways, the stress at the critical point was 199 MPa, and on uneven roads it increased to 304 MPa, which is comparable to the yield point. It was found that the second form of vibration may cause stresses in the damage area, but the excitation frequency would have to be quite high. The probability of such a load and failure event occurring is low under operating conditions.

**Keywords:** failure analysis; FEM analysis; fracture mechanics; macroscopic research; semi-trailers

**Citation:** Decker, Ž.; Tretjakovas, J.; Drozd, K.; Rudzinskas, V.; Walczak, M.; Kilikevičius, A.; Matijosius, J.; Boretska, I. Material's Strength Analysis of the Coupling Node of Axle of the Truck Trailer. *Materials* **2023**, *16*, 3399. https://doi.org/10.3390/ma16093399

Academic Editors: Grzegorz Lesiuk and Dariusz Rozumek

Received: 9 March 2023
Revised: 15 April 2023
Accepted: 20 April 2023
Published: 26 April 2023



## 1. Introduction

Trucks with semi-trailers are the main means of transporting goods by road all over the world. Transport companies have one economic goal—to transport as many goods as possible with the least number of journeys and at the lowest possible cost. In order to meet these expectations, these vehicles must be characterized by high quality construction, durability and reliability. In addition to the construction, suspension, drive and safety systems play an important role. In turn, the reliable operation of these systems has a direct impact on road safety [1–10]. On the other hand, the safety issues related to the transport of chemical products in road transport are presented in [11–13]. The durability of vehicle suspension components refers to the duration of the onset of fatigue, defined as the number of cycles to a specific length of component failure under cyclic loads [6,14,15]. Therefore, the axles of trailers and semi-trailers are one of the main and most important elements that carry the greatest loads during the transport of goods. The axles of semi-trailers are a structural element that withstands both the full weight of the semi-trailer and the load, as well as the reactions of the road surface. Proper construction of the vehicle and its performance in accordance with the requirements of the vehicle approval are key factors in ensuring the durability and reliability of the semi-trailer. Even if these requirements are met, there are cases where the vehicle is no longer used due to structural damage. Due to the measurable costs of decommissioning the vehicle and the threat to traffic safety, each case of damage should be investigated. This is the only way to identify the causes of damage and reduce transport costs.

Potential damage, whatever its cause, is modeled and estimated in modeling. The finite element method (FEM), sometimes referred to as finite element analysis (FEA), is widely used in the literature [16–21] to achieve essentially two purposes. One of the aims of this study is to analyze the stresses of the current semi-trailer design or to optimize its design. The other concerns the identification and verification of possible causes of injury. FEM is a computational technique used to obtain approximate solutions to limit value problems in engineering. In other words, a boundary value problem is a mathematical problem where one or more dependent variables must satisfy a differential equation in a known domain of independent variables and satisfy the specific boundary conditions of the domain. Simply put, a model of the object under study is created, it is divided into finite elements, the forces acting on the object are evaluated and the displacements, stresses, reactions and deformations of the elements are calculated. When performing FEM strength calculations, the main task is to calculate the load forces and torques and to model the road conditions so that the numerically modeled conditions reflect the real road conditions [17,22–25].

During operation, any trailer is subjected to loads that cause stresses, vibrations and noise in various parts of its structure. In order to withstand these loads, the strength, stiffness and fatigue properties of the respective components are required. In addition, the quality of the trailer as a system, which includes energy efficiency, safety and consumer comfort, is highly desirable [23].

Virtually every trailer and semi-trailer has a built-in data collection "black box". These data are very important in the event of a truck accident. A black box is a general term that can refer to several different elements of computerized systems typically installed in a commercial motor vehicle. In the specific case under analysis, the semi-trailer has an operational data logger integrated in the TEBS modulator, called ODR-Tracker. The ODR-Tracker program records the conditions under which the trailer was operated. These data are used to analyze the use of the vehicle and to evaluate the towing vehicle. During the analysis, it is possible to see the total mileage of the semi-trailer, the number of trips, the working hours of the driver, and so on. Importantly, this program records the average and maximum or exceeded driving speed of the semi-trailer, the air pressure, the frequency of the brake pedal, the braking time and force, the number of actuations of the ABS system and many other parameters [26]. Due to these advantages, the data obtained from this software are often the basis for assessing the efficiency of vehicle use (usually published) and real values of units and part load (most often unpublished). The last mentioned parameter is crucial if it is necessary to recreate the actual operating conditions under which the vehicle components have been damaged.

Very often the literature sources are limited to the analysis of FEM, but no less important is the real analysis of the axis design. When designing and manufacturing semi-trailer axles and other fasteners, it is very important to choose the right design solution. It is also important to choose the right materials from which the structure will be made. Properly designed constructions do not require technological adjustments and are durable. However, in the event of repeated axle fractures of semi-trailers, it is necessary to analyze the structural components, cross-sections and assess the reliability of the overall system [14,24,27,28].

The subject of the research presented in this paper was a damaged axle of a truck semi-trailer. The need to conduct the test arose from the fact that the transport company had several vehicles of the same type, all intensively used, and it was not the first case of damage of this character. Due to the fact that access to the ODR-Tracker data was obtained for the tested case of axle damage, it was decided to use them for analytical calculations and simulations. Thus, not only the geometry of the elements might be mapped, but also the values of the actual load. At the same time, the need to rely solely on nominal data was avoided, which distinguishes the approach presented here from those previously published.

## 2. Materials and Methods

The object of the research presented in this paper was the axle of a semi-trailer damaged during the operation of the vehicle. The damage to its material occurred in an unusual place. This did not happen near the wheel of the vehicle, where the greatest stresses due to road input seem to occur, but at the rocker arm, towards the centre of the vehicle. Therefore, it was decided to look for structural, technological, operational and dynamic factors that could explain the damage in this particular place.

The analysis of the structure and possible defects in the technology of making the elements according to the state after the damage was carried out using macroscopic metallography methods. The vehicle load and traction conditions were checked on the basis of recorded data on the courses of transport. Simulations of the axis dynamics were performed using FEM and based on analytical calculations. All test results complement each other and allow us to confirm or exclude possible causes of cracks.

As the damage to the axle assembly occurred in an unusual place, we decided to start the analysis by checking the frequency and form of natural vibrations. Although the mere occurrence of vibrations does not have to be the cause of damage in the joint, the fatigue strength of the materials is lower than the static one and the number of cycles that can be transferred is always limited.

### 2.1. Finite Element Method Simulations

Numerical simulations were carried out to check the location of potential stress concentration points. The model described in publication [29] was used for the simulation (Figure 1). The axis model (tube) contains 10,120 elements (20,424 nodes) type C3D4R with reduced integration. The mesh elements of the tube were intentionally denser in the area of their interaction with both bushing parts. A similar number of elements (9528) and nodes (15,052) consist of two bushings. For both parts, the mesh of C3D4R elements had been refined at the area of contact with the welds. Welds (for tying bushings and tube) were meshed with C3D4 elements (at least 10,785). The wishbones were modelled as a rigid body (not shown in Figure 1) between points (LeftHinge and RightHinge) and appropriate bushing.

Due to the complex system of interactions (tie, coupling, rigid body) between the geometry elements, the modelled parts were divided into partitions. The largest number of partitions, 16, were separated in the axis part. The quality of the mesh of all parts was checked in terms of the occurrence of geometrical errors, the time of a stable time increment and the consistency of the simulation results with the analytically calculated values.

The stiffness of the linear spring elements installed between axle (RP-3, RP-4) and left and right longerons was taken equal to 4.5 kN/mm. Shock absorbers with a damping coefficient of 10 N s/mm each were included between the same nodes like springs. The model omitted the stiffness and damping of wheels with tires. The wheel weight of 120 kg was simulated as hooked at the axle ends in points RP-1 and RP-2. Additionally, points RP-5 and RP-6 were assigned a mass of 25 kg each, which resulted from taking into account this part of the mass of the spring, the shock absorber and the wishbone, which are displaced together with the solid elements visible in Figure 1. The quality of the mesh of all parts was verified in terms of the occurrence of geometrical errors, the stable time increment and the consistency of the simulation results with the analytically calculated values. Simulations were performed using the Abaqus Explicit software.

The numerical model prepared in the described way was used to analyze the stress distribution in the axle material. The location of the zone of greatest stresses and their values were analyzed in terms of the possibility of damage observed on the real object. Additionally, the model was used to check the mode and frequency of the natural vibrations of the axis. This analysis was used to verify the values taken for the simulation and to check whether the generation of vibrations was related to the load on the axle material in the area of cooperation with the sleeve, where the origin of fracture usually takes place.

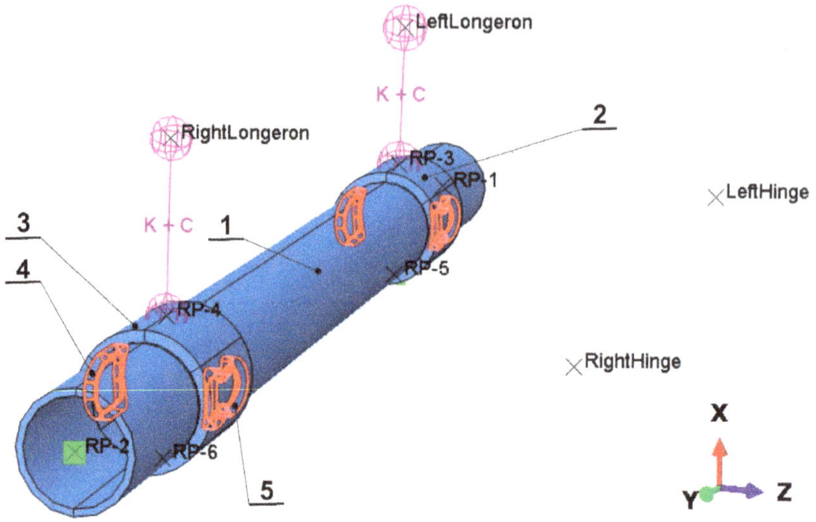


**Figure 1.** Assembly of the axle with constraint points on longerons and hinges: 1—axis; 2, 3—bushings; 4, 5—angle welds. RightLongeron, LeftLongeron, RightHinge and LeftHinge points have a rotational degree of freedom round the *Y* axis only. Between points RightHinge and RP-4, as well as RP-6, is rigid body type connection to the wishbone and a similar part is in model at the left side. Vehicle front at right.

### 2.2. Semi-Trailer ODR-Tracker Data Analysis

Trip data is recorded in the ODR-Tracker in the TEBS modulator of the semi-trailer and stores the last 200 trip data. Recording starts when the distance traveled is at least 5 km and the vehicle speed is at least 30 km/h.

According to the Register of Legislation, speed limiters for freight transport with a gross vehicle weight of more than 12 Mg must be set at a maximum speed of 85 km/h.

In Figure 2, we can see that the maximum speed was exceeded during almost every trip, with a speed of 90–110 km/h. In the load chart, we see journeys when the semi-trailer was running unloaded or with a small load and journeys when the semi-trailer was fully loaded. Comparing the speed and semi-trailer load diagrams, we see that the speeds were exceeded even when the semi-trailer was fully loaded.

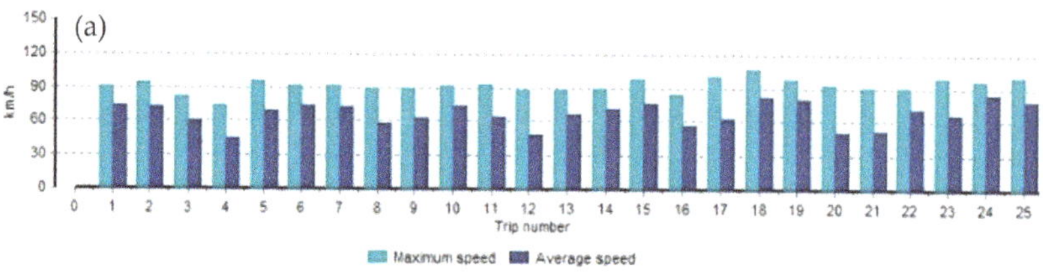


**Figure 2.** *Cont.*

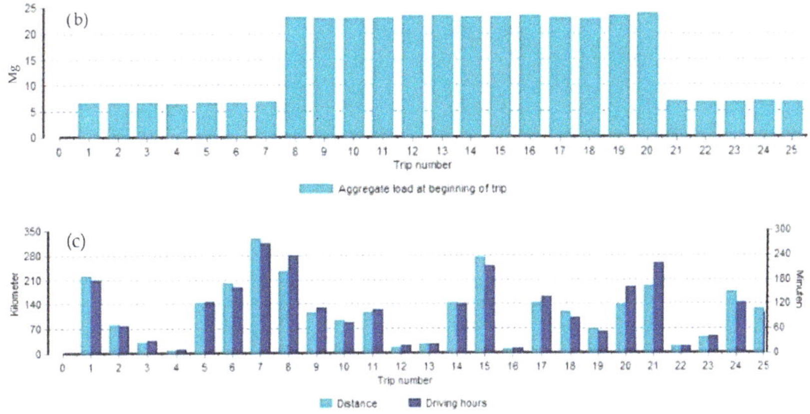


**Figure 2.** Characteristics of transportation parameters: speed (**a**), load (**b**), distance and time (**c**).

In the event of road irregularities or potholes, a fully loaded semi-trailer is more likely to break under chassis components or components at maximum or close speeds.

We observe that all the numerical values in red are the limits of the set parameters (Table 1). Data in the table are ordered in max speed. In 35% of all trips, we see speeding and exceeding the maximum load on the axle. These are factors that can affect the critical deformations of the axles.

**Table 1.** Vehicle parameters registered during full-loaded transport.

| Distance, km | Driving Time, h | Max Speed, km/h | Pressure, MPa | Break | | Max Load, Mg |
|---|---|---|---|---|---|---|
| | | | | Actuations, - | Frequency, 1/km | |
| 10.9 | 00:12 | 78 | **0.285** | 2 | 0.18 | 24.1 |
| 10.9 | 00:11 | 85 | **0.410** | 2 | 0.18 | 24.1 |
| 49.7 | 01:10 | 85 | 0.200 | 44 | 0.89 | 24.1 |
| 16.5 | 00:20 | 88 | **0.225** | 12 | 0.73 | 24.1 |
| 234.2 | 04:00 | 89 | 0.195 | 117 | 0.50 | 24.1 |
| 116.3 | 01:51 | 90 | 0.185 | 52 | 0.45 | 24.1 |
| 68.7 | 00:56 | 92 | **0.270** | 20 | 0.29 | 25.1 |
| 34.1 | 00:37 | 92 | 0.155 | 37 | 1.09 | 24.1 |
| 116.0 | 01:47 | 93 | **0.205** | 43 | 0.37 | 24.1 |
| 137.9 | 02:41 | 93 | 0.165 | 83 | 0.60 | 24.1 |
| 274.3 | 03:41 | 94 | 0.175 | 76 | 0.28 | 24.6 |
| 197.0 | 02:45 | 95 | **0.240** | 59 | 0.30 | 25.6 |
| 132.9 | 01:19 | 97 | **0.210** | 14 | 0.11 | 24.1 |
| 262.3 | 03:59 | 97 | 0.180 | 180 | 0.69 | 24.6 |
| 70.5 | 00:53 | 98 | 0.185 | 22 | 0.31 | 24.1 |
| 17.9 | 00:17 | 99 | 0.190 | 12 | 0.67 | 24.1 |
| 109.7 | 01:33 | 101 | **0.215** | 45 | 0.41 | 24.1 |
| 179.0 | 02:35 | 101 | 0.195 | 85 | 0.47 | 24.1 |
| 258.1 | 03:08 | 101 | 0.185 | 94 | 0.36 | 24.1 |

A jump in air pressure gauges has been recorded. Such significant changes in the parameters of the air system (2 times higher than the permissible ones) are possible only in the presence of strong, short-term effects of external factors, such as the wheel entering a sufficiently deep pit or entering an obstacle [21,30]. We can conclude that there were such externalities [24].

### 2.3. *Analytical Calculations of an Axle Load*

Analytical calculations were carried out for the middle part of the connection of the axle with the sleeve, in the place where the cross-section is the smallest. The geometric moments of inertia of the cross-section, axials and polar were determined. The axle load was assumed on the basis of data published in the literature. The location of the most loaded cross-section and the values of normal and shear stresses were determined. The calculations were carried out when driving on a good quality road and on a bumpy road.

## 3. Results and Discussion

### 3.1. *Effects of Numerical Simulations*

Mode 1 and 2 take place in a vertical plane (Figure 3). The first mode of vibration has the character of uniform deflection of the complete axis. In addition to the mass of the axle, the frequency of the vibrations is therefore affected by the stiffness of the suspension spring elements. At an excitation frequency of 24 Hz, with a deflection of less than 2 mm, the entire axis vibrates in the vertical direction. The figure shows the same deflection of the suspension springs for mode 1. For this reason, this mode of vibration cannot generate stresses in the material. The vibration period in this case is about 41 ms. Therefore, if the vehicle was moving at a speed of 90 km/h (a value close to the maximum speed in Figure 1), this mode of vibration would occur when the wheel passes over bumps on the road spaced approximately every 1 m.

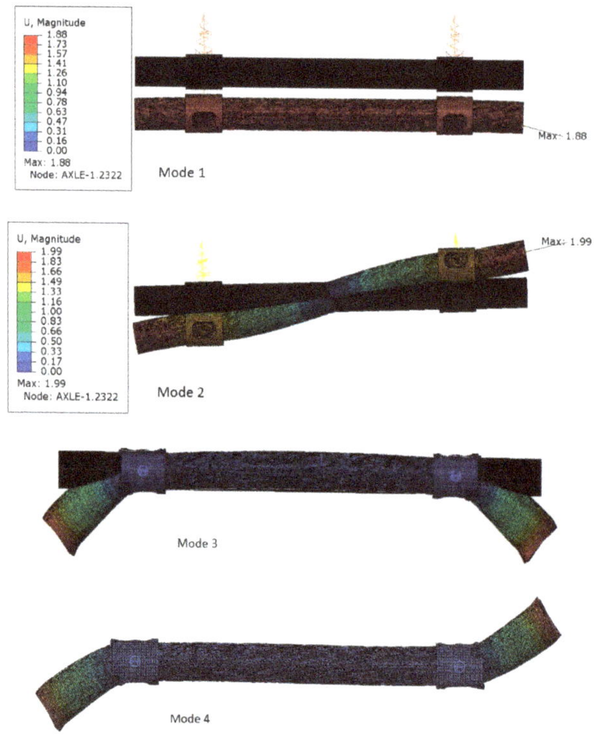


**Figure 3.** Modes of natural vibrations of the axle. Modes 1 and 2—view from the back side, modes 3 and 4—top view. Deformation scale factor 130.

The second mode of vibration also depends on the stiffness of the elastic elements of the suspension. Unlike the first form, this one involves the generation of stresses in the material, because the right and left sides of the axis deform in opposite directions—up and down. The greatest deflection, about 2 mm, should be expected at the end of the axle, where the wheels are mounted. This mode of vibration can occur at an excitation frequency of about 56 Hz, which is more than twice the first one. The greatest stresses should be generated in the area of the smallest deflection, i.e., about half the length of the axle, as well as from the outside, where the axle and bushing are connected by a weld cooperate. It may be concluded that vibrations of this type will not have a significant impact on the fatigue load of the central part of the axle, where good quality protective coating was observed. A greater risk may occur in the area of axle and bushing contact. Due to the impact of sand, stones and water thrown by the wheels, the protective coating is subject to faster wear there (see Figure 4). In addition, the mutual cooperation of the two parts that make up the axle may cause abrasion of the paint coating even in places invisible from the outside.

For the second form of vibration to occur, the excitation existing on the road should be approximately every 443 mm, if the vehicle is moving at a speed of 90 km/h. This distance is shorter than the dynamic radius of the wheel (about 497 mm), so in addition to the elastic characteristics of the suspension, the stiffness of the tires would be important here; this was neglected in the simulation.

The first form of vibration is the most likely, i.e., it can most often occur when the vehicle is moving at a speed of 90 km/h.

Mode 3 and mode 4 (Figure 3) are only in a horizontal plane. In this case, the maximum axle deflection of 2.16 mm occurs only on the outside of the wishbones. The excitation frequency necessary to induce these vibration modes is about 149 Hz, which is almost 3 times higher than for mode 2. The occurrence of such an excitation frequency is unlikely under operating conditions. The difference in the natural frequency for modes 3 and 4 does not exceed 0.32 Hz, which is less than 3%.

None of the four modes of vibration described can be caused by forces from rotating suspension components. In a vehicle traveling at a speed of 90 km/h, a wheel with a dynamic radius of 497 mm rotates without slipping at a frequency of less than 6 Hz. So, it is much lower than the frequency 24 Hz calculated for mode 1.

### 3.2. Axle Construction Analysis

The test object consists of round profile axle, couplings with welded brackets, shock absorbers, airbags, rubber bushings, wheel hubs with disc brake system and other fasteners. Mounting points for brackets and axle couplings are shown in Figure 4.

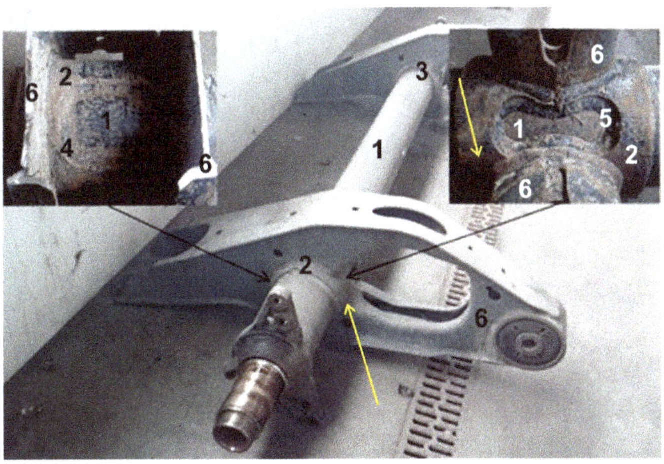


**Figure 4.** Mounting points for brackets and couplings: 1—axis; 2, 3—bushings; 4, 5—welds; 6—arm. The fracture origin area is pointed out by yellow arrows.

The brackets are connected to the shaft with sleeve joints with no thermal effect and are welded in four places. In the coupling joint, two technological holes of different geometries are cut from opposite sides. In those places, the couplings are welded to the shaft (Figure 5). From the results, we can see that the geometry of the shaft and the coupling changes are due to the welding process (there is a gap between the shaft and the coupling) (Figure 6).

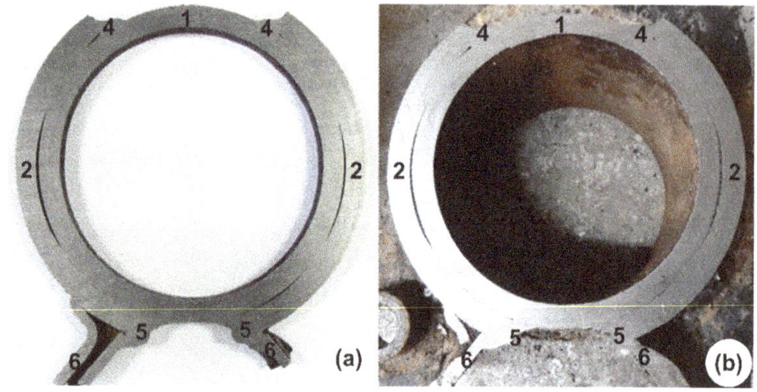


**Figure 5.** Two sides of the same cross-section at welds of the axis unit: (**a**)—broken side; (**b**)—unbroken side. The numerals mean: 1—axis; 2—bushing; 4, 5—angle welds; 6—arm.

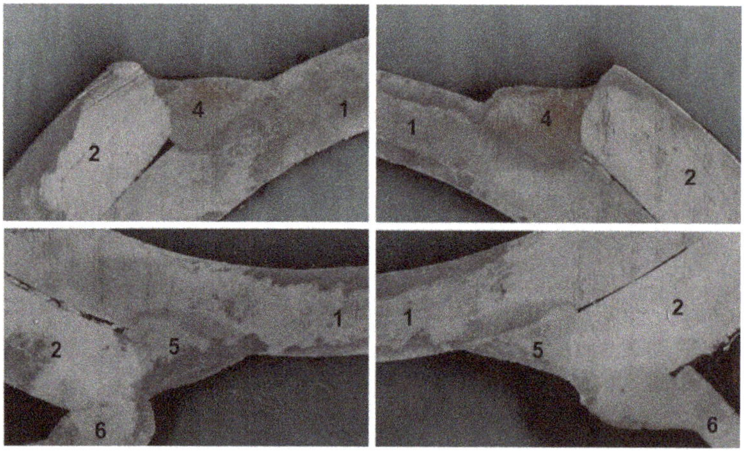


**Figure 6.** Cross-section of angle welds: 1—axis; 2—bushing; 4, 5—angle welds; 6—arm.

Examining the fastening of the brackets and couplings (Figure 4) we see that the total area of contact of the coupling with the axis is 84,811 mm$^2$. The area of the welds on which the sleeve joint with the shaft is reinforced is 8628 mm$^2$. During the destructive studies and the examination of the fracture (Figure 5), we notice that the contact area has changed due to welding. The coupling is detached from the axis and we can say that the axis is "held" only by the welds, which make up only 10% of the total contact area in the project [31–33].

We see that there is a gap between the shaft and the coupling connection, in places up to 1.5 mm (Figure 5a). To make sure that this is not the result of a fracture, but a construction, we also cut the part of the shaft without the fracture. In Figure 5b), we see that there is also a gap in the cross-section of the fractured side of the shaft between the shaft and the coupling joint. This confirms the assumption that the gap is not a consequence of a fracture [34].

The measurement of the welded joints shall assess the conformity of the joints, their arrangement and the amounts of defects observed during the visual inspection: inclusions, accumulations, cuts, indentations, convexities or depressions in the seam surface, fractures of the seam surface size and deviations in geometrical parameters.

Seam defects and damage are identified according to [35,36] of the literature.

The integrity of the weld seam was inspected and no cracks were observed. After the section of the axle at the welded joints (Figure 6), the cross-sections of the welds were inspected. Defects observed: 402—Non-weld (insufficient weld), 500—Shape defects (irregular geometry of the weld, imperfect shape of the outer surface of the weld, different geometries of the cross-sectional area, cross-sectional areas of the weld vary from 6 to 20%), 505—Insufficient weld edges, 509—Depression (formed by gravity; the amount of the weld metal on the axle metal is higher than on the coupling metal; significantly greater thermal impact area), large angular seam asymmetry (excessive structural unevenness), 513—Uneven width (large seam width tolerances), 514—Uneven surface, 602—Splashes (added metal droplets spattered by welding and adhering to the base and hardened seam metal).

For welded joints, the area of thermal exposure close to the weld limit is of great importance. The range of thermal effects of the test seam is very different: from 1 mm to 3.5 mm. The metal heats the seam almost to the melting point. At that point, the metal is overheated, so this area usually determines the strength of the welded joint.

It can be seen that there is a significant difference in the load on the axle material depending on whether the considered point is located on the outside of the wishbone (between the wishbone and the wheel hub journal) or in the place between the wishbones. During driving and acceleration, if friction in rolling bearings is omitted, no torsional stresses are generated in this section of the axle which is closer to the wheel. The torsional load of this part of the axle occurs only during braking. Apart from that, bending stresses prevail; the source of which is the weight on one side of the axle (normally half of the whole axle load), and geometric extortions caused by unevenness in the road [37,38]. In the middle part of the axle, between the wishbones, bending and torsion loads occur simultaneously. In addition, this load is greater than the difference in the deflection (rotation) of the wishbones of the same axle.

In the case of analytical calculations, which are shown below, all mentioned load types are included.

### 3.3. *Analytical Calculations*

#### 3.3.1. The Moments of Inertia of the Cross-Section

The geometry of the cross-section is calculated as a complex figure consisting of an annular tube with two external brackets (Figure 7). Additional areas in the cross-section of the welds and corbels (Figure 5) are unvalued. The moments of inertia of the cross-section of the axle are determined as a sum of the central annular tube and the two external brackets.

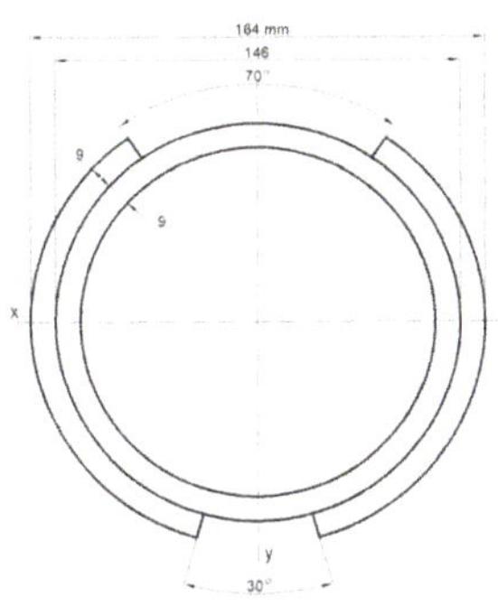


**Figure 7.** Axis cross-section geometry for solution of moments of inertia.

The shaft is deformed by two bending moments and torque and so, two moments of inertia and a polar moment of inertia are determined.

The polar moment of inertia in general cases [39] is:

$$I_p = \iint_A \rho^2 \mathrm{d}A = \iint_A \rho^2 \rho \, \mathrm{d}\varphi \mathrm{d}\rho \tag{1}$$

Due to symmetry, only half of the area is used as a domain of integration (Figure 8) for a solution of the polar moment of inertia.

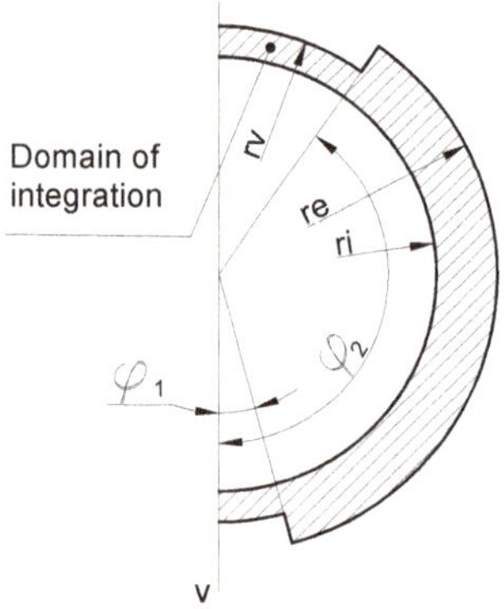


**Figure 8.** Domain of integration.

Using integration Formula (1) for a defined domain (Figure 8) from vertical axis $v$:

$$I_p = \int_0^{\varphi_1} \mathrm{d}\varphi \int_{r_i}^{r_v} \rho^3 \, \mathrm{d}\rho + \int_{\varphi_1}^{\varphi_2} \mathrm{d}\varphi \int_{r_i}^{r_e} \rho^3 \, \mathrm{d}\rho + \int_{\varphi_2}^{\pi} \mathrm{d}\varphi \int_{r_i}^{r_v} \rho^3 \, \mathrm{d}\rho \, , \tag{2}$$

In cases when (according to Figure 7) values of particular angles and radiuses are equal:

$$\varphi_1 = \frac{30^\circ}{2} = 15^\circ \; ; \; \varphi_2 = 180^\circ - \frac{70^\circ}{2} = 145^\circ, \tag{3}$$

$$r_i = \frac{146 \text{ mm} - 2 * 9 \text{ mm}}{2} = 64 \text{ mm} \; ; \; r_v = \frac{146 \text{ mm}}{2} = 73 \text{ mm} \; ; \; r_v = \frac{164 \text{ mm}}{2} = 82 \text{ mm} \, , \tag{4}$$

The polar moment of inertia $I_p$ is equal $37.3 \cdot 10^6$ mm$^4$.

The area of the cross-section and moments of inertia around the x and y axes (Figure 9) are solved and checked with CAD program.

The geometrical parameters of cross-section: area $A$ = 7030 mm$^4$; moments of inertia: $I_x = 15.6 \cdot 10^6$ mm$^4$; $I_y = 21.7 \cdot 10^6$ mm$^4$.

The real orientation of the cross-section of the axle is rotated with an angle of 74 degrees (Figure 10).

The moments of inertia around the horizontal and vertical axis: $I_{x1} = 21.2 \cdot 10^6$ mm$^4$; $I_{y1} = 16.2 \cdot 10^6$ mm$^4$. The polar moment of inertia and section modules will be used to determine maximal stresses in the cross-section and the finding of critical points.

### 3.3.2. Analytical Normal and Shear Stresses in Dangerous Zone of Cross-Section

Bending and torsion are described in [40,41] and the internal forces of the axle are taken from [42] and will be used in this chapter. The investigation quarter of the cross-section is in left and down with both positive stresses due to bending moments (Figure 11). In this quarter exists welding of internal and external tubes; consequently, welding is a stress concentrator factor. It follows that critical point A with maximal stresses is in the dangerous zone of weld or near this zone.

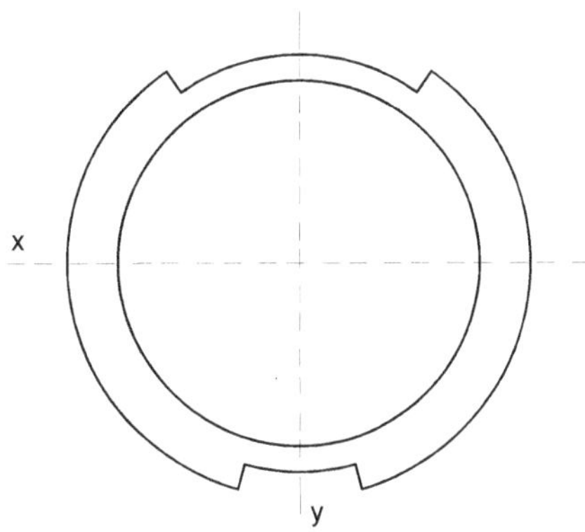


**Figure 9.** Area for determine moments of inertia.

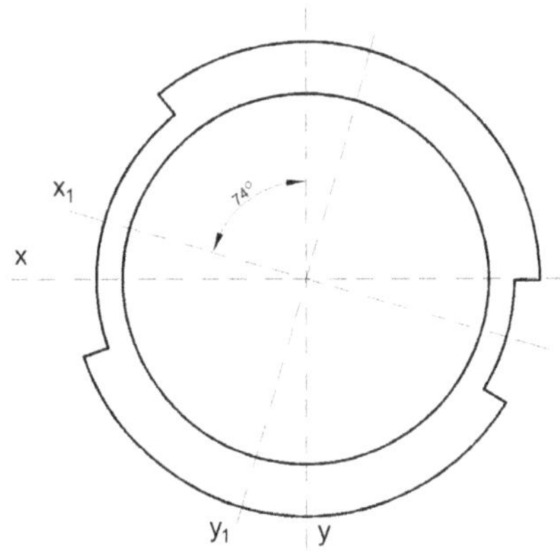


**Figure 10.** Area to determine moments of inertia with rotation.

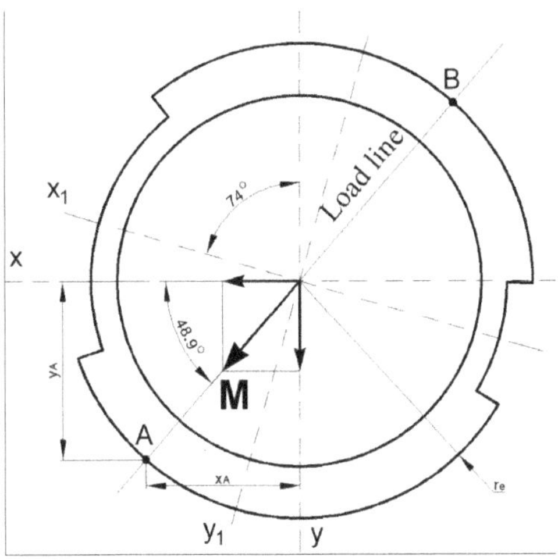


**Figure 11.** Sum moments and load line direction.

Two types of heavy loadings of axle are used to determine the stresses in point A [21,39–41]:

1. Driving of the truck on highways without appreciable path holes (the maximal depth is 50 mm);
2. Diving of the truck on highways or roads with path holes with a depth of up to 100 mm.

3.3.3. Driving of the Truck on Highways

Internal forces from [29] are:

- Bending moment $M_x$ = 31.1 kNm;
- Bending moment $M_y$ = 27.3 kNm;
- Torque $T$ = 9.0 kNm.

Both bending moments can be replaced with resultant vector sum **M** (Figure 11). This sum moment is acting on a plane which is termed as a load plane. The intersection of the load plane with the cross-sectional plane is the load line.

The direction of the load line is described by angle $\varphi$:

$$\varphi = \text{atan}\left(\frac{M_x}{M_y}\right) = \text{atan}\left(\frac{31.1\text{ kNm}}{27.3\text{ kNm}}\right) = 0.85\text{ rad} = 48.7^{\circ} \tag{5}$$

Points A and B are on the plane where sum moment **M** acting and normal stresses are extreme. From previous investigations [39], the position of the dangerous quarter has been found in the cross-section—front and down quarter. As a result, the normal stresses in point A are determined:

$$\sigma_A = \frac{M_x}{I_{x1}} \cdot y_A + \frac{M_y}{I_{y1}} \cdot x_A, \tag{6}$$

where, $y_A$ and $x_A$ are coordinates of the point A in Figure 11.

The real values of coordinates are:

$$x_A = r_e \cdot cos\varphi = 82 \cdot cos48.7^{\circ} = 54.1\text{ mm}$$

$$y_A = r_e \cdot sin\varphi = 82 \cdot sin48.7^{\circ} = 61.6\text{ mm}$$

The normal stress in point A due to bending moments acting about the horizontal and vertical axes is:

$$\sigma_A = \frac{31.1 \cdot 10^6\text{ Nmm}}{21.2 \cdot 10^{-6}\text{ Nmm}} \cdot 61.6\text{ mm} + \frac{27.3 \cdot 10^6\text{ Nmm}}{16.2 \cdot 10^{-6}\text{ Nmm}} \cdot 54.1\text{ mm} = 161\text{ MPa} \tag{7}$$

Shearing stresses in point A:

$$\tau_A = \frac{T}{I_p} \cdot \rho_A = \frac{9.00 \cdot 10^6\text{ Nmm}}{37.3 \cdot 10^6\text{ mm}^4} \cdot 82\text{ mm} = 19.8\text{ MPa} \tag{8}$$

Compound influence of bending and torsion is calculated below:

$$\sigma_{d,A} = \sqrt{\sigma_A^2 + 3 \cdot \tau_A^2} = \sqrt{161^2\text{ MPa} + 3 \cdot 19.8^2\text{ MPa}} = 165\text{ MPa} \tag{9}$$

The value of stress when the truck drives on highways proves that the construction is enough strong. The dominant type of stress here is normal and its value is less than 30 % of the yield strength (581 MPa) for axis material.

3.3.4. Driving of the Truck on Roads with Roughness

For analyzing the case of driving of the track on pavement roads, values of loading forces of the axle was taken from publication [29]. They are:

- Axial force caused, e.g., by truck turning $N$ = 44.1 kN;
- Bending moment which is generated by the force acting between the tire and the road surface $M_x$ = 53.3 kNm and $M_y$ = 27.3 kNm;
- The torque of the axle which is a sum of the braking force acting at the dynamic radius of the tire $T$ = 12.4 kNm.

The load line changed direction due to increasing the bending moment about the $x$ axis (Figure 12). The angle of the moment vector was calculated according to the equation:

$$\varphi = \text{atan}\left(\frac{M_x}{M_y}\right) = \text{atan}\left(\frac{53.3\ \text{kNm}}{27.3\ \text{kNm}}\right) = 1.098\ \text{rad} = 62.9^{\circ} \tag{10}$$

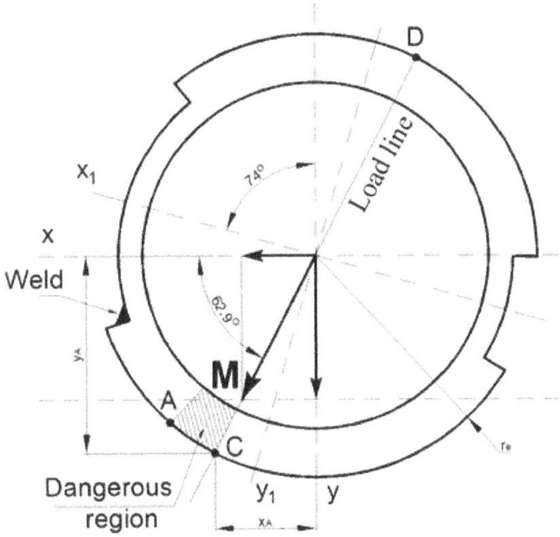


**Figure 12.** Load line direction and rotation of coordination system.

The normal stresses have additional influence due to the axial force increasing due to bending moments and expressed as follows:

$$\sigma_C = \frac{N}{A} + \frac{M_x}{I_x}\cdot y_C + \frac{M_y}{I_y}\cdot x_C \tag{11}$$

The real values of coordinates are:

$$x_C = r_e\cdot cos\varphi = 82\cdot cos62.9^{\circ} = 37.4\ \text{mm}$$

$$y_C = r_e\cdot sin\varphi = 82\cdot sin62.9^{\circ} = 73.0\ \text{mm}$$

With values:

$$\sigma_C = \frac{44.1\cdot 10^3\ \text{N}}{7030\ \text{mm}^2} + \frac{53.3\cdot 10^6\ \text{Nmm}}{21.2\cdot 10^6\ \text{mm}^4}\cdot 73\ \text{mm} + \frac{27.3\cdot 10^6\ \text{Nmm}}{16.2\cdot 10^6\ \text{mm}^4}\cdot 37.4\ \text{mm} = 253\ \text{MPa} \tag{12}$$

Shearing stresses with road roughness influence:

$$\tau_C = \frac{T}{I_p}\cdot r_e = \frac{12.4\cdot 10^9\ \text{Nmm}}{37.3\cdot 10^6\ \text{mm}^4}\cdot 82\ \text{mm} = 4.05\ \text{MPa} \tag{13}$$

Compound determining stress in weld region:

$$\sigma_{d,C} = \sqrt{\sigma_C^2 + 3\cdot\tau_C^2} = \sqrt{253^2 + 3\cdot 4.05^2} = 253\ \text{MPa} \tag{14}$$

Advancing load strength determined by the dynamics coefficient 1.2 [35,42–44].

On highways, the maximal stress in a dangerous zone will increase due to a stress concentrator: from $\sigma_{d,A}$ = 161·1.2 = 193 MPa, until $\sigma_{d,A}$ = 165·1.2 = 198 MPa.

During driving on low quality roads, the actual stresses in the axle increase to: $\sigma_{d,C}$ = 253·1.2 = 304 MPa.

Between points A and C, there is a dangerous region of cross-sections with extreme possible stresses. This region has a stable position in plane due to bending moments. When the wheel meets roughness in the road, due to the rotating of the cross-section counter clockwise, the weld between the external and internal tubes with the entire section also rotates to a dangerous zone (Figure 12).

Another stress concentrator factor is weld in the cross-section. Since the position of the weld seam is different, such as critical points with extreme stresses in the load line, the stresses in the weld cross-sectional area will be analyzed in the next paper.

#### 3.3.5. Fatigue of Coupling Node

In order to specify a strength for axle under repetitive loading, it is necessary to determine fatigue limit $\sigma_{-1}$ for steel of axle. It could be performed experimentally and theoretically. Endurance experiments are expensive and need special laboratory equipment.

In theory [36], fatigue limit $\sigma_{-1}$ for the symmetrical cycle described as function on ultimate tensile stress $\sigma_u$. For tension, this function is $\sigma_{-1} = 0.28 \cdot \sigma_u$, for bending $\sigma_{-1} = 0.40 \cdot \sigma_u$, respectively, for torsion $\tau_{-1} = 0.22 \cdot \sigma_u$. Stresses of axle vary according to non-symmetrical function, depend on road roughness and could be similar due to distribution in Figure 12.

Is it known that non symmetrical cycle loading has a lesser endurance limit [45]. It is enough to know that for those analyses, that in bending and torsion fatigue limit is maximum 50% of ultimate limit. From previous investigations [29,42], it is known that ultimate tensile stress is 663 MPa, so a theoretical maximal fatigue limit of steel will not exceed 332 MPa; accordingly, this limit will be decreased in reality. If the value of stresses generated in the material of the vehicle axle (304 MPa) increased by about 9%, they would reach the calculated tensile stress limit at fatigue load.

It can be seen that the second term of the sum in Equation (12) has the greatest influence on the stress value. With unchanged values of other parameters, a 20-degree left rotation of the system shown in Figure 12 could cause such an increase in stresses that they would reach the fatigue stress limit. The same effect could be caused by changing the direction of action of the unchanged value of the total moment M by about 20 degrees to the right. In this case, the angle would not be 62.8 but 42.8 degrees.

It is worth paying attention to an additional aspect, structural or technological. It was shown that the elements of the tested axle did not touch each other over the maximum surface area. This means that the entire cross-section could not be active in the load transfer. When driving on a poor quality road in this case, damage may occur in the observed place. In the literature [5], one can find an example of simulating a similar system of the whole vehicle in which the authors indicate the same area as having a critically low fatigue strength.

## 4. Conclusions

1. From analyzing the ODR-Tracker data, it was found that 35% of all trips were speeding and exceeding the maximum axle load.
2. In destructive studies, it was observed that the contact area between the two elements changed due to welding. The coupling is in contact with the shaft only at the points where the welds are located, which is only 10% of the total contact area of the joint.
3. The measurement of welded joints shall assess the conformity of the joints, their arrangement and the amounts of defects observed during the visual inspection: inclusions, accumulations, cuts, indentations, convexities or depressions in the seam surface, fractures of the seam surface size and deviations in geometrical parameters.

4. After the section of the axle at the welded joints, the cross-sections of the welds were inspected. Defects observed: non-weld, shape defects, insufficient weld edges, depression, large angular seam asymmetry, uneven width, uneven surface and splashes.
5. Maximal analytical stresses were calculated in the dangerous region for two types of driving. On highways stresses in coupling, the node critical point increase until 198 MPa, on roads with roughness—304 MPa. Overload by 9% would cause a fatigue limit of steel.
6. The extreme stresses occur in the defined dangerous region of the cross-section and in rotating the cross-section, the weld is closer to this region with increasing value of stresses.
7. It is shown, that in the dangerous quarter of the cross-section exists weld and the weld position does not coincide with the load line. Due to welding, additional stress concentration in the weld region appears and results from this investigation will be used for weld structure analyses.
8. The first vibration mode takes place at the excitation frequency of 24 Hz and can most often occur when the vehicle is moving at a speed of 90 km/h. However, the deformation of the system for this mode is such that it does not generate stresses in the material. The deformation of the axis in the hazardous area, where the damage was observed, may occur as a result of natural vibrations of the second mode. The excitation frequency in this case would have to be 56 Hz, which is less likely under operating conditions.

**Author Contributions:** Conceptualization, Ž.D., V.R., J.T., K.D., M.W., A.K., J.M. and I.B.; methodology, Ž.D., V.R., M.W., A.K. and I.B.; software, K.D. and I.B.; validation, M.W., A.K. and J.M.; formal analysis, K.D. and M.W.; investigation, Ž.D., V.R., J.T., K.D., M.W., A.K., J.M. and I.B.; resources, Ž.D., V.R., J.T., M.W. and J.M.; data curation, J.T., K.D. and V.R.; writing—original draft preparation, Ž.D., V.R., J.T., K.D., M.W., A.K., J.M. and I.B.; writing—review and editing, Ž.D., J.T., K.D. and J.M.; visualization, K.D. and I.B.; supervision, K.D. and J.M.; project administration, J.M.; funding acquisition, K.D. and M.W. All authors have read and agreed to the published version of the manuscript.

**Funding:** This research received no external funding.

**Institutional Review Board Statement:** Not applicable.

**Informed Consent Statement:** Not applicable.

**Data Availability Statement:** Not applicable.

**Acknowledgments:** This work was partly prepared as part of the Kazimierz Drozd scientific internship at the Institute of Mechanical Science of Vilnius Gediminas Technical University which took place from 26 July to 6 August 2021; and the Mariusz Walczak scientific internship at the Institute of Mechanical Science of Vilnius Gediminas Technical University which took place from 11 July to 8 August 2022.

**Conflicts of Interest:** The authors declare no conflict of interest.

## References

1. Gajek, A. Directions for the development of periodic technical inspection for motor vehicles safety systems. *Arch. Automot. Eng. Arch. Motoryz.* **2018**, *80*, 37–51. [CrossRef]
2. Gogola, M.; Ondruš, J.; Kubalak, S.; Turiak, P. Comparison of braking properties of selected vehicle with different methods. *Arch. Automot. Eng. Arch. Motoryz.* **2022**, *95*, 5–17. [CrossRef]
3. Hudec, J.; Šarkan, B.; Czödörová, R.; Caban, J. The Influence of Quality Management System on the Operation of Periodical Technical Inspection Stations. *Appl. Sci.* **2021**, *11*, 4854. [CrossRef]
4. Jagelčák, J.; Kiktová, M.; Frančák, M. The Analysis of Maneuverability of Semi-trailer Vehicle Combination. *Transp. Res. Procedia* **2020**, *44*, 176–181. [CrossRef]
5. Sejkorova, M.; Verner, J.; Sejkora, F.; Hurtova, I.; Senkyr, J. Analysis of operation wear of brake fluid used in a Volvo car. In *Transport Means—Proceedings of the International Conference, Trakai, Lithuania, 3–5 October 2018*; Kaunas University of Technology: Kaunas, Lithuania, 2018; pp. 592–596.

6. Vdovin, D.; Levenkov, Y.; Chichekin, I. Prediction of fatigue life of suspension parts of the semi-trailer in the early stages of design. *IOP Conf. Ser. Mater. Sci. Eng.* **2020**, *820*, 012002. [CrossRef]
7. Vlkovský, M.; Veselík, P. Cargo Securing—Comparison of Different Quality Roads. *Acta Univ. Agric. Silvic. Mendel. Brun.* **2019**, *67*, 1015–1023. [CrossRef]
8. Vrabel, J.; Skrucany, T.; Bartuska, L.; Koprna, J. Movement analysis of the semitrailer with the tank-container at hard braking—The case study. *IOP Conf. Ser. Mater. Sci. Eng.* **2019**, *710*, 012025. [CrossRef]
9. Zhao, X.; Kang, J.; Lei, T.; Li, X.; Cao, Z.; Wang, Y. Research on air cushion suspension support system with low speed and heavy load. *J. Phys. Conf. Ser.* **2020**, *1549*, 052020. [CrossRef]
10. Zuska, A.; Kurczyński, D.; Jackowski, J.T. Study of Loads Acting on the Load during the Sudden Braking of a Vehicle. *Appl. Sci.* **2023**, *13*, 1559. [CrossRef]
11. Bartnik, G.; Krzysiak, Z.; Samociuk, W.; Łysiak, G.; Plizga, K.; Szmigielski, M.; Nieoczym, A.; Kaliniewicz, Z.; Brumerčik, F. Dokumentowanie spełniania wymagań w obszarze bezpieczeństwa technicznego na przykładzie dystrybucji paliw ciekłych. *Przem. Chem.* **2017**, *96*, 1039–1041. [CrossRef]
12. Stoma, M.; Maj, G.; Dudziak, A.; Slaska-Grzywna, B.; Piekarski, W.; Andrejko, D. Rotation of chemical products in compliance with the Safety and Quality Research SQAS. *Przem. Chem.* **2015**, *94*, 1883–1886.
13. Vasalic, D.; Masonicic, Z.; Milojević, S.; Ivkovic, I.; Pesic, R. Some aspects concerning management of road transport of dangerous goods using contemporary information systems. *Mobil. Veh. Mech.* **2021**, *47*, 25–34.
14. Jilek, P.; Berg, J.; Tchuigwa, B.S.S. Influence of the Weld Joint Position on the Mechanical Stress Concentration in the Construction of the Alternative Skid Car System's Skid Chassis. *Appl. Sci.* **2022**, *12*, 397. [CrossRef]
15. Nešić, N.; Simonović, J.; Blagojević, M.; Milojević, S.; Jović, S. Vehicle Suspension System with Integrated Inerter—Extended Analysis. *IOP Conf. Ser. Mater. Sci. Eng.* **2022**, *1271*, 012030. [CrossRef]
16. Caban, J.; Nieoczym, A.; Gardyński, L. Strength analysis of a container semi-truck frame. *Eng. Fail. Anal.* **2021**, *127*, 105487. [CrossRef]
17. Dębski, H.; Koszałka, G.; Ferdynus, M. Application of FEM in the analysis of the structure of a trailer supporting frame with variable operation parameters. *Eksploat. Niezawodn.* **2012**, *14*, 107–114.
18. Gorbunov, M.; Dižo, J.; Blatnický, M.; Kravchenko, K.; Semenov, S.; Mikhailov, E. Proposal of a method for detection of a damaged hydraulic shock absorber in a vehicle's suspension system. *Commun. Sci. Lett. Univ. Zilina* **2022**, *24*, B41–B48. [CrossRef]
19. Kaiwu, C.; Cheng, W.; Lu, J. Finite element analysis on the bendig condition of truck frame before and after. *AIP Conf. Proc.* **2018**, *1967*, 030004. [CrossRef]
20. Podkowski, K.; Barszcz, Z.; Senko, J. FEM (Finite Element Method) Numeric Analyses of the Syrenka S201 Car Model. In Proceedings of the 1st Renewable Energy Sources-Research and Business (RESRB-2016), Wrocław, Poland, 22–24 June 2016; Springer: Cham, Switzerland, 2017; pp. 415–421.
21. Tarkowski, S.; Nieoczym, A.; Caban, J.; Jilek, P.; Sejkorová, M. The Analysis of Pneumatic Wheel Rim Deformation While Hitting an Obstacle. *Appl. Sci.* **2022**, *12*, 6371. [CrossRef]
22. Aloni, S.; Khedkar, S. Comparative evaluation of tractor trolley axle by using Finite Element analysis approach. *Int. J. Eng. Sci. Technol.* **2012**, *4*, 135–1360.
23. Chaudhari, P.C.; Sonara, V.D.; Rathod, P.P. Analysis and Design of Tractor Rear Axle using Finite Element Method-A review. *Int. J. Adv. Eng. Res. Dev.* **2015**, *2*, 105–109.
24. Čepukė, Ž. Puspriekabės Saf Ašies Lūžio Priežasties Tyrimas. Doctoral Dissertation, VGTU, Vilnius, Lithuania, 2017.
25. Sojka, M.; Cornák, Š.; Droppa, P. Selected problems of tracked vehicle movement modelling. In Proceedings of the 21st International Scientific Conference Transport Means 2017, Juodkrante, Klaipeda, Lithuania, 20–22 September 2017; pp. 493–498.
26. Krenevičius, A.; Leonavičius, M.K. *Eksperimentinė Medžiagų Mechanika*; Vilnius Tech: Vilnius, Lithuania, 2007; pp. 98–106.
27. Ambiger, H.; Reddy, B. Design Optimization of Automobile Rear Axle Housing for Fatigue Loads using Finite Element Analysis. *IJSRD Int. J. Sci. Res. Dev.* **2016**, *4*, 633–636.
28. Asi, O. Fatigue failure of a rear axle shaft of an automobile. *Eng. Fail. Anal.* **2006**, *13*, 1293–1302. [CrossRef]
29. Decker, Ž.; Rudzinskas, V.; Drozd, K.; Caban, J.; Tretjakovas, J.; Nieoczym, A.; Matijošius, J. Analysis of the Vehicle Chassis Axle Fractures. *Materials* **2023**, *16*, 806. [CrossRef] [PubMed]
30. Barta, D.; Dižo, J.; Blatnický, M.; Molnár, D. Experimental Research of Vibrational Properties of a Single-Axle Trailer when Crossing an Individual Road Obstacle. *Stroj. Cas.* **2022**, *72*, 19–26. [CrossRef]
31. Andrzejczak, K.; Młyńczak, M.; Selech, J. Poisson-Distributed Failures in the Predicting of the Cost of Corrective Maintenance. *Eksploat. Niezawodn. Mintenance Reliab.* **2018**, *20*, 602–609. [CrossRef]
32. Čepukė, Ž.; Tretjakovas, J.; Rudzinskas, V. Review of fracturing axles of semi-trailers and solutions methods. In Proceedings of the 21th International Conference Mechanika, Kaunas, Lithuania, 12–13 May 2016.
33. Kilikevičius, A.; Kasparaitis, A. Dynamic Research of Multi-Body Mechanical Systems of Angle Measurement. *Int. J. Precis. Eng. Manuf.* **2017**, *18*, 1065–1073. [CrossRef]
34. Ratov, B.T.; Mechnik, V.A.; Gevorkyan, E.S.; Matijosius, J.; Kolodnitskyi, V.M.; Chishkala, V.A.; Kuzin, N.O.; Siemiatkowski, Z.; Rucki, M. Influence of CrB2 Additive on the Morphology, Structure, Microhardness and Fracture Resistance of Diamond Composites Based on WC–Co Matrix. *Materialia* **2022**, *25*, 101546. [CrossRef]

35. *EN 12195-1:2010/AC:2014;* Load Restraining on Road Vehicles—Safety—Part 1: Calculation of Securing Forces. Institute for Reference Materials and Measurements: Geel, Belgium, 2014.
36. Małecka, J.; Łagoda, T. Fatigue and Fractures of RG7 Bronze after Cyclic Torsion and Bending. *Int. J. Fatigue* **2023**, *168*, 107475. [CrossRef]
37. Dižo, J.; Blatnický, M.; Sága, M.; Harušinec, J.; Gerlici, J.; Legutko, S. Development of a New System for Attaching the Wheels of the Front Axle in the Cross-Country Vehicle. *Symmetry* **2020**, *12*, 1156. [CrossRef]
38. Matijošius, J.; Vasiliauskas, A.V.; Vasilienė-Vasiliauskienė, V.; Krasodomskis, Ž. The Assess-ment of Importance of the Factors That Predeter-mine the Quality of a Service of Transportation by Road Vehicles. *Procedia Eng.* **2016**, *134*, 422–429. [CrossRef]
39. Meidutė-Kavaliauskienė, I.; Stanujkic, D.; Vasiliauskas, A.V.; Vasilienė-Vasiliauskienė, V. Significance of Criteria and Resulting Significance of Factors Affecting Quality of Services Provided by Lithuanian Road Freight Carriers. *Procedia Eng.* **2017**, *187*, 513–519. [CrossRef]
40. Vrabel, J.; Stopka, O.; Rievaj, V.; Šarkan, B.; Pruskova, K.; Michalk, P. Measuring the resistance of tires for passenger vehicle against the rolling and sliding on loading area of the flatbed truck when providing the transport services. *Commun. Sci. Lett. Univ. Zilina* **2016**, *18*, 124–128. [CrossRef]
41. Buhari, R.; Rohani, M.M.; Abdullah, M.E. Dynamic load coefficient of tyre forces from truck axles. *Appl. Mech. Mater.* **2013**, *405–408*, 1900–1911. [CrossRef]
42. Kilikevičlenė, K.; Skeivalas, J.; Kilikevičlus, A.; PečEliūNas, R.; Bureika, G. The Analysis of Bus Air Spring Condition Influence upon the Vibration Signals at Bus Frame. *Eksploat. Niezawodn. Maint. Reliab.* **2015**, *17*, 463–469. [CrossRef]
43. Vrabel, J.; Jagelcak, J.; Zamecnik, J.; Caban, J. Influence of Emergency Braking on Changes of the Axle Load of Vehicles Transporting Solid Bulk Substrates. *Procedia Eng.* **2017**, *187*, 89–99. [CrossRef]
44. Singh, J.; Singh, S.P.; Joneson, E. Measurement and analysis of US truck vibration for leaf spring and air ride suspensions, and development of tests to simulate these conditions. *Packag. Technol. Sci.* **2006**, *19*, 309–323. [CrossRef]
45. Cebon, D. Heavy vehicle vibration—A case study. *Veh. Syst. Dyn.* **1986**, *15* (Suppl. 1), 30–43. [CrossRef]

materials

MDPI

*Article*

# A Phase Field Approach to Two-Dimensional Quasicrystals with Mixed Mode Cracks

**Tong Li [1], Zhenting Yang [1], Chenghui Xu [2], Xinsheng Xu [1] and Zhenhuan Zhou [1,*]**

[1] State Key Laboratory of Structural Analysis, Optimization and CAE Software for Industrial Equipment, Department of Engineering Mechanics, International Center for Computational Mechanics, Dalian University of Technology, Dalian 116024, China
[2] School of Mechanics, Civil Engineering and Architecture, Northwestern Polytechnical University, Xi'an 710072, China
* Correspondence: zhouzh@dlut.edu.cn; Tel.: +86-1384-2853-461

**Abstract:** Quasicrystals (QCs) are representatives of a novel kind of material exhibiting a large number of remarkable specific properties. However, QCs are usually brittle, and crack propagation inevitably occurs in such materials. Therefore, it is of great significance to study the crack growth behaviors in QCs. In this work, the crack propagation of two-dimensional (2D) decagonal QCs is investigated by a fracture phase field method. In this method, a phase field variable is introduced to evaluate the damage of QCs near the crack. Thus, the crack topology is described by the phase field variable and its gradient. In this manner, it is unnecessary to track the crack tip, and therefore remeshing is avoided during the crack propagation. In the numerical examples, the crack propagation paths of 2D QCs are simulated by the proposed method, and the effects of the phason field on the crack growth behaviors of QCs are studied in detail. Furthermore, the interaction of the double cracks in QCs is also discussed.

**Keywords:** phase field model; decagonal quasicrystal; crack propagation; brittle fracture; mixed mode crack; finite element method

**Citation:** Li, T.; Yang, Z.; Xu, C.; Xu, X.; Zhou, Z. A Phase Field Approach to Two-Dimensional Quasicrystals with Mixed Mode Cracks. *Materials* **2023**, *16*, 3628. https://doi.org/10.3390/ma16103628

Academic Editors: Grzegorz Lesiuk and Dariusz Rozumek

Received: 31 March 2023
Revised: 5 May 2023
Accepted: 8 May 2023
Published: 9 May 2023



## 1. Introduction

Quasicrystals (QCs) are a new kind of material with perfect long-range order but no periodicity. Due to their unique atomic structures, QCs exhibit many excellent physical properties, such as high hardness, a low friction coefficient, and high resistance. Therefore, QCs are very promising materials for potential applications in corrosion-resistant coatings, hydrogen storage, photovoltaic solar cells, etc. [1]. However, QCs are usually brittle, and consequently, cracks, holes, and other defects will inevitably occur during daily use. If the crack propagates, the QC will fail, which may lead to a catastrophic accident. Therefore, the fracture analysis of QCs is of great importance, and it is very necessary to investigate crack propagation behaviors in cracked QCs.

Plenty of research work has been carried out on the fracture problems of QCs. Li et al. [2] extended the classical linear elasticity fracture mechanics to investigate a decagonal QC with a Griffith crack. The results indicated the phonon and phason stresses at the crack tip exhibit the well-known square root singularity, and the strain energy release rate was obtained. Zhou and Fan [3] developed the plane elasticity theory of two-dimensional (2D) octagonal QCs and obtained the exact analytic solution of a Mode I Griffith crack. Guo and Fan [4] studied the fracture problem of a Mode II Griffith crack in decagonal quasicrystals and obtained the corresponding stress intensity factors and strain energy release rate. Shen and Fan [5] calculated the stress intensity factors for an infinitely long strip of finite height containing two straight, semi-infinite collinear cracks. Li and Fan [6] obtained exact solutions for two semi-infinite collinear cracks in a strip of 1D hexagonal QCs. After that, they [7] further obtained an analytic solution for the elliptic notch problem of the material in icosahedral QCs by using the complex variable method. The solution can be reduced to that of a Griffith crack problem. Fan et al. [8] presented linear, nonlinear, and dynamic fracture problems for different QCs. Li and Liu [9] obtained closed-form expressions for the elastic displacement and stress fields induced by a dislocation in icosahedral QCs. Sladek et al. [10] developed a meshless method based on the local Petrov-Galerkin approach for fracture analysis of decagonal QCs; both static and transient dynamic boundary value problems were considered. After that, they [11] present path-independent integrals for accurate evaluation of energy release and stress intensity factors in decagonal QCs. Wang et al. [12] obtained the phonon-phason coupling field in the half-space, which can be expressed in terms of elementary functions. These solutions could have applications in 3D contact and crack problems in QCs. Li et al. [13,14] derived solutions for elliptical crack and planar crack problems in 2D hexagonal QCs. Li and Shi [15] employed the method of potential function theory to solve plane defect problems originating from two-dimensional decagonal QCs. Zhao et al. [16] derived the fundamental solutions for interface cracks in 1D hexagonal QC coatings under in-plane loads.

For crack propagation in QCs, some investigations have been conducted based on the atomistic model [17–29]. However, it is inconvenient to apply atomistic simulation to engineering. Therefore, Wang and Ricoeur [30] adopted the finite element method (FEM) to simulate the crack growth in 1D QCs and predicted the crack pattern for different boundary conditions. Fan et al. [31–39] investigated crack propagation based on the elastodynamic/hydrodynamic model.

From the above literature review, it is found that increasing attention has been paid to the fracture of QCs. However, most of them were concentrated on the derivation of exact solutions and determination of fracture parameters of cracked QCs, and the investigations on crack propagation were mainly based on the atomistic model [17–29]. The work on crack growth in QCs based on continuum mechanics is still insufficient. Additionally, in the traditional FEM, the crack topology is modeled by geometry. The remesh process is necessary when simulating crack propagation, which has a huge computational cost. In this paper, a phase field method is introduced to predict the crack propagation path in QCs. Unlike conventional discrete crack models (such as the FEM), the fracture phased field method employs diffusive cracks to avoid an explicit representation of kinematic discontinuities, and therefore the propagating cracks are tracked automatically without additional ad-hoc criteria in the classical Griffith fracture theory [40,41]. In this method, the fracture energy and degraded strain energy of QCs are formulated using the phase field variable. Subsequently, the total potential energy of QCs under the phase field framework is obtained, and the governing equations for the phase field model are derived by means of the Francfort-Marigo variational principle. The phase field evolution equation for QCs is constructed. Finally, the FEM is adopted to solve the phase field governing equations. The phase field variable and phonon/phason displacement of the entire model can be obtained, as well as the reaction force and the crack pattern.

This paper is organized as follows: The basic equations of 2D decagonal QCs are presented in Section 2.1. The phase field model for 2D QCs is formulated in Section 2.2. The finite element implementation for the phase field model is presented in Section 2.3. Several numerical examples are presented in Section 3. Conclusions are summarized in Section 4.

## 2. Phase Field Method for 2D Decagonal QCs

### 2.1. The Basic Equations

According to the elasticity of 2D decagonal QCs, the basic equations for the plane problem of QCs in the absence of body force and phason self-action are [42]:

$$\begin{cases} \frac{\partial \sigma_x}{\partial x} + \frac{\partial \tau_{xy}}{\partial y} = 0 \\ \frac{\partial \tau_{yx}}{\partial x} + \frac{\partial \sigma_y}{\partial y} = 0 \end{cases}, \begin{cases} \frac{\partial H_x}{\partial x} + \frac{\partial H_{xy}}{\partial y} = 0 \\ \frac{\partial H_{yx}}{\partial x} + \frac{\partial H_y}{\partial y} = 0 \end{cases} \tag{1}$$

$$\begin{Bmatrix} \varepsilon_x \\ \varepsilon_y \\ \gamma_{xy} \end{Bmatrix} = \begin{bmatrix} \frac{\partial}{\partial x} & \\ & \frac{\partial}{\partial y} \\ \frac{\partial}{\partial y} & \frac{\partial}{\partial x} \end{bmatrix} \begin{Bmatrix} u_x \\ u_y \end{Bmatrix}, \begin{Bmatrix} \omega_x \\ \omega_y \\ \omega_{xy} \\ \omega_{yx} \end{Bmatrix} = \begin{bmatrix} \frac{\partial}{\partial x} & \\ & \frac{\partial}{\partial y} \\ \frac{\partial}{\partial y} & \\ & \frac{\partial}{\partial x} \end{bmatrix} \begin{Bmatrix} w_x \\ w_y \end{Bmatrix} \tag{2}$$

$$\begin{Bmatrix} \sigma_x \\ \sigma_y \\ \tau_{xy} \end{Bmatrix} = \begin{bmatrix} C_{11} & C_{12} & \\ C_{12} & C_{22} & \\ & & C_{66} \end{bmatrix} \begin{Bmatrix} \varepsilon_x \\ \varepsilon_y \\ \gamma_{xy} \end{Bmatrix} + \begin{bmatrix} R_1 & R_1 & R_2 & -R_2 \\ -R_1 & -R_1 & -R_2 & R_2 \\ R_2 & R_2 & -R_1 & R_1 \end{bmatrix} \begin{Bmatrix} \omega_x \\ \omega_y \\ \omega_{xy} \\ \omega_{yx} \end{Bmatrix} \tag{3}$$

$$\begin{Bmatrix} H_x \\ H_y \\ H_{xy} \\ H_{yx} \end{Bmatrix} = \begin{bmatrix} R_1 & -R_1 & R_2 \\ R_1 & -R_1 & R_2 \\ R_2 & -R_2 & -R_1 \\ -R_2 & R_2 & R_1 \end{bmatrix} \begin{Bmatrix} \varepsilon_x \\ \varepsilon_y \\ \gamma_{xy} \end{Bmatrix} + \begin{bmatrix} K_1 & K_2 & & \\ K_2 & K_1 & & \\ & & K_1 & -K_2 \\ & & -K_2 & K_1 \end{bmatrix} \begin{Bmatrix} \omega_x \\ \omega_y \\ \omega_{xy} \\ \omega_{yx} \end{Bmatrix} \tag{4}$$

where $u_x$ and $u_y$ are the phonon displacements; $w_x$ and $w_y$ are the phason displacements; $\varepsilon_x$, $\varepsilon_y$, and $\gamma_{xy}$ are the phonon strains; $\omega_x$, $\omega_y$, $\omega_{xy}$, and $\omega_{yx}$ are the phason strains; $\sigma_x$, $\sigma_y$, and $\sigma_{xy}$ are the phonon stresses; $H_x$, $H_y$, $H_{xy}$, and $H_{yx}$ are the phason stresses; $C_{11}$, $C_{12}$, $C_{22}$, and $C_{66}$ are the phonon moduli; $K_1$ and $K_2$ are the phason moduli; and $R_1$ and $R_2$ are the phonon-phason coupling coefficients.

The strain energy of QCs is:

$$\Pi_s = \int_{\Omega} \psi(\boldsymbol{\varepsilon},\ \boldsymbol{\omega})\ \mathrm{d}V = \int_{\Omega} \frac{1}{2}\boldsymbol{\varepsilon}^{\mathrm{T}}\boldsymbol{\sigma} + \frac{1}{2}\boldsymbol{\omega}^{\mathrm{T}}\mathbf{H}\ \mathrm{d}V \tag{5}$$

where $\boldsymbol{\sigma} = \{\sigma_x,\ \sigma_y,\ \sigma_{xy}\}^{\mathrm{T}}$ and $\mathbf{H} = \{H_x,\ H_y,\ H_{xy},\ H_{yx}\}^{\mathrm{T}}$ are the stress vectors; $\boldsymbol{\varepsilon} = \{\varepsilon_x,\ \varepsilon_y,\ \gamma_{xy}\}^{\mathrm{T}}$ and $\boldsymbol{\varepsilon} = \{\omega_x,\ \omega_y,\ \omega_{xy},\ \omega_{yx}\}^{\mathrm{T}}$ are the strain vectors; and $\psi(\boldsymbol{\varepsilon},\ \boldsymbol{\omega})$ is the strain energy density.

### 2.2. Phase Field Method

Consider a 1D QC strip with a centered crack, as shown in Figure 1. The fracture energy is:

$$\Pi_c = \int_{\partial\Omega_c} G_c\ \mathrm{d}S = G_c A_c \tag{6}$$

where $G_c$ is the critical energy release rate (CERR); $\partial\Omega_c$ is the crack surface; and $A_c$ is the cross-sectional area of the strip.

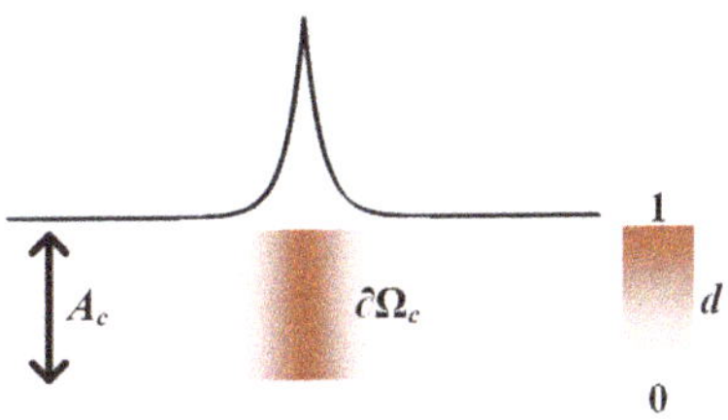


**Figure 1.** A QC strip with a centered crack.

In a phase field model, the crack is supposed to be surrounded by a diffusive degraded zone, and a phase field variable $d$ is introduced to describe the damage in the diffusive degraded zone:

$$d = e^{-\frac{|x|}{l_c}} \tag{7}$$

where $l_c$ is an internal length scale that controls the width of the diffusive zone. Here, it is noted that, when $x$ approaches infinity ($\pm\infty$), $d$ converges to zero, which indicates that the material is intact; when $x$ is zero, $d$ equals one, which indicates that the material is totally broken and the crack surface is produced. Define a crack's surface density as functional [43]:

$$\gamma(d,\ d') = \frac{1}{2l_c}\left(d^2 + l_c^2 d'^2\right) \tag{8}$$

The crack topology can then be described by the phase field variable.

Substituting Equation (8) into Equation (6) yields the fracture energy in the phase field model:

$$\Pi_c = \int_\Omega G_c \gamma(d,\ d')\ \mathrm{d}V \tag{9}$$

where $\Omega$ is the domain of the overall model. As observed in Equation (9), the integral over the crack path is transformed into the integral over the model. The crack topology is implicitly expressed in the framework of the phase field method. Therefore, it is unnecessary to track the crack's path while it is propagating.

Similarly, the 2D crack surface density functional can be defined as:

$$\gamma(d,\ \nabla d) = \frac{1}{2l_c}\left(d^2 + l_c^2|\nabla d|^2\right) \tag{10}$$

Therefore, the fracture energy of QCs in a plane problem is:

$$\Pi_c = \int_\Omega G_c \gamma(d,\ \nabla d)\ \mathrm{d}V \tag{11}$$

In the phase field model, the material at the crack is assumed to be softened. Therefore, a degradation function is introduced to evaluate the damage to the material in the diffusive degraded zone [43]:

$$\omega(d) = (1-d)^2 + k \tag{12}$$

where $k$ is a small positive number to ensure the non-singularity of the matrix. It can be observed that $\omega(d)$ satisfies $\omega(0) = 1$ and $\omega(1) = 0$.

The strain energy for QCs is modified as follows:

$$\Pi_s = \int_\Omega \omega(d)\psi(\boldsymbol{\varepsilon},\ \boldsymbol{\omega})\ \mathrm{d}V \tag{13}$$

It should be pointed out that the crack will not propagate if the crack face is under compression. Therefore, the strain energy density $\psi$ should be decomposed into a tensile part and a compressed part, and the energy degradation only occurs on the tensile part [43]:

$$\Pi_s = \int_\Omega \omega(d)\psi_+ + \psi_-\ \mathrm{d}V \tag{14}$$

where $\psi_+ = \lambda\langle\varepsilon_1 + \varepsilon_2 + \varepsilon_3\rangle_+^2/2 + \mu\left[\langle\varepsilon_1\rangle_+^2 + \langle\varepsilon_2\rangle_+^2 + \langle\varepsilon_3\rangle_+^2\right]$ and $\psi_- = \psi - \psi_+$ are the tensile and compressed strain energy densities, respectively; $\langle\rangle_+$ is defined as $\langle x\rangle_+ = (|x| + x)/2$; and $\lambda$ and $\mu$ are the Lamé constants.

The potential energy of the overall model contains the strain energy, fracture energy, and potential energy of the external force:

$$\begin{aligned}\Pi_p &= \Pi_s + \Pi_c - \Pi_e \\ &= \int_\Omega \omega(d)\psi_+ + \psi_-\ \mathrm{d}V + \int_\Omega G_c\gamma(d,\ \nabla d)\ \mathrm{d}V \\ &-\int_{\partial\Omega} \mathbf{u}^{\mathrm{T}}\mathbf{t}^u\ \mathrm{d}S - \int_{\partial\Omega} \mathbf{w}^{\mathrm{T}}\mathbf{t}^w\ \mathrm{d}S\end{aligned} \tag{15}$$

where $\mathbf{u} = \{u_x,\ u_y\}^{\mathrm{T}}$; $\mathbf{w} = \{w_x,\ w_y\}^{\mathrm{T}}$; and $\mathbf{t}^u$ and $\mathbf{t}^w$ are the external forces in the phonon and phason fields, respectively.

The Francfort-Marigo variational principle states that the real displacement **u** and the phase field variable $d$ will minimize the potential energy.

$$\begin{aligned}\delta \Pi_p &= \int_{\partial\Omega_u} \omega(d)\sigma_{ij}^{+} n_i \delta u_j + \sigma_{ij}^{-} n_i \delta u_j \, \mathrm{d}S + \int_{\partial\Omega_w} \omega(d) H_{ij}^{+} n_i \delta w_j + H_{ij}^{-} n_i \delta w_j \, \mathrm{d}S \\ &\quad - \int_{\Omega} \frac{\partial\left[\omega(d)\sigma_{ij}^{+}\right]}{\partial x_i} \delta u_j + \frac{\partial \sigma_{ij}^{-}}{\partial x_i} \delta u_j \, \mathrm{d}V - \int_{\Omega} \frac{\partial\left[\omega(d) H_{ij}^{+}\right]}{\partial x_i} \delta w_j + \frac{\partial H_{ij}^{-}}{\partial x_i} \delta w_j \, \mathrm{d}V \\ &\quad \int_{\Omega} \omega'(d)\psi_{+} \delta d \, \mathrm{d}V + \int_{\Omega} \frac{G_c}{l_c} d \delta d \, \mathrm{d}V + \int_{\partial\Omega} G_c l_c \frac{\partial d}{\partial x_i} n_i \delta d \, \mathrm{d}S - \int_{\Omega} G_c l_c \frac{\partial^2 d}{\partial x_i^2} \delta d \, \mathrm{d}V \\ &\quad - \int_{\partial\Omega} \delta u_j t_j^u \, \mathrm{d}S - \int_{\partial\Omega} \delta w_j t_j^w \, \mathrm{d}S \\ &= 0 \end{aligned} \tag{16}$$

where $\sigma_{ij}^{+}$ and $H_{ij}^{+}$ are the stresses induced by stretch, while $\sigma_{ij}^{-}$ and $H_{ij}^{-}$ are induced by compression. Equation (16) is valid for arbitrary $\delta u_i$, $\delta w_i$, and $\delta d$. Hence, the governing equations for the phase field model of QCs are:

$$\nabla\left[\omega(d)\sigma_{ij}^{+} + \sigma_{ij}^{-}\right] = 0, \Omega \tag{17}$$

$$\nabla\left[\omega(d)H_{ij}^{+} + \partial H_{ij}^{-}\right] = 0, \Omega \tag{18}$$

$$(2d-2)\psi_{+} + \frac{G_c}{l_c} d - G_c l_c \Delta d = 0, \Omega \tag{19}$$

$$\left[\omega(d)\sigma_{ij}^{+} + \sigma_{ij}^{-}\right] n_i = t_j^u, \ \partial\Omega \tag{20}$$

$$\left[\omega(d)H_{ij}^{+} + H_{ij}^{-}\right] n_i = t_j^w, \ \partial\Omega \tag{21}$$

$$d_{,i} n_i = 0, \ \partial\Omega \tag{22}$$

It should be mentioned that crack growth is an irreversible process. Therefore, Equation (19) should be modified by considering the history of the load [44]:

$$(2d-2)H + \frac{G_c}{l_c} d - G_c l_c \Delta d = 0 \tag{23}$$

where $H = \max\limits_{[0,\ t]}\{\psi_{+}\}$ is a history variable that is the maximum strain energy during the crack propagation. This history variable ensures that the crack face does not close under compression. Equation (23) is the evolution law of the crack phase field for QCs. Cracks grow only if this equation is valid.

### 2.3. Finite Element Implementation

Due to the strong nonlinearity of the governing equations in the phase field model, the FEM is often adopted to solve the problem. In the FEM, the phonon/phason displacements and the phase field variable are approximated by the shape functions:

$$\{u_x,\ u_y\}^{\mathrm{T}} = \sum_{i=1}^{4} \mathbf{N}_i^u \mathbf{u}_i \tag{24}$$

$$\{w_x,\ w_y\}^{\mathrm{T}} = \sum_{i=1}^{4} \mathbf{N}_i^w \mathbf{w}_i \tag{25}$$

$$d = \sum_{i=1}^{4} N_i d_i \tag{26}$$

where $N_i$ is the shape function; $\mathbf{N}_i^u = \mathbf{N}_i^w = \mathrm{diag}(N_i,\ N_i)$; and $\mathbf{u}_i$, $\mathbf{w}_i$, and $d$ are the nodal phonon and phason displacements and phase field variables, respectively.

Therefore, the phonon and phason strains and the gradient of the phase field variable are, respectively:

$$\{\varepsilon_x,\ \varepsilon_y,\ \gamma_{xy}\}^{\mathrm{T}} = \sum_{i=1}^{4} \mathbf{B}_i^u \mathbf{u}_i \tag{27}$$

$$\{\omega_x,\ \omega_y,\ \omega_{xy},\ \omega_{yx}\}^{\mathrm{T}} = \sum_{i=1}^{4} \mathbf{B}_i^w \mathbf{w}_i \tag{28}$$

$$\nabla d = \sum_{i=1}^{4} \mathbf{B}_i^d \mathbf{d}_i \tag{29}$$

where $\mathbf{B}_i^u = \begin{bmatrix} \partial N_i/\partial x & & \partial N_i/\partial y \\ & \partial N_i/\partial y & \partial N_i/\partial x \end{bmatrix}^{\mathrm{T}}$, $\mathbf{B}_i^w = \begin{bmatrix} \partial N_i/\partial x & & \partial N_i/\partial y & \\ & \partial N_i/\partial y & & \partial N_i/\partial x \end{bmatrix}^{\mathrm{T}}$, and $\mathbf{B}_i^d = [\partial N_i/\partial x \quad \partial N_i/\partial y]^{\mathrm{T}}$. Substituting Equations (24)–(29) into Equation (15) yields the residuals:

$$\mathbf{r}^a = \int_{\partial\Omega} \mathbf{N}^{\mathrm{T}}\mathbf{t}\ \mathrm{d}S - \int_{\Omega} (1-d)^2 \mathbf{B}^{\mathrm{T}} \mathbf{D}_{QC} \mathbf{B} \mathbf{a}\ \mathrm{d}V = 0 \tag{30}$$

$$\mathbf{r}^d = -\int_{\Omega} 2(d-1) H^u \left(\mathbf{N}^d\right)^{\mathrm{T}} \mathrm{d}V - \int_{\Omega} \frac{G_c}{c_0} \left[\frac{2d}{l_c}\left(\mathbf{N}^d\right)^{\mathrm{T}} + 2l_c \left(\mathbf{B}^d\right)^{\mathrm{T}} \nabla d\right] \mathrm{d}V = 0 \tag{31}$$

where $\mathbf{a} = \{u_{x1},\ u_{y1},\ \ldots,\ u_{x4},\ u_{y4},\ w_{x1},\ w_{y1},\ \ldots,\ w_{x4},\ w_{y4}\}^{\mathrm{T}}$; $\mathbf{N} = \begin{bmatrix} \mathbf{N}_0 & \mathbf{0}_{2\times 8} \\ \mathbf{0}_{2\times 8} & \mathbf{N}_0 \end{bmatrix}$ is the shape function matrix where $\mathbf{N}_0 = \begin{bmatrix} N_1 & 0 & N_2 & 0 & N_3 & 0 & N_4 & 0 \\ 0 & N_1 & 0 & N_2 & 0 & N_3 & 0 & N_4 \end{bmatrix}$; $\mathbf{B} = \mathbf{L}\mathbf{N}$ is the strain matrix where $\mathbf{L} = \begin{bmatrix} \frac{\partial}{\partial x} & 0 & \frac{\partial}{\partial y} & 0 & 0 & 0 & 0 \\ 0 & \frac{\partial}{\partial y} & \frac{\partial}{\partial x} & 0 & 0 & 0 & 0 \\ 0 & 0 & 0 & \frac{\partial}{\partial x} & 0 & \frac{\partial}{\partial y} & 0 \\ 0 & 0 & 0 & 0 & \frac{\partial}{\partial y} & 0 & \frac{\partial}{\partial x} \end{bmatrix}^{\mathrm{T}}$; and $\mathbf{t} = \{t_x^u,\ t_y^u,\ t_x^w,\ t_y^w\}^{\mathrm{T}}$ is the load vector.

Equations (30) and (31) can be solved by this iteration method:

$$\begin{Bmatrix} \mathbf{a}_{n+1} \\ \mathbf{d}_{n+1} \end{Bmatrix} = \begin{Bmatrix} \mathbf{a}_n \\ \mathbf{d}_n \end{Bmatrix} + \begin{bmatrix} \mathbf{K}_n^{aa} & \mathbf{K}_n^{ad} \\ \mathbf{K}_n^{da} & \mathbf{K}_n^{dd} \end{bmatrix}^{-1} \begin{Bmatrix} \mathbf{r}_n^a \\ \mathbf{r}_n^d \end{Bmatrix} \tag{32}$$

where $\mathbf{K}^{aa} = \int_{\Omega} \omega(d) \mathbf{B}^{\mathrm{T}} \mathbf{D} \mathbf{B}\ \mathrm{d}V$, $\mathbf{K}^{ad} = \left(\mathbf{K}^{da}\right)^{\mathrm{T}} = \int_{\Omega} 2(d-1) \mathbf{B}^{\mathrm{T}} \boldsymbol{\sigma} \mathbf{N}\ \mathrm{d}V$, and $\mathbf{K}^{dd} = \int_{\Omega} \left(2H + \frac{G_c}{l}\right) \mathbf{N}^{dT} \mathbf{N}^d dV + \int_{\Omega} l G_c \mathbf{B}^{dT} \mathbf{B}^d dV$.

Due to the high nonlinearity of Equation (32), there is a convergence problem using the classic Newton iteration. A highly robust staggered algorithm is usually adopted to solve Equation (32) [41,44]. In this algorithm, one of the two unknowns (displacement and phase field variable) is assumed to be unchanged while solving the other unknown during one iteration, which yields:

$$\mathbf{a}_{n+1} = \mathbf{a}_n + \left(\mathbf{K}_n^{aa}\right)^{-1} \mathbf{r}_n^a \tag{33}$$

$$\mathbf{d}_{n+1} = \mathbf{d}_n + \left(\mathbf{K}_n^{dd}\right)^{-1} \mathbf{r}_n^d \tag{34}$$

Finally, the nodal phonon/phason displacements and the phase field variables can be calculated by Equations (33) and (34), respectively.

## 3. Numerical Results

In this section, a few numerical examples are presented to illustrate the application of the phase field method to the fracture of 2D decagonal QCs. The material parameters are tabulated in Table 1 [45–47].

**Table 1.** Material properties of 2D decagonal QCs.

| | $C_{11}$ | $C_{12}$ | $C_{66}$ | $K_1$ | $K_2$ | $R_1$ | $R_2$ |
|---|---|---|---|---|---|---|---|
| Al-Ni-Co (GPa) | 234.30 | 57.34 | 88.45 | 122 | 24 | −1.1 | 0.1 |

According to Fan's work [42], the expression of CERR is:

$$G_c = \frac{\lambda(K_1 + K_2) + 2(R_1^2 + R_2^2)}{8(\lambda + M)c} K_{IC}^2 \tag{35}$$

where $M = (C_{11} - C_{12})/2$ and $c = M(K_1 + K_2) - 2(R_1^2 + R_2^2)$. The fracture toughness of Al-Ni-Co QCs is $K_{IC} = 1\text{MPa}\sqrt{\text{m}}$. Therefore, the CERR in this paper is selected as $G_c = 5.56 \times 10^{-4}$ N/mm, according to the material constants in Table 1. Although no experimental method has been reported to apply a constant phason displacement on the surface of QCs, some investigations [48,49] reveal that some ways can cause the disorder of the phason field. Therefore, different phason displacement loads are considered in the following calculation to investigate the effect of the phason field on the fracture behavior of QCs.

### *3.1. The Rectangular QCs with Edge Crack*

A rectangular QC model is considered, as shown in Figure 2. The geometry of the model is width $W$, height $L$, and crack length $a$. The upper edge is constrained to have the same displacement in phonon and phason fields, and the lower edge is only constrained in the vertical direction. A concentrated phonon force $P_\sigma$ is applied to the upper edge. The angle between $P_\sigma$ and the horizon is $\varphi$.

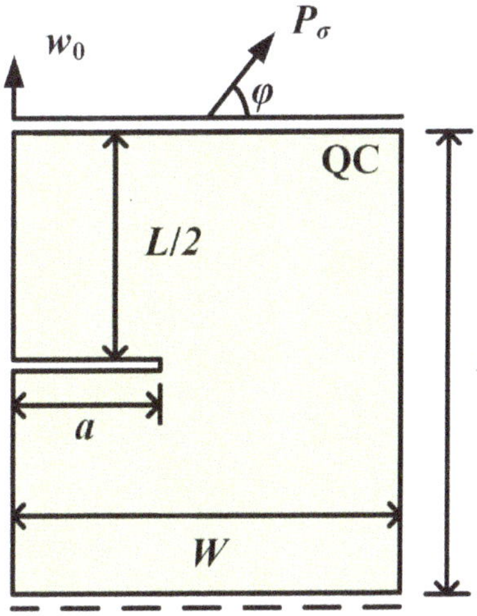


**Figure 2.** A rectangular QC with an edge crack.

At the current stage, the crack growth of 2D QCs has not been reported in the open literature. To demonstrate the accuracy of the present method, an elastic material is selected by degenerating all the material constants in phason field, i.e., $K_1 = K_2 = R_1 = R_2 = 0$. The geometrical parameters are taken as $a = 0.5$ mm and $W = L = 1$ mm. The elastic material constants come from Ref. [44]. The variation of the vertical concentrated force $P_\sigma$ ($\varphi = \pi/2$) versus the displacement $u$ is plotted in Figure 3. As observed, the present results are in excellent agreement with the reference data from Ref. [44]. Furthermore, it is noted that the peak force $P_\sigma$ shows an increasing trend as the length scale $l_c$ decreases, which indicates the crack grows slowly as the length scale declines. In the previous study, to ensure the high resolution of the crack topology, the minimum size of the element was required to meet the condition of $h < l_c/2$ [43]. Therefore, in the following calculations, the length scale is selected as $l_c = 1$ mm, and the element size is selected as $h < l_c/5$.

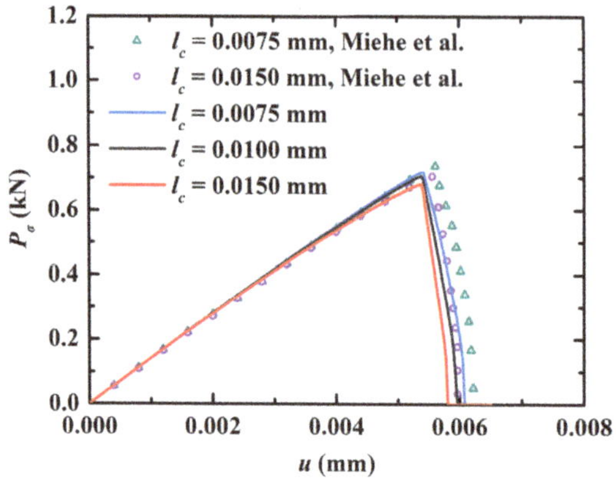


**Figure 3.** Variation of the force $P_\sigma$ versus the displacement $u$ [44].

Subsequently, the effect of the phason field on the crack growth is studied in Figure 2 by considering an initial phason displacement $w_0$ on the upper edge of the model. The parameters are selected as $a = 50$ mm and $W = L = 100$ mm. The variation of the applied force versus the displacement for different $w_0$ values is illustrated in Figure 4. The results show that the peak force decreases as the initial phason displacement increases, which indicates that the initial stretch phason load will undermine the strength of the model. In addition, as the initial phason load gets larger, the crack grows slower. The crack pattern with $w_0 = 6 \times 10^{-4}$ mm is plotted in Figure 5.

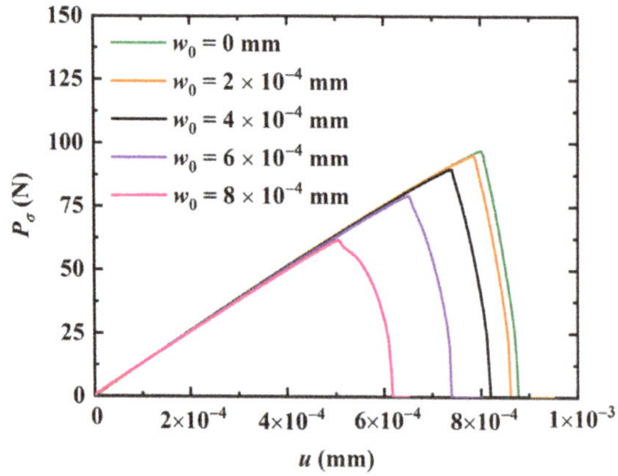


**Figure 4.** Force-displacement relations for different initial phason loads $w_0$.

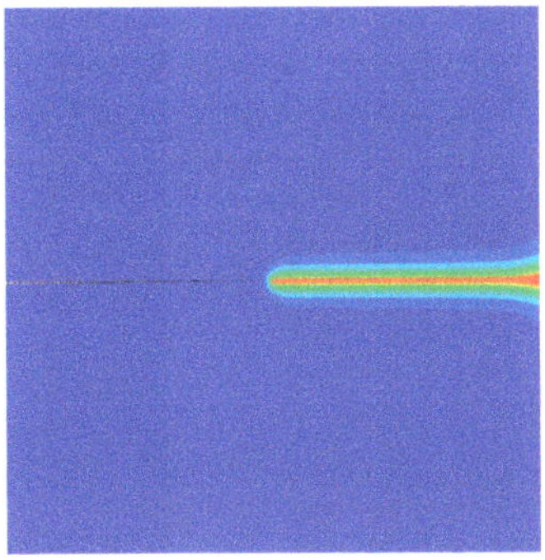

**Figure 5.** The crack pattern with $w_0 = 6 \times 10^{-4}$ mm.

Finally, the QC model subjected to a concentrated shear load at its upper edge is considered. The variation of the shear phonon force $P_\sigma$ ($\varphi = \pi$) versus the horizontal displacement at the upper edge is illustrated in Figure 6. Two peak forces were observed, and they decrease as the initial vertical phason displacement $w_0$ increases. The crack patterns with different initial phason displacements are shown in Figure 7. Clearly, the increasing initial phason displacement will lead to a significant crack deflection. Therefore, it is concluded that the phason load is a key influencing factor for the peak force and crack patterns under the applied shear loads.

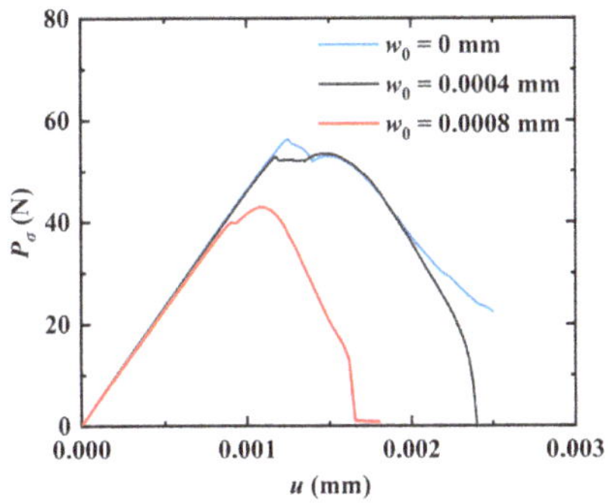


**Figure 6.** The shear force-displacement relations for different initial tensile phason loads.

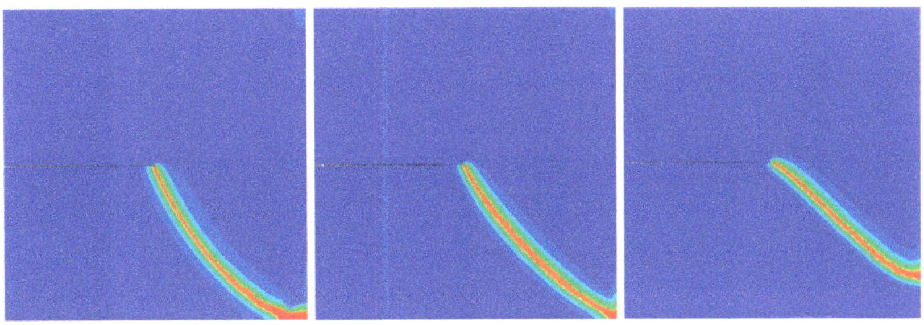

(a) (b) (c)

**Figure 7.** The crack pattern with shear load for different initial phason loads (edge crack): (**a**) $w_0 = 0$ mm; (**b**) $w_0 = 4 \times 10^{-4}$ mm; and (**c**) $w_0 = 8 \times 10^{-4}$ mm.

### 3.2. The Rectangular QCs with an Internal Crack

In this example, a rectangular QC with an internal crack is considered in Figure 8. The length of the crack is $a$, and the angle between the crack and the horizon is $\theta$. The middle of the crack is centered in the model. The boundary conditions of the QC are the same as those in Section 3.1. The parameters are selected as $a = 25$ mm and $W = L = 100$ mm. The QC model is subjected to a vertical initial phason displacement $w_0$ at the upper edge.

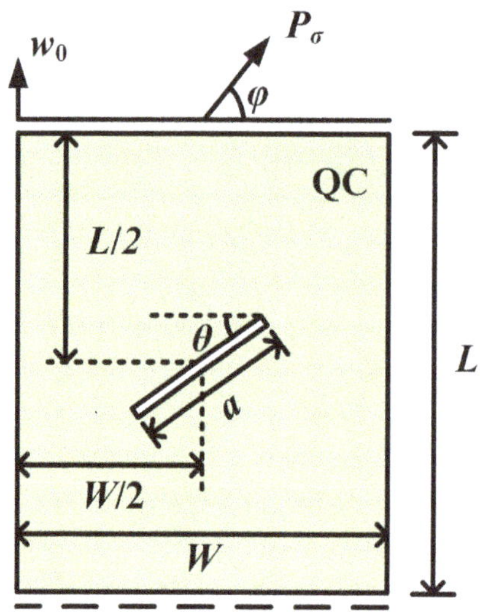


**Figure 8.** A rectangular QC with an internal crack.

Firstly, the QC model is subjected to a tensile load in the phonon field ($\varphi = \pi/2$). The peaks of the phonon forces for different angles $\theta$ are shown in Figure 9. It can be observed that the peak force monotonously increases with the increase in $\theta$, while it shows an opposite trend as the initial phason displacement increases. The force-displacement relation for different angles $\theta$ is shown in Figure 10. Clearly, the slope of the force-displacement curve increases as the angle $\theta$ increases. The crack patterns with different angles are plotted in Figure 11. As observed, the crack grows along a straight line to the edge of the model when subjected to a tensile load.

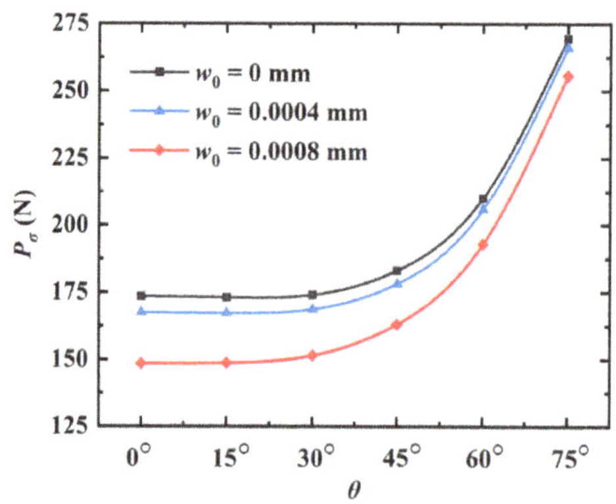


**Figure 9.** The peak phonon forces for different angles $\theta$ (edge crack).

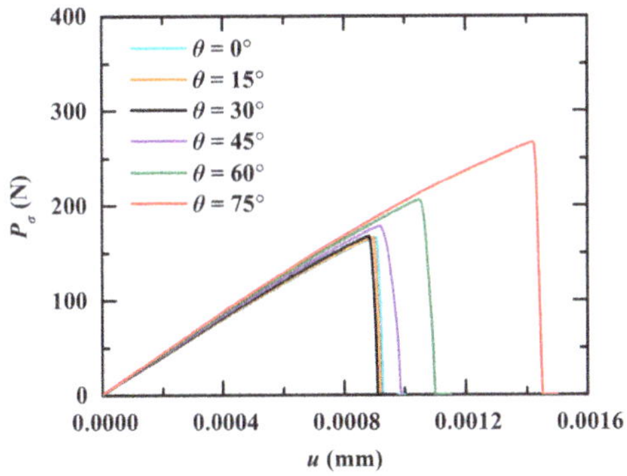


**Figure 10.** The force-displacement relation for different angles $\theta$.

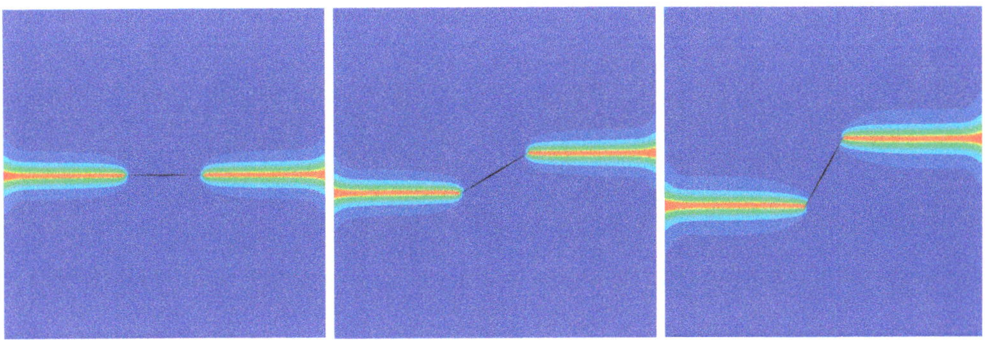

(a) (b) (c)

**Figure 11.** The crack patterns with $w_0 = 6 \times 10^{-4}$ mm for different angles $\theta$: (**a**) $\theta = 0$; (**b**) $\theta = \pi/6$; and (**c**) $\theta = \pi/3$.

Subsequently, the model is subjected to shear loads in the phonon field at the upper edge. The peak phonon forces for different angles $\theta$ are illustrated in Figure 12. As observed, the peak force first increases and then decreases as the angle $\theta$ increases. The crack patterns are plotted in Figure 13. Similar to the observation in Figure 7, the increasing initial phason displacement has a significant influence on the crack propagation path.

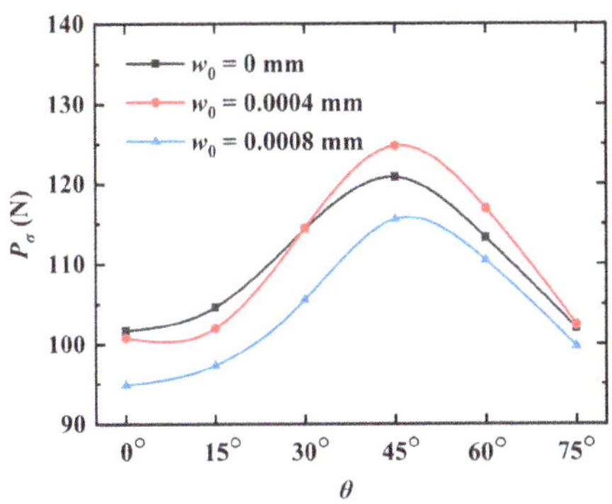


**Figure 12.** The peak forces for different angles $\theta$ (internal crack).

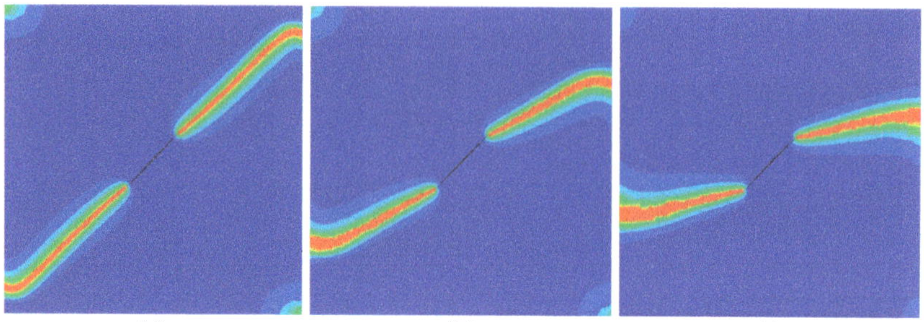

(**a**) (**b**) (**c**)

**Figure 13.** The crack pattern with shear load for different initial phason loads (internal crack): (**a**) $w_0 = 8 \times 10^{-4}$ mm; (**b**) $w_0 = 1.2 \times 10^{-3}$ mm; and (**c**) $w_0 = 1.6 \times 10^{-3}$ mm.

### 3.3. *The Rectangular QCs with Double Cracks*

As the last example, the QC model with double cracks is investigated in Figure 14. The parameters are selected as $a_1 = a_2 = 25$ mm and $W = L = 100$ mm. The left crack is horizontal. The angle between the right crack and the horizon is $\theta$. The lengths of the left and right cracks are $a_1$ and $a_2$, respectively. Point $B$ is located at the center of the right crack. The distance between the right edge of the model and point $B$ is $\Delta_1 = 27.5$ mm. The distance between the left crack and point $B$ is $\Delta_2$. The model is subjected to a tensile phonon/phason displacements. The peak forces for different angles $\theta$ with $\Delta_2 = 32.5$ mm are plotted in Figure 15. It can be found that, as $\theta$ increases, the peak force monotonously increases for $w_0 = 0$ mm and $w_0 = 0.0004$ mm, while it increases first and then decreases for $w_0 = 0.0008$ mm. The crack patterns for different angles $\theta$ and distances $\Delta_2$ are illustrated in Figures 16 and 17. As observed, the interaction of two cracks has a big contribution to their crack propagation path. The two cracks eventually merge together as the applied load increases.

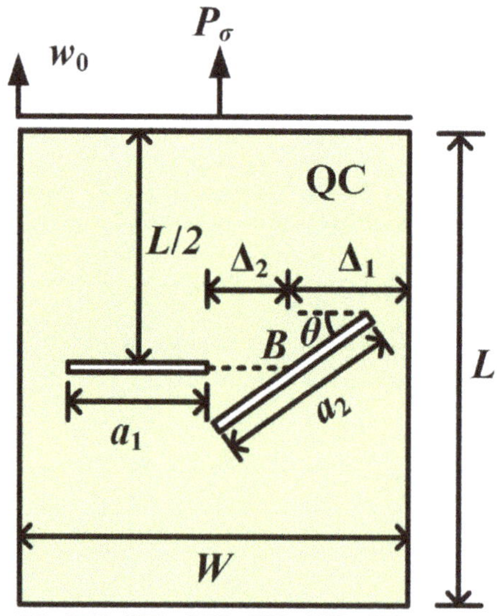


**Figure 14.** A rectangular QC with double internal cracks.

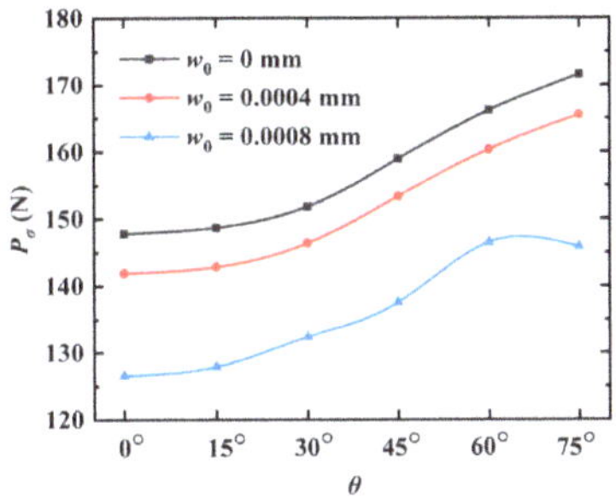


**Figure 15.** The peak forces for different angles $\theta$ (double cracks).

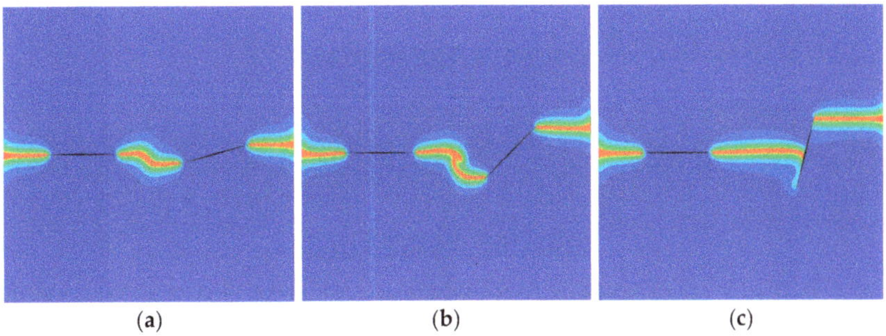

**Figure 16.** The crack pattern with $\Delta_2 = 32.5$ mm for different angles (double cracks): (**a**) $\theta = \pi/12$; (**b**) $\theta = \pi/4$; (**c**) $\theta = 5\pi/12$.

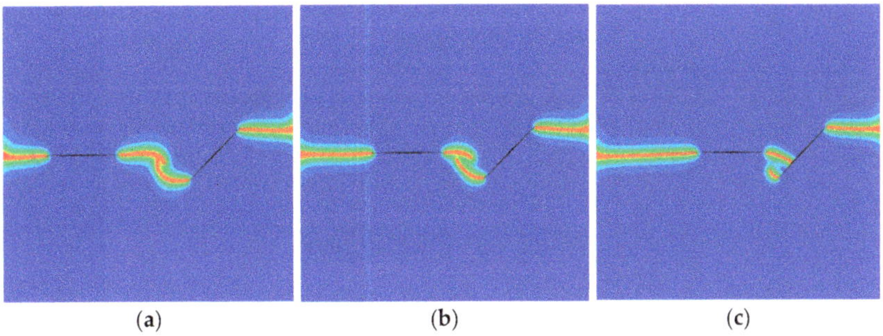

**Figure 17.** The crack pattern with $\theta = \pi/4$ for different distances $\Delta_2$: (**a**) $\Delta_2 = 32.5$ mm; (**b**) $\Delta_2 = 22.5$ mm; and (**c**) $\Delta_2 = 12.5$ mm.

## 4. Conclusions

In this paper, a phase field model is developed to predict crack propagation in 2D decagonal QCs. The contribution of the phason field to the potential energy is considered, and the evolution law of the crack phase field for QCs is established. Therefore, the crack topology in QCs is described by the phase field variable, which can be solved by the FEM. In this manner, the crack propagation in QCs can be accurately simulated without remeshing, and the evolution of the crack can be vividly illustrated. Numerical examples illustrate that the proposed model can predict accurate results for the crack propagation of QCs, and the phason field has a big contribution to both the force-displacement relation and the crack pattern.

**Author Contributions:** Conceptualization, Z.Z. and Z.Y.; methodology, Z.Y. and T.L.; data curation, T.L.; writing—preparation of original draft, T.L., Z.Y. and C.X.; writing—proofreading and editing, Z.Z. and X.X. All authors have read and agreed to the published version of the manuscript.

**Funding:** This research was funded by the Fundamental Research Funds for the Central Universities, [grant number DUT21LK35, DUT22LK16]; State Key Laboratory of Structural Analysis, Optimization and CAE Software for Industrial Equipment, [grant number GZ22109].

**Institutional Review Board Statement:** Not applicable.

**Informed Consent Statement:** Not applicable.

**Data Availability Statement:** The data presented in this study are available on request from the corresponding authors.

**Conflicts of Interest:** The authors declare no conflict of interest.

## References

1. Maciá Barber, E. *Quasicrystals: Fundamentals and Applications*, 1st ed.; CRC Press: Boca Raton, MA, USA, 2020.
2. Li, X.F.; Fan, T.Y.; Sun, Y.F. A decagonal quasicrystal with a Griffith crack. *Philos. Mag.* **1999**, *79*, 1943–1952.
3. Zhou, W.M.; Fan, T.Y. Plane elasticity problem of two-dimensional octagonal quasicrystals and crack problem. *Chin. Phys.* **2001**, *10*, 743–747.
4. Guo, Y.C.; Fan, T.Y. A mode-II Griffith crack in decagonal quasicrystals. *Appl. Math. Mech.* **2001**, *22*, 1311–1317. [CrossRef]
5. Shen, D.W.; Fan, T.Y. Exact solutions of two semi-infinite collinear cracks in a strip. *Eng. Fract. Mech.* **2003**, *70*, 813–822. [CrossRef]
6. Li, L.H.; Fan, T.Y. Exact solutions of two semi-infinite collinear cracks in a strip of one dimensional hexagonal quasicrystal. *Appl. Math. Comput.* **2008**, *196*, 1–5. [CrossRef]
7. Li, L.H.; Fan, T.Y. Complex variable method for plane elasticity of icosahedral quasicrystals and elliptic notch problem. *Sci. China Phys. Mech.* **2008**, *51*, 773–780. [CrossRef]
8. Fan, T.Y.; Tang, Z.Y.; Chen, W.Q. Theory of linear, nonlinear and dynamic fracture for quasicrystals. *Eng. Fract. Mech.* **2012**, *82*, 185–194. [CrossRef]
9. Li, L.H.; Liu, G.T. Stroh formalism for icosahedral quasicrystal and its application. *Phys. Lett. A* **2012**, *376*, 987–990. [CrossRef]
10. Sladek, J.; Sladek, V.; Krahulec, S.; Zhang, C.; Wünsche, M. Crack analysis in decagonal quasicrystals by the MLPG. *Int. J. Fract.* **2013**, *181*, 115–126. [CrossRef]
11. Sladek, J.; Sladek, V.; Atluri, S.N. Path-independent integral in fracture mechanics of quasicrystals. *Eng. Fract. Mech.* **2015**, *140*, 61–71. [CrossRef]
12. Wang, T.; Li, X.Y.; Zhang, X.; Müller, R. Fundamental solutions in a half space of two-dimensional hexagonal quasicrystal and their applications. *J. Appl. Phys.* **2015**, *117*, 154904. [CrossRef]
13. Li, Y.; Fan, C.; Qin, Q.H.; Zhao, M. Closed-form solutions of an elliptical crack subjected to coupled phonon–phason loadings in two-dimensional hexagonal quasicrystal media. *Math. Mech. Solids* **2019**, *24*, 1821–1848. [CrossRef]
14. Li, Y.; Zhao, M.; Qin, Q.H.; Fan, C. Analysis solution method for 3D planar crack problems of two-dimensional hexagonal quasicrystals with thermal effects. *Appl. Math. Model.* **2019**, *69*, 648–664. [CrossRef]
15. Li, W.; Shi, Y.Q. Extension of elastic models to decagonal quasicrystals. *Crystals* **2020**, *10*, 469. [CrossRef]
16. Zhao, M.H.; Fan, C.Y.; Lu, C.S.; Dang, H.Y. Analysis of interface cracks in one-dimensional hexagonal quasi-crystal coating under in-plane loads. *Eng. Fract. Mech.* **2021**, *243*, 107534. [CrossRef]
17. Stadler, J.; Mikulla, R.; Trebin, H.R. IMD: A software package for molecular dynamics studies on parallel computers. *Int. J. Mod. Phys. C* **1997**, *8*, 1131–1140. [CrossRef]
18. Trebin, H.R.; Mikulla, R.; Stadler, J.; Schaaf, G.; Gumbsch, P. Molecular dynamics simulations of crack propagation in quasicrystals. *Comput. Phys. Commun.* **1999**, *121–122*, 536–539. [CrossRef]
19. Schaaf, G.D.; Roth, J.; Trebin, H.R.; Mikulla, R. Numerical simulation of dislocation motion in three-dimensional icosahedral quasicrystals. *Philos. Mag. A* **2000**, *80*, 1657–1668. [CrossRef]
20. Krdzalic, G.; Brunelli, M.; Trebin, H.R. Temperature dependence of dislocation motion and crack propagation in a two-dimensional binary model quasicrystal. *MRS Online Proc. Libr.* **2001**, *643*, K7.1.1-6. [CrossRef]
21. Rudhart, C.; Gumbsch, P.; Trebin, H.R. From crystalline to glassy: Crack propagation modes in decagonal quasicrystals. *MRS Online Proc. Libr.* **2003**, *805*, 312–317. [CrossRef]
22. Rudhart, C.; Rösch, F.; Gähler, F.; Roth, J.; Trebin, H.R. *Crack Propagation in Icosahedral Model Quasicrystals*; Springer: Berlin/Heidelberg, Germany, 2003; pp. 107–116.
23. Schaaf, G.D.; Roth, J.; Trebin, H.R. Dislocation motion in icosahedral quasicrystals at elevated temperatures: Numerical simulation. *Philos. Mag.* **2003**, *83*, 2449–2465. [CrossRef]
24. Rosch, F.; Rudhart, C.; Gumbsch, P.; Trebin, H.R. Cleavage planes of icosahedral quasicrystals: A molecular dynamics study. *Mat. Res. Soc. Symp. Proc.* **2004**, *805*, 329–334. [CrossRef]

25. Rudhart, C.; Trebin, H.R.; Gumbsch, P. Crack propagation in perfectly ordered and random tiling quasicrystals. *J. Non-Cryst. Solids* **2004**, *334–335*, 453–456. [CrossRef]
26. Rösch, F.; Rudhart, C.; Roth, J.; Trebin, H.R.; Gumbsch, P. Dynamic fracture of icosahedral model quasicrystals: A molecular dynamics study. *Phys. Rev. B* **2005**, *72*, 014128. [CrossRef]
27. Rudhart, C.; Gumbsch, P.; Trebin, H.R. Temperature dependence of crack propagation in a two-dimensional model quasicrystal. *Philos. Mag.* **2005**, *85*, 3259–3272. [CrossRef]
28. Engel, M.; Umezaki, M.; Trebin, H.R.; Odagaki, T. Dynamics of particle flips in two-dimensional quasicrystals. *Phys. Rev. B* **2010**, *82*, 134206. [CrossRef]
29. Lipp, H.; Engel, M.; Sonntag, S.; Trebin, H.R. Phason dynamics in one-dimensional lattices. *Phys. Rev. B* **2010**, *81*, 064302. [CrossRef]
30. Wang, Z.; Ricoeur, A. Numerical crack path prediction under mixed-mode loading in 1D quasicrystals. *Theor. Appl. Fract. Mech.* **2017**, *90*, 122–132. [CrossRef]
31. Sun, Z.F.; Fan, T.Y.; Wu, X.F. Convolution of the impact three-dimensional elasto-dynamics and dynamic stress intensity factor for an elliptic crack. *Acta. Mech. Sin.* **2002**, *18*, 302–308.
32. Wang, X.F.; Fan, T.Y.; Zhu, A.Y. Dynamic behaviour of the icosahedral Al-Pd-Mn quasicrystal with a Griffith crack. *Chin. Phys. B* **2009**, *18*, 709–714.
33. Zhu, A.Y.; Fan, T.Y. Dynamic crack propagation in decagonal Al-Ni-Co quasicrystal. *J. Phys.-Condes. Matter* **2008**, *20*, 295217. [CrossRef]
34. Yin, Z.H.; Fan, T.Y.; Zhu, A.Y. Dynamic crack propagation in five-fold symmetry quasicrystals. *Mod. Phys. Lett. B* **2009**, *23*, 1509–1518. [CrossRef]
35. Qiao, L.P.; Wu, L.; Fan, T.Y. Dynamic response of an icosahedral quasi-crystalline medium with a Griffith crack under mechanical loadings. *Adv. Mech. Eng.* **2017**, *9*, 1–12. [CrossRef]
36. Li, W.; Fan, T.Y. Elasto-dynamics of quasicrystals. *Crystals* **2016**, *6*, 152. [CrossRef]
37. Cheng, H.; Fan, T.Y.; Wei, H. Phonon–phason dynamics and hydrodynamics of fivefold and tenfold symmetry quasicrystals. *Acta Mech.* **2017**, *228*, 136–1372. [CrossRef]
38. Tang, Z.Y.; Fan, T.Y. Three-dimensional equations of generalized dynamics of 18-fold symmetry soft-matter quasicrystals. *Mod. Phys. Lett. B* **2020**, *34*, 2050109. [CrossRef]
39. Fan, T.Y.; Tang, Z.Y. Three-dimensional generalized dynamics of soft-matter quasicrystals. *Adv. Mater. Sci. Eng.* **2020**, *2020*, 4875854. [CrossRef]
40. Ambati, M.; Gerasimov, T.; De Lorenzis, L. A review on phase-field models of brittle fracture and a new fast hybrid formulation. *Comput. Mech.* **2015**, *55*, 383–405. [CrossRef]
41. Bui, T.Q.; Hu, X.F. A review of phase-field models, fundamentals and their applications to composite laminates. *Eng. Fract. Mech.* **2021**, *248*, 107705. [CrossRef]
42. Fan, T.Y. *Mathematical Theory of Elasticity of Quasicrystals and Its Applications*; Springer: Singapore, 2016.
43. Miehe, C.; Welschinger, F.; Hofacker, M. Thermodynamically consistent phase-field models of fracture: Variational principles and multi-field FE implementations. *Int. J. Numer. Methods Eng.* **2010**, *83*, 1273–1311. [CrossRef]
44. Miehe, C.; Hofacker, M.; Welschinger, F. A phase field model for rate-independent crack propagation: Robust algorithmic implementation based on operator splits. *Comput. Meth. Appl. Mech. Eng.* **2010**, *199*, 2765–2778. [CrossRef]
45. Chernikov, M.A.; Ott, H.R.; Bianchi, A.; Migliori, A.; Darling, T.W. Elastic moduli of a single quasicrystal of decagonal Al-Ni-Co: Evidence for transverse elastic isotropy. *Phys. Rev. Lett.* **1998**, *80*, 321–324. [CrossRef]
46. Jeong, H.C.; Steinhardt, P.J. Finite-temperature elasticity phase transition in decagonal quasicrystals. *Phys. Rev. B* **1993**, *48*, 9394–9403. [CrossRef]
47. Edagawa, K. Phonon–phason coupling in decagonal quasicrystals. *Philos. Mag.* **2007**, *87*, 2789–2798. [CrossRef]
48. Jaric, M.V.; Nelson, D.R. Diffuse scattering from quasicrystals. *Phys. Rev. B* **1988**, *37*, 4458–4472. [CrossRef] [PubMed]
49. Coddens, G.; Bellissent, R.; Calvayrac, Y.; Ambroise, J.P. Evidence for phason hopping in icosahedral AlFeCu quasi-crystals. *Europhys. Lett.* **1991**, *16*, 271–276. [CrossRef]

MDPI
St. Alban-Anlage 66
4052 Basel
Switzerland
Tel. +41 61 683 77 34
Fax +41 61 302 89 18
www.mdpi.com

*Materials* Editorial Office
E-mail: materials@mdpi.com
www.mdpi.com/journal/materials

www.ingramcontent.com/pod-product-compliance
Lightning Source LLC
Chambersburg PA
CBHW040746250926
51850CB00007BB/1422
*9783036581248*